T0122942

The Evolution of Obesity

The Evolution of Obesity

MICHAEL L. POWER AND JAY SCHULKIN

The Johns Hopkins University Press

Baltimore

© 2009 The Johns Hopkins University Press
All rights reserved. Published 2009
Printed in the United States of America on acid-free paper

Johns Hopkins Paperback edition, 2013
9 8 7 6 5 4 3 2 1

The Johns Hopkins University Press
2715 North Charles Street
Baltimore, Maryland 21218-4363
www.press.jhu.edu

The Library of Congress has cataloged the hardcover edition of this book as follows:

Power, Michael L.
    The evolution of obesity / Michael L. Power and Jay Schulkin.
        p. ; cm.
    Includes bibliographical references and index.
    ISBN-10: 0-8018-9262-7 (hardcover : alk. paper)
    ISBN-13: 978-0-8018-9262-2 (hardcover : alk. paper)
    1. Obesity. 2. Adipose tissues—Evolution. 3. Human evolution. I. Schulkin, Jay. II. Title.
[DNLM: 1. Obesity. 2. Adaptation, Physiological—physiology. 3. Adipose Tissue—
metabolism. 4. Adipose Tissue—physiology. 5. Diet. 6. Evolution. WD 210 P887m 2009]
    RC628.P65 2009
    616.3'98—dc22   2008033977

A catalog record for this book is available from the British Library.

ISBN-13: 978-1-4214-0960-3
ISBN-10: 1-4214-0960-7

*Special discounts are available for bulk purchases of this book. For more information, please
contact Special Sales at 410-516-6936 or specialsales@press.jhu.edu.*

The Johns Hopkins University Press uses environmentally friendly book materials, including re-
cycled text paper that is composed of at least 30 percent post-consumer waste, whenever possible.

Contents

....

Preface

....

Humanity is changing in size and shape. There has been a remarkable upsurge in the number of people considered overweight or obese. This change has occurred far too fast to represent population-level genetic change, but there must be biological, and thus genetic, factors involved with the obesity epidemic. Not all people are becoming fat. How can we explain both the sudden change in human mean weight and the variability among people in essentially the same environments?

This book focuses on aspects of human metabolism, physiology, and behavior relevant to obesity—evolved, possibly adaptive, traits that interact with the modern human environment to produce a vulnerability to weight gain. Of course, biology is not the only way to explore and understand human obesity. For example, cultural, socioeconomic, and technological factors influence the eating and the physical activity patterns that underlie our propensity to gain weight. These topics might well deserve books of their own. But in the end, people are biological organisms, and it is the interaction of our environment with our biology that leads to obesity. And as we, the authors, are biologists, our interests and our relevant expertise center on biological aspects of the obesity epidemic.

Fundamentally, this is a book about human biology. Biological knowledge is advancing at an amazing rate. Entire genomes are being sequenced; the basic building blocks of life are becoming understood. From this expansion of knowledge comes an appreciation of both the diversity of life and its common underpinnings. There is a significant amount of conservatism regarding the molecules and molecular pathways of life. Human beings and honeybees share a molecular heritage. At the same time there has been a tremendous diversification of function among the molecules of life. Most molecules produced by our bodies perform multiple functions; they perform diverse roles in the diverse tissues that make up a whole organism. To understand our biology it is necessary to integrate the different levels of biological investigation, from DNA to hormones to metabolic pathways and eventually to whole-body physiology and behavior.

In this book we examine the physiological regulation of energy metabolism in the broadest sense, from appetite and regulation of food intake to reorganization of energy stores in the body. We approach these issues from anatomical, neuroendocrine, physiological, and ecological perspectives, and at the level of the cell, organs, and the organism. Our approach is comparative and firmly anchored in an evolutionary perspective. Although the book is primarily about human biology we believe that a comparative perspective is important. We can learn much that is relevant to ourselves from the study of nonhuman species, especially our closest relatives, the nonhuman primates. Therefore, we include discussions of nonhuman biology where they seem to be illuminating or enjoyable examples of the concepts and principles we are exploring. For example, we discuss chimpanzee feeding behavior, the role of stored fat in black bear reproduction, and the energetic advantages and challenges of being a mouse or an elephant or something in between, like a human.

The biology of obesity is very broad and quite complex. This book cannot be all-encompassing. We have focused on a systems approach, attempting to integrate the many levels of biology into a comprehensible whole. We hope that this book will have a broad readership and thus have strived to make the text as accessible as possible without losing academic rigor; to focus on important concepts and at the same time to provide a sufficient grounding in the empirical data. For any given topic there will undoubtedly be readers who think we went into too much detail and others who wish we had elaborated more. Our hope is that all readers will come away with an appreciation of the core ideas. Of primary importance: evolved systems differ from engineered ones in fundamental ways; we, like all other creatures, carry a legacy from our past. Our biology reflects evolutionary responses to the challenges that our ancestors faced. The important signaling (information) molecules that regulate physiology and behavior are ancient and have been co-opted to perform multiple functions that vary with tissue, stage of development, and the conditions of the internal milieu. There is redundancy and overlap in function between molecules and metabolic pathways. Biology is inherently regulatory; from genes to metabolism to behavior, it is regulation that allows the flexibility and diversity that characterizes life.

The thoughts expressed in this book are ours, but many colleagues have inspired and informed us during our careers. We thank all of them collectively and a few specifically. Suzette Tardif has been a career-long colleague of MLP; she has been invaluable for the many informal conversations about biology and science as well as for the productive collaborations. JS thanks Mark Friedman, Lauren Hill, and Tim Moran and dedicates the book to them.

The Evolution of Obesity

Human Biology, Evolution, and Obesity

....

Daniel Lambert was born in Leicester, England, on March 13, 1770. During his relatively short life (he died at the age of 39), he became moderately famous. He met the king of England and other noblemen. People paid money to see him (Bondeson, 2000). He is still relatively famous today. There are exhibits of his clothes and other personal effects in museums in both Leicester where he was born and Stamford where he died. His portrait hangs in the mayor's office in Stamford Town Hall and was reproduced on the cover of the *Quarterly Journal of Medicine* (Figure I.1). What was the source of his fame? At the time of his death on June 21, 1809, Daniel weighed more than 700 pounds (Table I.1).

What is rare and unusual can often be perceived as valuable or desired. Consider the word *portly*. The modern definition in the dictionary is "corpulent" or "stout." However, the archaic definition is "stately" or "imposing." A portly gentleman was a prosperous gentleman, a person who had succeeded in life. "Portly" was a compliment when obesity was not common. It's not very likely to be taken as a compliment today.

Obese people are not a uniquely modern occurrence; it is the *prevalence* of obesity that has changed, not its existence. The evidence for human obesity goes back over 20,000 years. The Venus of Willendorf, found in an archeological site in Willendorf, Germany, has been dated to 20,000 BCE or earlier (Figure I.2). We do not know if this work of art is a true representation of an individual, but the level of detail and realism in the figurine implies that the artist had seen an extremely obese woman.

There are many accounts of extreme obesity in history, some better documented than others. In Europe in the 1700s and 1800s, obese people

1

exhibited themselves as curiosities. There were also so-called human skeletons—people exhibiting extreme thinness. In U.S. circuses the male "human skeleton" often married the circus fat woman, largely for marketing purposes no doubt. Extremes of body form have long been known among humans. In many of the examples of extreme obesity in history, the facts suggest a strong genetic or pathological component. These individuals often were obese from early life and as young children weighed hundreds of pounds.

Why did Daniel Lambert become obese? We do not know. As a young man he was considered large, but in the sense of being tall (5 foot 11 inches, tall for that time in England, but not exceptionally so) and stout, but not necessarily fat. He was a vigorous young man who was quite

FIGURE I.1. Daniel Lambert was the fattest man in England in his day. He was popular and well liked.

TABLE I.1   *Daniel Lambert's physical statistics at time of death*

Height: 5 ft 11 in (180 cm)
Waist: 9 ft 4 in (284 cm)
Calf circumference: 3 ft 1 in (94 cm)
Weight: 739 lbs (52 stone 11 lbs [335 kg])

strong, reputed to have been able to lift over 500 pounds. He was an excellent swimmer and gave swimming lessons to the youth of Leicester (Bondeson, 2000). At the age of 21 Daniel took over his father's job as the keeper of the Bridewell (or House of Corrections) of Leicester County. He effectively took a desk job. Although he remained active in hunting pursuits his actual occupation required little exertion. He spent most of

FIGURE I.2.   The Venus of Willendorf is more than 22,000 years old, implying that human obesity existed at least that long ago.

his workday sitting in front of the building smoking his pipe. It was at that point that his weight began to steadily increase (Bondeson, 2000).

During his life Daniel Lambert was known as the fattest man in England. That sobriquet was not pejorative. There was a great deal of curiosity concerning his immense size and amount of body fat, but the general feeling toward him seemed to be quite positive; he was a human wonder. His meeting with the elderly count Josef Boruwlaski was a celebrity event: an encounter between the world's largest and smallest man. His image was used in political cartoons, usually in a positive manner representing England's greatness. There are references to him in works of literature, such as William Makepeace Thackeray's *Barry Lyndon* and *Vanity Fair* and Charles Dickens's *Nicholas Nickleby*. His name came to imply colossal, again in a positive sense. There are still numerous pubs and taverns named after him, presumably associating his immense size with good food and drink. Ironically, Daniel Lambert apparently was quite moderate in his eating, and did not even drink beer (Bondeson, 2000).

His friends certainly held him in great affection. Upon his death they paid for a handsome tombstone inscribed with a tribute to his good character, as well as his remarkable physical characteristics:

> In Remembrance, of that PRODIGY in NATURE DANIEL LAMBERT a Native of LEICESTER who was possessed of an exalted and convivial Mind and in personal Greatness had no COMPETITOR. He measured three Feet one Inch round the LEG nine Feet four Inches round the BODY and weighed FIFTY-TWO STONE ELEVEN POUNDS! He departed this Life on the 21st of June 1809 AGED 39 YEARS.
>
> As a Testimony of Respect this Stone is erected by his Friends in Leicester.

Modern-day sentiments appear not to be as charitable; his tombstone has been spray-painted with the word "FATTY" (Bondeson, 2000).

Human Biology

This is a book about human biology. That is, of course, too large a topic for a single book. It would take a multivolume set to explore all the fascinating aspects of our biology. But we begin with this statement to emphasize that this is not a typical book about obesity. Yes, we will examine the "obesity epidemic" evident in the dramatic increase in the proportion of humanity that is overweight and obese; yes, we will discuss the health consequences of excess fat, and yes, we will explore the relevant biology underlying appetite, energy balance, and feeding behavior. However, obesity itself is less the focus and more the example used to illuminate human biology and the interactions between biology and environment. We are not trying to give advice on how to prevent or "cure" obesity; we are trying to understand the how and why of human obesity. Without that understanding attempts to modify obesity within individuals will be problematic, if not doomed to failure.

Our thesis is that much of the increase in human obesity is due to a mismatch between adaptive biological characteristics of our species and the modern environment, which has changed dramatically from the one under which we evolved. Whether we like it or not we carry our species' biological past with us, and this affects how we react to our environment, regardless of how changed that environment has become or how good we as a species are at changing it. We believe this concept is critically important in understanding human obesity, but it is also important for understanding many other health and well-being issues of the modern world. In our opinion many modern diseases have, at least in part, underlying connections to adaptive physiology inappropriately expressed due to confounding environmental conditions.

Human obesity is a fascinating example of this concept. Storing fat on the body is adaptive. Fat is essential. Human beings have evolved to be very good at storing fat; fat appears to have been very important in our evolution. For example, human babies are among the fattest of all mammals (Kuzawa, 1998). This neonatal fatness may have been a key element in humankind's survival. Being too thin increases the risks of morbidity and mortality. It was common knowledge that having a little extra weight could offer protection from the ravages of disease. Indeed, the latest epidemiological research appears to confirm that old lore: for certain diseases, being overweight reduces mortality (Flegal et al., 2007). However, being overweight increases other mortality risks (Adams et al., 2006; Flegal et al., 2007). Extra fat on our bodies appears to be a double-edged sword.

In the past the ability of most people to lay down increasing amounts of adipose tissue (fat) was largely constrained by the external environment. Life was hard; food was often scarce. This is the ecological view of the regulation of food intake. It is interesting to note that within the broad science of biology there are very different conceptions of feeding biology. The ecological perspective often posits that animals are intake maximizers (Stubbs and Tolkamp, 2006). Evolution has acted to favor adaptations that increase the efficiency of foraging behavior and the motivation of animals to feed. The limits on food intake are considered to be largely external, imposed on the individuals by the environment (Stubbs and Tolkamp, 2006). This differs quite dramatically from the laboratory physiologists' perspective; for them, food intake is regulated by internal mechanisms that act to defend homeostasis. Under this paradigm, animals will not be selected to be intake maximizers; the motivation to feed will vary with the internal state, and there will be times when feeding is actually aversive. Of course it makes the most sense to combine these two perspectives; both contain elements of truth.

If animals must constantly struggle to achieve an energy intake that matches their energy expenditure, the adaptations that arise relevant to the regulation of food intake, energy expenditure, and energy storage will likely differ significantly from those that would be favored if energy intake routinely could exceed even high energy expenditures. Under conditions where food intake is usually externally limited, the general prediction would be that adaptations such as high motivation to feed, the ability to regulate energy expenditure, and the ability to store significant energy on the body during the rare occasions the environment provides excess food would be favored. Seasonal or episodic instances where food will not be limiting for a species occur in virtually all habitats. In addition, even in food-restrictive environments there will be time scales, usually short, during which many animals will be faced with excess food, for example, a carnivore feeding on a kill that weighs more than its own body weight. Internal regulation of food intake will always be important. There are internal as well as external limits on eating.

Modern life looks very different to a large proportion of humanity than the past did to their ancestors. Food is plentiful, and it doesn't require extreme or prolonged exertion to obtain. This is not true for all people; there are still areas of the world where people work very hard for barely enough food. Indeed, the dramatic increase in the price of staple foods throughout the world has caused intense concern and even riots. Some economists warn that the era of cheap food may be over, with possibly dire consequences for poor people in many areas of the world. Nonetheless, over the last few decades more and more human beings have managed to lay down excessive fat in the modern environment. Obesity is not new, but to have countries where up to a third of the population is obese is a very recent occurrence. That is the phenomenon we are interested in understanding; the science of biology is our tool to achieve that understanding.

Biology of Fat

We spend a good proportion of the book examining the biology of fat. After all, the obesity epidemic is not about excess weight, it is about excess fat. Our understanding of the roles fat plays on our bodies and the metabolic consequences of too much fat have changed substantially. This is an exciting area of biological research.

Most fat is stored in adipose tissue. Under our previous conception of fat and adipose tissue, *stored* was indeed the operative word. Adipose tissue was a means to store fat, and fat was an efficient way to store energy. Excess energy consumed at one time was stored as fat in adipose tissue to be mobilized and used at a later time when food supply was inadequate; a very adaptive system. Of course there are other biological roles for fat; fat is an essential part of our bodies. Fat is important to all animals, in ways similar and dissimilar to the role of fat in humans. For example, among marine mammals fat stored in the blubber is crucial as an insulator. Water is an excellent conductor of heat; it removes heat from an object about 25 times faster than air does. Fat in marine mammals performs a critical energy-conserving function via its insulating ability.

Fat was always thought to have adaptive purpose, but the previous conception of fat and adipose tissue was to a large extent passive. It was a consequence of positive energy balance (eating more calories than are expended) or deliberately deposited to perform important but essentially static functions. Adipose tissue was not considered to be very metabolically active. It stored excess ingested energy and thus provided metabolic fuel when needed if future food intake was not sufficient.

Our conception of fat and adipose has changed dramatically. We now know that adipose tissue is an active regulator of physiology and metabolism and not just a passive result of positive energy balance. Adipose tissue is actually an endocrine organ (Kershaw and Flier, 2004), producing and metabolizing a number of peptides and steroids, as well as immune-function molecules (Fain, 2006). Many of the health consequences of obesity are due to the metabolic effects of this endocrine and immunological organ becoming "oversized"; physiology becomes out of balance.

Why do we get fat? The simple answer is that to become obese people must eat more calories than they expend over a sustained period of time. The simple answer is correct, but hides a great deal of complexity behind its simplicity. Understanding energy metabolism is fundamental to understanding obesity, and both energy and metabolism are complex and profound concepts. How metabolism and behavior become connected adds even more complexity. Yes, accumulating fat requires that food intake must exceed energy expenditure, but there are many ways that can happen.

The Age of Biology

In the early to mid-1900s, physics dominated scientific discovery. Relativity, quantum mechanics, $E = mc^2$, the atomic bomb, landing on the moon, the expanding universe all dramatically changed the way we think about the world. In the late 1900s, information technology and material science may have had the greatest impact. Portable computers with greater computing power than the old mainframes, memory storage materials that can contain entire libraries, plastics and other synthetic materials, and of course the Internet have revolutionized the way we work and play. Now, at the beginning of the 21st century, biological knowledge appears to be ready to explode in a similar fashion with a comparable effect. Entire genomes are being sequenced; animals are being cloned; and the very building blocks of life are being investigated with tools more varied and powerful than were imaginable 50 years ago. Along with the discoveries has come a humbling understanding of the complexity of living systems. Exciting and fascinating discoveries often bring the realization that much of what we thought was true was actually naive, simplistic, or at best only a small part of the story. The more we learn, the more we realize we have so much more to learn.

The tools in our modern-day biological research toolbox are amazing. Consider knockout models: laboratory animals with specific genes deleted or inactivated, either globally or restricted to specific tissues. These animals are unable to produce the specific peptide coded by the inactivated or deleted gene. These knockout models have produced fascinating but often counterintuitive results.

For example, an oxytocin knockout mouse model was produced with the expectation that the lack of oxytocin would compromise parturition. After all, oxytocin is well known to stimulate uterine contractions; synthetic oxytocin is given to women to induce labor. A pregnant mammal without the ability to synthesize oxytocin should be in trouble. However, these female mice gave birth quite competently, with no difference in mean gestation length compared to control subjects (Young et al., 1996; Russell and Leng, 1998; Muglia, 2000). The pups were born healthy but died soon after birth. Although the dams exhibited (reasonably) normal maternal care-giving behavior, and the lack of oxytocin had not significantly affected the gestation and parturition process, it did result in a failure of milk let-down. The females could not nurse their pups and the pups starved to death (Young et al., 1996; Russell and Leng, 1998; Muglia, 2000). Thus of the three instances (parturition, parental behavior, and milk let-down) where oxytocin was presumed to perform a necessary and even essential reproductive role across mammalian species, only for milk let-down was the lack of oxytocin critical in the mouse.

What does oxytocin have to do with obesity? Not much really. We use oxytocin as an example of a principle that will form a recurring theme in this book. Living organisms are evolved systems. Evolution often leads to redundancy and multiplicity of function. Animals can sometimes compensate for the lack of a supposedly essential factor. Most metabolic pathways are complex; many have alternatives.

Evolution has produced potent information molecules such as oxytocin that shape and guide the physiology and behavior of organisms. There are a large but finite number of these molecules. Most are very ancient, being found in all vertebrates and some invertebrates. These molecules have been adapted and co-opted to perform multiple and diverse functions in the body. Their genes have often been duplicated; the duplicate genes have then undergone an evolutionary journey of their own, changing and adapting. These molecules and their receptors will have multiple and often overlapping functions.

Scientists are constantly discovering new information molecules. They also discover new functions and complexities regarding the known functions for information molecules. Based on evolutionary principles, and the lessons of the past, we make the following predictions. Whenever a new molecule is discovered and a function assigned to it by the investigating scientists, it is a good bet that as the molecule becomes better known more functions will be discovered for it; its functions will vary by tissue and in context with other molecules of physiology, endocrinology, and metabolism; and there will be other molecules that appear to be able to perform some or all of its functions, at least in certain contexts.

Human Obesity from an Evolutionary Perspective

One purpose of this book is to explore the possible evolutionary, adaptive origins of the physiology and behavior of modern humans that might predispose us to obesity under current conditions. This is by no means a novel concept; it has been postulated by many authors in different contexts. The terms *thrifty genotype* and *thrifty phenotype* have been used to capture the notion that for much of our past, food insecurity (to use the latest bureaucratic phrase for hunger) was common, and we evolved to respond and adapt to food shortages (e.g., Neel, 1962). The original conception of a thrifty genotype came from a consideration of insulin resistance from an adaptive perspective (Neel, 1962). Both predictable (e.g., seasonal) and unpredictable decreases in food supply could favor metabolic adaptations to buffer the organism from hunger by storing fat. Interestingly, metabolic changes by animals that increase energy stores before winter are similar to the changes brought about by type 2 diabetes (Scott and Grant, 2006). Insulin resistance leads to a net flux of fatty acids into adipose tissue, and thus the accumulation of fat. It is reasonable to hypothesize that adaptive responses that increased our ancestors' ability to exploit times of food plenty in order to survive times of food scarcity would have been favored. The ability to store significant amounts of fat on our bodies would be an example of this type of adaptation.

We are not arguing that obesity per se was adaptive. Absolutely not! A central theme of this book is that human obesity is an inappropriate adaptive response to modern living conditions. The ancestral genus *Homo* evolved from scavenger-gatherers to become hunter-gathers and eventually agriculturalists. Their descendants are now patrons of fast-food restaurants. There is a mismatch between our evolved biology and our modern life. The advantages of fat storage in the past have become significant disadvantages today.

How does the evolutionary perspective inform our efforts to reduce obesity on individual and societal levels? We argue that there is a significant difference between viewing obesity as pathology versus obesity as inappropriate adaptation. Both perspectives will provide insights, both can provide useful strategies to allow people to manage their weight, and both can inform policy decision makers regarding public health strategies and interventions that may or may not be effective.

Homeostasis, Allostasis, and Allostatic Load

The concept of homeostasis is a central principle in regulatory physiology, especially in biomedicine (Bernard, 1865; Cannon, 1932, 1935; Richter, 1953). Obesity can be conceived of as a failure to maintain weight homeostasis. Because obesity results from a sustained positive energy balance, perhaps a more accurate conception would be a failure of total body energy homeostasis. Weight by itself is likely poorly monitored by internal biological mechanisms.

Thus a refinement of the weight homeostasis theory is that the amount of stored energy on the body is the parameter defended. Most energy stores are in the form of lipids (fats) stored in adipose tissue—thus the lipostatic theory of feeding behavior. A quick and simplistic interpretation of the lipostatic theory is that the total amount of adipose tissue is defended by physiological and behavioral adaptations that affect appetite (food intake) and energy expenditure. The bottom line is that obesity ultimately derives from the accumulation of fat. The homeostatic models define obesity as the failure of lipostatic mechanisms, and thus as pathology.

The lipostatic theory fits well within the homeostatic paradigm, but it makes less sense when viewed from the evolutionary perspective. In the wild, most animals routinely experience fluctuations in body mass, especially fat mass. It is not clear to what extent our ancestors experienced fluctuations in body weight due to variations in food supply, but it is a reasonable hypothesis that this was not an infrequent occurrence. In the wild, animals rarely are homeostatic in the strictest sense. Rather they change states to match challenges, sometimes even to anticipate challenges, for example seasonal changes (Wingfield, 2004). This aspect of physiological regulation has been termed *allostasis* (Sterling and Eyer, 1988; Schulkin, 2003). The key concept is that animals regulate their physiology in order to achieve viability, defined as the ability to pass on their genetics to the next generation (Power, 2004). In other words, physiological regulation serves evolutionary fitness, not homeostasis. *Allostasis* is regulation to achieve viability through changes of state; *homeostasis* is regulation to achieve viability through resistance to change (Power, 2004). The two terms merely describe different ways in which physiological regulation achieves its adaptive purpose.

The homeostatic paradigm works well for critical parameters that must be maintained within tight limits for the animal to remain viable. But many nutrients do not fit this model, at least not in all ways, over all time scales. Body fat would appear to be a case where a minimal level is crucial for a human being to be a viable organism (capable of passing its genes on to the next generation), but where the upper limit is far less constrained. This is not to say that weight homeostasis doesn't exist or that the concept that fat stores are defended to a certain extent is not true. But apparently the lipostatic mechanisms of human beings are more successful at resisting fat loss than they are at resisting fat gain.

We argue that for most of our evolutionary history there was an asymmetry between the fitness consequences of physiology and metabolism that were conducive to fat accumulation versus those that maintained leanness. The advantages of (moderate) fat accumulation generally outweighed the disadvantages; the disadvantages of excessive fat accumulation that we see today were largely invisible due to external constraints presented by the environment. In other words, for most of our past as a species we lacked the modern ability to dramatically lower our energy expenditure and raise our calorie intake to be able to become obese. The constraints of energy intake and expenditure were external as well as internal.

Due to technological and social advances the external constraints have become greatly relaxed. We are now aghast at how poorly our internal constraints appear to work. Not for everyone, however. There are significant numbers of people who maintain their adiposity within the healthy range, with little change over time. Humans vary; another characteristic of evolved species.

The pathologies associated with obesity reflect, in part, a metabolic imbalance caused by excess adipose tissue. The concept of allostatic load is that physiological systems are finite, and extended up-regulation of physiological adaptations can eventually wear down the system (McEwen, 2000; Schulkin, 2003). Normal adaptive responses can be stretched beyond function into pathology. Obesity is associated with markers of inflammation (Fain, 2006) and hormone imbalances. These metabolic dysregulations derive directly and indirectly from the normal adaptive function of adipose tissue that is out of balance with the other organ systems due to the great increase in adiposity. Adipose tissue is an endocrine organ whose natural function allows it to greatly increase and decrease in size; adipose tissue is meant to be variable. However, the extent of adiposity that is possible in today's world exceeds the normal adaptive range of endocrine and immune function.

The Structure of the Book

There are many ways to examine the issue of human obesity. Nutrition, energetics, endocrinology, reproductive biology, gut physiology, neuroendocrinology, and psychology are all relevant for understanding how a person might be vulnerable to obesity. And those are just some of the biological categories. We cannot hope to satisfactorily address all issues in this book. We have done our best to present those that we do cover in as logical and coherent an order as possible, given that there really is no beginning, middle, or end to this story.

The first chapter examines the question of whether there really is a human obesity epidemic. In other words, has the proportion of humanity that has become overweight or obese really increased so dramatically, and if so, are the health consequences associated with this change serious enough to warrant using the word *epidemic*. We lay out the evidence and explain why we are convinced the term is appropriate, at least in its broadest sense.

We do see the value in warnings against a "crisis mentality" toward this health issue and that there are possible unintended consequences that could harm certain individuals (e.g., people vulnerable to eating disorders or poor body image). However, the evidence supports the hypothesis that diseases related to excess adipose tissue have increased and are now among the most significant factors in human morbidity and mortality worldwide. The evidence further supports the contention that the prevalence of obesity and its associated comorbidities is likely to continue to increase in the near future. Regardless of whether obesity represents a health crisis or not, it is an issue that needs better understanding.

In chapter 2 we begin our evolutionary journey. We explore the evolution of the genus *Homo* from its inception approximately 2 million years ago until the recent past. We examine our morphology and how it has and has not changed over that time. We look at body size, guts, and of course, brain. This chapter is not meant to be an introductory course in paleoanthropology. Indeed, the challenge of this book is that almost every chapter could be expanded into a book on its own. We restrict our exploration of our evolution to key areas that we judge were likely to have significant effects on our metabolism and our feeding biology, and likely evolved in response to challenges that are greatly reduced today.

Regardless of what we eat, a fascinating and unusual aspect of human feeding is that we eat in meals. And by that we are not just talking about the dictionary definition of a meal: "the food served and eaten in one sitting or a customary time or occasion of eating." That definition does have some interesting and significant implications for human feeding behavior and how it varies from most other mammals. That behavior is certainly different from how our closest relatives, the nonhuman primates, feed. In the wild most primates are grazers rather than meal feeders. But the key aspect of human meals is that they have social meaning and purpose that is often even more important than the nutritional purpose. Food is laden with more than just calories. For humans, eating and social behavior are inextricably intertwined.

In chapter 3 we examine the evolution of meals and argue that the "meal" is one of the most significant evolutionary changes in our behavior. A meal-eating feeding strategy, with all its social, political, and even sexual connotations, provided significant selective pressures that favored enhanced intelligence and thus helped drive the increase in brain size that characterizes the genus *Homo*. It also separates eating from nutrition to a certain extent; motivations to eat no longer come from a strictly nutritional perspective. We eat for psychosocial reasons as well as nutritional ones.

In chapter 4 we discuss homeostasis, allostasis, allostatic load, and the mismatch paradigm. Human beings have a remarkable ability to exist in environments and under conditions that are far removed from those under which we evolved. Because of this, human beings are frequently confronted with a mismatch between environmental conditions and evolved adaptive responses. This is a prominent component of evolutionary medicine (Williams and Nesse, 1991; Trevathan et al., 1999, 2007) called the mismatch paradigm (Gluckman and Hanson, 2006) that is highly relevant to human obesity. Adaptive responses are successful solutions to past challenges; evolution doe not anticipate or predict the future. In this chapter we lay out the basic theory, with some examples of how past adaptations can cause modern pathology.

In chapter 5 we examine our modern environment in light of the information and theory reviewed in chapters 2 through 4. Where are the mismatches between our evolution and the modern world? What are the possible proximate causes of the modern obesity epidemic? We examine our food, our habitual level of activity, the built environment, and our sleep patterns. We also discuss our gut microbes.

Chapter 6 is about energy and metabolism. Obesity occurs when, over a sustained period of time, more energy is taken into the body than is expended in metabolism. Energy and metabolism are complex, powerful, and often confusing concepts. In this chapter we explore the thermodynamic principles on which living systems work. We examine different kinds of biological energy, how they are used by living things, and how they are measured by scientists. A key aspect of this chapter is that metabolism is regulated. This seems intuitively obvious, but the implications for energy metabolism are sometimes overlooked. Energy intake and energy expenditure are simple concepts in principle but very complex in actual physiology. The simple solution for weight loss, eat fewer calories and expend more, can be very difficult to achieve for good, metabolically adaptive reasons.

In chapter 7 we examine gut-brain peptides, the hormones that connect the gut and the brain to regulate appetite and satiety, and to coordinate behavior with digestion. We also discuss the concept of information molecules: molecules produced by cells that convey information to other cells or end-organ systems to link, regulate, and coordinate metabolism. Peptide and steroid hormones are excellent examples. Many if not most of these potent information molecules appear to be extremely ancient. They are found within all vertebrates and some are found in invertebrates. We discuss the implications of the evolutionary perspective on the expectations we should have regarding the function and regulation of these molecules. We examine leptin, a peptide produced by adipose tissue and secreted in proportion to the amount of body fat, in some detail.

There are metabolic and endocrine signals that affect food intake and energy expenditure. Energy intake and energy expenditure are regulated. In chapter 8 we examine some of the endocrine and metabolic signals and their peripheral and central circuits that influence food intake and energy expenditure. Variation in metabolism among individuals affects the extent to which ingested fat is oxidized or deposited in adipose tissue. Variation in the ability to switch between primarily glucose metabolism and primarily fatty acid metabolism may play a significant role in the susceptibility to obesity. However, behavior also plays a role, and that means neural circuits that mediate appetite and satiety are key to our understanding of human obesity. We briefly explore the neural circuits, distributed throughout the brain, that are important in eating behaviors.

Animals anticipate. Physiology is not just reactive; physiology changes in anticipation of need as well as in response to need. This is certainly true for adaptations relevant to feeding biology. From the work of Pavlov came the concept of cephalic-phase responses: centrally based responses to the sight, smell, taste, and other aspects of food that prepare animals to ingest, digest, absorb, and metabolize food. Chapter 9 explores these anticipatory responses in feeding biology, and related topics such as taste biology.

We must eat to survive. In addition, there are critical parameters of our internal milieu that must remain within constrained limits or health will suffer. Eating provides the materials to sustain life, but also results in significant homeostatic challenges that must be met. Eating is both protective and challenging to homeostasis. One function of satiety may be to protect homeostasis by restraining eating. In chapter 10 we examine the paradox of feeding to see how it might affect appetite and satiety, and the extent to which the control of eating may not always be about energy balance, at least in the short term.

In chapter 11 we examine the biology of fat. Obesity is not excess weight; it is excess fat. In general, the health consequences of obesity are not due to being heavy, although there are some (e.g., osteoarthritis). Rather, the health consequences are the result of excess adipose tissue. Adipose tissue is an active regulator of metabolism; excess adipose tissue may disrupt the regulation of other aspects of physiology and metabolism. The exaggeration of normal physiology can lead to pathology, a concept called allostatic load (McEwen and Stellar, 1993; McEwen, 2000, 2005; Schulkin, 2003). For example, obesity is associated with an inflammatory state, as measured by the concentrations of proinflammatory hormones and cytokines. Many of these are produced and released by adipose tissue. The normal functions of adipose tissue may become maladaptive when it represents an abnormally large proportion of the body.

Men and women differ. There are good biological reasons why they must. Vive la différence! In chapter 12 we examine sex differences in the biology of fat. We propose that many of these differences represent adaptive changes that support female reproduction. Fat is more important for female reproduction than it is for male reproduction. For example, maternal body condition is correlated with infant body composition. An interesting fact is that human babies are among the fattest of mammalian neonates, much fatter, on average, than any of our nonhuman primate relatives. This increased neonatal fatness may have been important to support the extensive postnatal brain growth necessary due to our larger brains. It may have also required a corresponding increase in maternal fatness. We examine the associations among fat, leptin (an important hormone of adipose tissue), and reproduction, focusing on the effects of obesity on female reproduction, but including male fertility as well. We do not argue that obesity confers reproductive advantage. Obesity has detrimental reproductive consequences for both men and women, but so does excessive leanness. In our past the latter was probably more prevalent. We examine the associations between adiposity, leptin, sexual maturation, reproductive function, and birth outcome.

A vulnerability to obesity has a genetic component. The heritability of size is well established. However, except for some rare, single-gene mutations, the genetic risks for obesity are not yet well defined. The number of candidate genes for a susceptibility to obesity is large, and complex interactions between genes and the environment would appear to be the most likely cause of obesity. In chapter 13 we examine genetic correlates of obesity and differences among human populations in the vulnerability to obesity and its associated diseases. Much of the variation among human geographic populations may represent the accumulation of random mutations. Some might reflect local adaptations to regional food types. Regardless, these variations among people that affect susceptibility to obesity were likely invisible in our past due to external constraints that kept people from obtaining excess food and required high energy expenditures to survive.

The data by which these issues can be explored are scant. Nonetheless, we believe understanding variations in human vulnerability to obesity is important. There are probably many genes and polymorphisms, nonrandomly distributed throughout humanity, that affect people's tendency to gain weight. There are racial and ethnic variations in the susceptibility to obesity, the health consequences of different levels of adiposity, the level of adiposity for any given BMI (body mass index), the distribution of fat, and the ability to use fat as a metabolic fuel. All of these may reflect genetic variation within the human genome, although race is a poor marker for genetic differences.

In chapter 13 we also examine the phenomenon of in utero programming of metabolism. There is a U-shaped distribution to the risk of adult obesity due to birth weight; both small and large babies are at risk of future obesity. We critique the thrifty genotype and phenotype hypotheses that have been proposed as evolutionary explanations for the link between low birth weight and later obesity. There are weaknesses in the various hypotheses proposed, but the link between the uterine environment and later physiology, metabolism, and health is real.

In the conclusion we encapsulate our general evolutionary hypotheses. The major adaptive changes in our lineage related to the importance of fat, both in our diet and on our bodies, likely arose to a large extent to support our larger brain. This hypothesis explains our fat babies, which explains the tendency for women to put on more fat than men do. It is not the only factor that likely affected our ancestors' selective pressures that encouraged a fat-accumulating physiology. An increase in pathogen density may also have played a role in favoring fatness, especially in infants. For many diseases, especially of the gastrointestinal tract, carrying extra fat and thus being able to go longer without food would have been adaptive. Increased population density, a tendency to occupy the same site for an extended period of time (especially after the advent of agriculture), and the domestication of animals, which brought novel pathogens into close contact with people, all probably increased the net pathogen load an individual accumulated through life compared to a low-population-density, hunter-gatherer existence. Supporting evidence for this theory is that human milk has the highest concentrations of antimicrobial function molecules of any milk so far examined. In the past our babies may have been exposed to far more pathogens than most other infants.

And there was likely a strong asymmetry due to external constraints on food intake and a required high energy expenditure that would have made polymorphisms that encouraged fatness to be incompletely expressed but would have allowed the full expression of genes that predisposed individuals to remain lean. We argue that such an environment would have, on balance, preferentially selected against the leanness genes and favored those that predisposed individuals to accumulate fat during the rare occasions it was possible. This would have encouraged the accumulation of numerous polymorphisms that, in the modern era, predispose us to obesity.

We also examine various strategies to halt, possibly even reverse, the obesity epidemic. The human obesity syndrome arises from many factors interacting in often complex and nonintuitive ways. There are no easy solutions. To view obesity as a simple problem of eating too much and exercising too little is both true and confounding. We evolved as a species that likes high energy density food. In the past it required substantial effort and exertion to acquire such foods; evolution has given us adaptive responses that motivate us to make the effort. As a species we have proven very talented at modifying our environment in ways that increase our access to food, especially high-calorie food, and at the same time reduce our necessary energy output. That ability arguably was the prime reason why our ancestors were successful and colonized the world. It was necessary for their survival. We are now paying the price for their success.

However, we also have the ability to modify our environment. We are capable of understanding ourselves and the world around us. We have economic, social, and political strategies that can and do modify the way we see the world and ourselves. The problem is difficult, but not, we think, insurmountable. Furthermore, the goal is health. Although obesity has health risks, excess fat does not explain all the variation in health among people. Such factors as physical fitness and psychosocial characteristics are also important. There are ways that people can improve their health even without dramatic weight loss. Physical activity may be one of the most important. As a species we are well adapted to sustained moderate exertion; it was necessary for survival for most of our existence as a species. Consistent evidence from epidemiological studies supports that moderate but consistent physical activity improves health. Strategies to increase physical activity, for individuals, communities, and society as a whole, may have a greater effect on health than all of the obsessions about our diet and our weight.

Finally, we consider that any quick-fix pharmacological interventions are more likely to fail than to succeed. The metabolic signals that underlie our propensity to become obese or to remain lean are complex. Many factors interact and integrate in a dynamic equilibrium. Change one factor and it is difficult to predict how many others will be affected, but it is a good bet that there will be unanticipated and unintended consequences. The health consequences of obesity can be very serious, and need to be addressed. However, to simply force a few selected parameters that are indicative of health back to a "normal" level will not necessarily return the whole body to health. This approach is often fraught with unintended, and unpredictable, consequences. Ameliorating the health consequences of obesity will probably require a whole-body approach.

ONE

Humanity on the Fat Track

H umanity is getting fat. Not everyone, but many of us in every quarter of the globe: men and women, young and old, rich and poor, from every race and ethnicity. There is a worldwide obesity epidemic that shows little sign of slowing, let alone reversing.

What is amazing, and frightening, is how quickly this change in human body weight is occurring. Within a few generations the bell curve of human weight distribution has shifted and skewed toward greater weight. The median-weight person of today would have been considered to be significantly heavier than average only a short time ago.

And this trend appears to be continuing. In the United States, the good news is that the proportion of the population that is underweight has decreased over the last 20 years. Morbidity and mortality due to hunger has mostly been eliminated. Unfortunately this has not translated into an increased proportion of the U.S. population being at a healthy weight. Instead, the proportions of overweight and obese people have increased—and continue to increase at an alarming rate. The proportion of the population considered extremely obese has proportionately increased the most (Freedman et al., 2002). The number of people considered extremely obese in the United States has more than tripled since 1960.

Why is this happening? Or perhaps more to the point, why is it happening now? Why are so many human beings susceptible to sustained weight gain in the modern environment? There have been obese people throughout history, but for most of that time obesity has been rare and unusual. In the past, being fat was a mark of wealth and status in many

FIGURE 1.1. The Tuscan general Alessandro del Borro painted by Charles Mellin (ca. 1635). This gentleman was successful and proud of it. He made sure his portrait displayed his stout build. Photo: Bildarchiv Preussischer Kultur-besitz / Art Resource, NY.

cultures (Figure 1.1). It was difficult to become obese. Nowadays, being thin is rare and unusual, and the mark of celebrity and wealth.

The rapidity with which the incidence of obesity has increased world-wide suggests that genetic change on a population level is an unlikely cause. We haven't suddenly become genetically more obesity prone. The underlying genetic and biological factors that are contributing to the large proportion of people that become obese in the modern environment likely have been extant in our species for a considerable time. Our evolved genome is interacting with our dramatically changed environment; for many of us the result is sustained weight gain.

Although technological, economic, and cultural factors have created the circumstances allowing the increased incidence of overweight and obese people in the world, to understand why these factors lead to obesity requires understanding the underlying biology. And in our opinion, a complete understanding of our biology relevant to obesity requires a careful consideration of the evolutionary events and pressures that have shaped our adaptive responses to hunger, food, exertion, and energy stores that, in today's world, may not be as appropriate as in the past. All living things carry the past with them. What they are depends on who their ancestors were. We don't live like our ancestors did, but we carry their biology. The evolutionary past both enables and constrains. To understand ourselves now we need to examine our evolutionary past.

People who follow "traditional ways of life" are becoming rare. Most people do not live like their ancestors did. There are, of course, many good things associated with this change: longer life spans and lower infant mortalities, for example. Let there be no mythology about the past. Traditional lifestyles represent successful strategies for the human species but can be brutally hard on individuals. We are not interested in making value judgments regarding lifestyles in this book. Rather, we are exploring the value of understanding the adaptations to past life in order to understand the challenges and consequences of modern circumstances. We carry biology that evolved to solve challenges of our ancestors 5,000 years, 50,000 years, 500,000 years, and even longer ago. We see the value of understanding this past in order to understand the present. Neither of us wishes to go back to living as our ancestors did.

In this book we explore human biology as it pertains to obesity. We examine feeding biology, digestion, energy metabolism, and the physiology and endocrinology of fat. We explore our bodies from the brain to the gut. And always underlying and guiding our exploration are the principles of evolution.

But first we turn to the science of epidemiology. If our thesis that a significant contributor to modern human obesity is the mismatch between our current environment and our evolved adaptive responses is correct, then human obesity must indeed be dramatically more prevalent now than in our past. We present what is known about the recent spread of obesity through the human race to examine whether this is true.

Measuring Obesity

How is obesity measured? This is a key question. There are many factors that affect body weight: height, sex, age, body build, bone density, and muscle mass to name some of the major sources of weight variation among people. A nation's population could increase in weight over generations without any change in fatness. Indeed, there was a general increase in height in the United States until fairly recently. Because of better nutrition more people were achieving their genetic growth potential. This was accompanied by an increase in weight simply because taller people were on average appropriately heavier. But that is not what has happened over the last 25 years. The increase in height has slowed or stopped, but the increase in weight has not (Ogden et al., 2004; Figure 1.2).

Among developed nations the United States has the dubious distinction of being a leader in obesity prevalence. The United States also has had consistent data collection protocols in place and has been able to track the changes in its population over the last 20 to 30 years. The United States is an excellent place to start our investigation.

The primary source of national data in the United States on obesity and overweight is the National Health and Nutrition Examination Survey (NHANES) an initiative that spends about $30 million per year to accumulate data nationwide. NHANES includes both an extensive take-home questionnaire (2 to 3 hours) and a physical examination in the mobile examination center. A key feature of NHANES is that these studies allow for standardized measurements of height and weight. This allows the calculation of the body mass index (BMI), which is defined as the weight of a person in kilograms divided by the square of their height in meters. Although not perfect, BMI is considered the best indicator of weight relative to height, and is strongly associated with the proportion of body mass that is fat among all races and ethnic groups so far examined (e.g., Norgan and Ferro-Luzzi, 1982; Norgan, 1990; Gallagher et al., 2000). BMI has become the preferred index to assess overweight and obesity (Ogden et

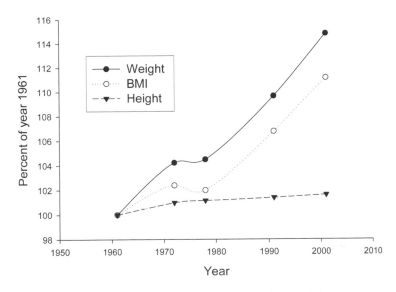

FIGURE 1.2. NHANES data indicate that the mean height of adults ages 20 to 74 years in the United States changed little (about 1%), while mean weight increased substantially (about 15% higher in 2002). As a result, mean BMI has also shown a substantial increase (about 11%). Data from Ogden et al., 2004.

al., 2006). The mean BMI for adults in the United States has dramatically increased since 1980 (see Figure 1.2).

Different BMI ranges have different levels of health risks. Both very low BMI (extreme thinness) and high BMI (overweight and obesity) are associated with greater morbidity and mortality. There is still discussion over what are the best cutoff points to define the upper and lower limits of BMI that represent the optimal health condition. There appear to be racial differences in how health risks vary by BMI (Araneta et al., 2002; Yajnik, 2004), and other parameters, such as waist circumference and waist-to-hip ratio (Iacobellis and Sharma, 2007), modify the expected risks within BMI categories. BMI by itself can be a blunt tool to assess individual health risk; individual characteristics change the health risks associated with any given BMI. For example, a high BMI due to a large muscle mass does not imply the same risks as one due to a large fat mass. Similarly, a supposedly healthy BMI may be misleading if the individual has low muscle mass and hence a greater proportion of fat than would be

TABLE 1.1   *International classification of adult underweight, overweight, and obesity according to BMI (body mass index)*

Classification	BMI (kg/m²)	
	Principal cutoff points	Additional cutoff points
Underweight	<18.50	<18.50
Severe thinness	<16.00	<16.00
Moderate thinness	16.00–16.99	16.00–16.99
Mild thinness	17.00–18.49	17.00–18.49
Normal range	18.50–24.99	18.50–22.99
		23.00–24.99
Overweight	≥25.00	≥25.00
Preobese	25.00–29.99	25.00–27.49
		27.50–29.99
Obese	≥30.00	≥30.00
Obese class I	30.00–34.99	30.00–32.49
		32.50–34.99
Obese class II	35.00–39.99	35.00–37.49
		37.50–39.99
Obese class III	≥40.00	≥40.00

Source: WHO, accessed at www.who.int/bmi/index.jsp?introPage=intro_3.html.

predicted by BMI. In most cases the health effects of overweight and obesity are due to the metabolic consequences of excess fat, not excess weight per se.

However, BMI is still a useful tool to examine health risks among a population. It is easy and inexpensive to obtain and it is minimally invasive to individuals. Of course the health implications for any given BMI need to be considered in the broader context of the person's personal circumstances, but BMI provides a screening tool to examine population-level variation in health related to adiposity. In general, for Caucasian populations the range of BMI from 18.5 to 25 kg/m² is classified as normal weight; below 18.5 kg/m² is considered underweight; and over 25 kg/m² is considered overweight. Within the overweight category, a BMI of 30 kg/m² or above is considered to define obese. Extreme or morbid obesity is defined as a BMI of 40 kg/m² or above. The current more extensive BMI risk categories as used by the World Health Organization (WHO) are given in Table 1.1.

Is There Really an Obesity Epidemic?

The word *epidemic* conjures up visions of substantial numbers of healthy people suddenly being stricken with some affliction. Epidemics are usually associated with communicable diseases. Indeed, one dictionary definition of the word *epidemic* is "a rapidly spreading outbreak of a contagious disease." Certainly human obesity does not fit that definition. However, another definition is "a rapid spread, growth, or development," with the example given being "an epidemic of robberies." So epidemics are not restricted to infectious diseases, or even health and disease. Flegal (2006) carefully reviews definitions of the word *epidemic*, including the one given in the *Dictionary of Epidemiology,* and concludes that the recent changes in human obesity prevalence do indeed have characteristics of an epidemic. She relies mainly on the epidemiological definition of epidemic, in which the salient point is that an epidemic is an occurrence of "health-related events clearly in excess of normal expectancy." She concludes that the extent of the increase in the prevalence of obesity over the last 25 years was not predictable from the obesity prevalence data prior to 1980.

Others have criticized the use of the word *epidemic* in regards to human obesity (e.g., Campos et al., 2006), arguing that it inappropriately raises the level of concern regarding this health issue. Such critics argue that the rise in overweight and obesity is not rapid and dramatic enough and the health consequences not so dire as to justify the notion of an obesity epidemic. Campos and colleagues (2006) caution that many economic interests (e.g., the diet industry, health food industry, and even biomedical researchers) have a vested interest in an exaggerated concern over humanity's increase in body weight. Others (e.g., Kim and Popkin, 2006) have pointed out that there are other economically powerful groups that would benefit from a lack of concern (e.g., fast-food and soft drink companies).

Still, the word *epidemic* does suggest a crisis, and scientists especially should be careful about contributing to the crisis mentality that seems to permeate our culture these days. The norms of body form and adiposity have changed many times over history. Some of the changes, especially as they have related to women, are not biologically sound. Corsets to produce 18-inch waists were based on fashion and socially accepted ideas of women's physiques. They were not health-based—and health-based guidelines for human adiposity are what are needed.

There is no exact, quantitative definition of epidemic. This is partially because the phenomena that can become epidemic can differ substantially in time scale and mode of increase. A cholera epidemic is different from the epidemic of HIV infection, and both are substantially different from the epidemic of obesity. Changes in the prevalence of obesity occur on the time scale of human generations. On that time scale the increase in obese people over the last 20 years is dramatic and would not have been predicted from an examination of data from previous centuries. This is not an isolated or regional phenomenon. For example, in the United States, the percentage of obese people has increased in every state. Something has changed, and humanity is getting fatter.

Does this change have important health consequences? Obesity is associated with many chronic diseases, including diabetes, hypertension, cardiovascular disease, and osteoarthritis. Countries with a high prevalence of obesity also have a high prevalence of adult-onset (type 2) diabetes. As modern medicine succeeds in curing, preventing, and even eradicating many of the most prevalent diseases of the past, the incidence of obesity-related diseases of modern humans continues to increase. Our society is becoming burdened by health care expenditures and excess morbidity and mortality attributable to our expanding waistline. Direct U.S. medical costs associated with obesity are estimated to exceed $61 billion per year (Stein and Colditz, 2004). High BMI is associated with higher health care costs and greater job absenteeism (Bungum et al., 2003). In most South Pacific island nations roughly three of four deaths are due to noncommunicable diseases (Table 1.2), with the leading causes being cardiovascular disease and diabetes. According to the World Health Organization (WHO), the cost of obesity-associated diseases represents close to half of health care expenditures in some of these countries.

In our opinion the answer to the question of whether there is an epidemic of obesity throughout the developed world that is rapidly spreading to the developing world is yes. The increase in human obesity is a "health-related event in excess of normal expectations." This conclusion does not negate or override the cautions and caveats brought up by researchers who disagree. Health is a multidimensional parameter. Body mass index and even percent body fat will only explain part of the variation in the

TABLE 1.2   *Age-standardized death rates among South Pacific island nations*

Country	All causes per 100,000	% from noncommunicable diseases
Cook Islands	817	75
Fiji	1065	77
Marshall Islands	1333	75
Nauru	1446	79
Palau	968	77
Samoa	1026	76
Tonga	888	77
Tuvalu	1428	73
Vanuatu	1033	75

health of people. What is indisputable is that humanity has changed quite dramatically in BMI over the last few decades, and this change appears likely to continue worldwide for the near future. This change has consequences that must be addressed. In the next sections we review the evidence for the increasing prevalence of obesity worldwide and its associated disease risks.

Global Obesity Prevalence

Based on BMI, the WHO has estimated that there are now over 1 billion overweight or obese people in the world, far more than the 800 million people suffering from malnutrition. There is consistent evidence that people of Asian descent have higher fat mass (especially visceral fat) for any given BMI, and this is true wherever they have grown up (see Araneta et al., 2002). In that case the BMI definitions of overweight and obese will be different for Asians, with lower cutoffs. That would raise the number of overweight people to approximately 1.3 billion, more than one of every five people alive today. No region of the world appears to be immune from this change in human weight. There are differences among nations, however.

Again, we start with the United States as the current obesity leader among developed nations. Since the 1980s, the distribution of BMI in the United States population has shifted to the right and has become skewed,

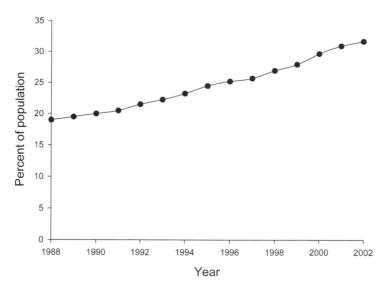

FIGURE 1.3. Prevalence of adult obesity (defined as BMI >30 kg/m²) in the United States from 1988 to 2002 using data from the Behavioral Risk Factor Surveillance System, corrected for self-report bias using NHANES data. Data from Ezzati et al., 2006.

with the heaviest getting heavier. The proportion of normal-weight Americans is at an all-time low. The proportion of people that are considered overweight has remained stable, while the proportion that is considered obese has increased (Figure 1.3). The largest percentage increase has been among those considered extremely obese (Freedman et al., 2002). It is estimated that there are more than 12 million people in the United States with a BMI greater than 40 kg/m². Approximately half of those have a BMI greater than 50 kg/m², and approximately a million Americans have a BMI greater than 70 kg/m².

Men and women and all age groups are affected. From 1999 to 2002, 62% of U.S. women age 20 years or older were overweight (defined as having a BMI >25 kg/m²); 33% were obese (BMI ≥ 30 kg/m²) (Hedley et al., 2004; Moore, 2004); and 15% of girls age 12 to 19 years were overweight (defined as having a BMI ≥ the 95th percentile-age according to the CDC growth charts) (Hedley et al., 2004). Under the current U.S. military's recommended enlistment standard for BMI, 40% of young women and 25% of young men in the United States are not eligible (National Academy of Sciences, 2006). They are too overweight.

Perhaps most worrisome is that this phenomenon is also occurring in children, a group that has experienced a 3.4-fold increase in overweight and a 7.8-fold increase in extreme overweight between the 1960s and 2000 (Hedley et al., 2004). Thus there is no indication that U.S. obesity prevalence will decline in the future, and indeed it may continue to increase. Data do, however, indicate a leveling off of the prevalence of obesity among women (Ogden et al., 2006), which is a hopeful sign. Or perhaps this result simply reflects that the proportion of people that are vulnerable to becoming obese in the modern Western environment is approximately one of three, and the U.S. population is approaching saturation.

But the United States does not have the highest rates of obesity. In the South Pacific the incidence of obesity has skyrocketed. Some of these small islands have populations with the highest prevalence of obesity in the world (Figure 1.4a). Over 70% of the people of the Pacific island country of Nauru are classified as obese, and 40% have type 2 diabetes. Several other Pacific island countries have comparable rates of overweight and obesity that are associated with high rates of comorbid diseases. The prevalence of diabetes is also dramatically increasing in these populations (Figure 1.4b).

Obesity rates in Europe are approaching those of the United States. In the United Kingdom adult obesity prevalence is approaching 30% and is expected to be about one of three adults by 2010 (Department of Health, 2006). The prevalence of overweight and obesity among 15-year-olds in Europe varies considerably by country but is disturbingly high. Interestingly, except for Ireland, the rate of overweight and obesity in 15-year-old European adolescents is higher in boys than in girls.

Even in Africa the rates of overweight and obesity are climbing, especially among women, and especially in the urban areas of the more economically advantaged countries. Rates in South Africa approach those of Europe: The rate of obesity is low in black men (8%), higher in white men (20%), and highest in black women (30.5%) (van der Merwe and Pepper, 2006). In many areas of sub-Saharan Africa malnutrition and obesity can be found in the same communities.

The prevalence of overweight and obesity is increasing in many Asian countries. Based on the standard BMI categories used for Caucasian populations, the prevalence of overweight and obesity is still lower in Asia compared with the United States (Figure 1.5a). However, among Asians

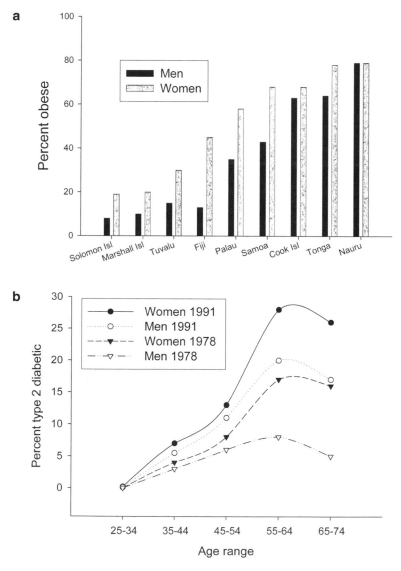

FIGURE 1.4. (a) Obesity rates for nine South Pacific island countries. (b) Combined diabetes rates for the countries by age and sex.

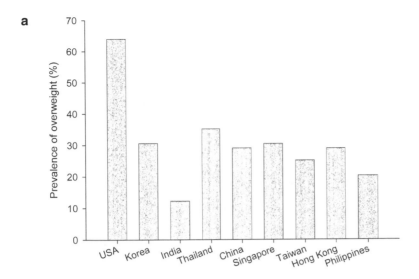

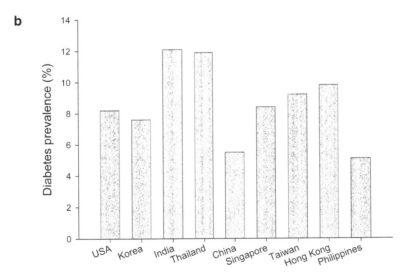

FIGURE 1.5. (a) Prevalence of overweight and obese people (BMI >25 kg/m^2) in the United States and selected Asian countries in the late 1990s. (b) Prevalence of diabetes in the late 1990s. Diabetes prevalence in Asian countries is greater than would be expected for their obesity prevalence, based on the data from the United States.

the health risks associated with adiposity appear to be elevated at lower BMI than for people of European descent. The prevalence of type 2 diabetes in Asian countries is similar to or even higher than that in the United States (Figure 1.5b). The increase in diabetes prevalence in Asian countries over the last 20 to 30 years is greater than the increase in the United States, and parallels the increase in BMI in those countries. There appear to be racial (and ethnic) differences in the vulnerability to obesity and its sequelae. In particular, Asians may be more likely to deposit fat in visceral depots, which increases their risks of diabetes and other obesity related diseases (Yajnik, 2004). The health-based BMI categories may need to be revised to account for differences among people of Asian, sub-Saharan African, and European descent (see chapter 13).

Health Consequences

The increases in the prevalence of obesity are associated with increases in the prevalence of many noncommunicable diseases. Obese people are at higher risk for a number of major and minor pathologies (Bray and Gray, 1988; Table 1.3). Chief among the most debilitating of these maladies are cardiovascular diseases and diabetes. Recent evidence has shown strong associations among many cancers and obesity (Renehan et al., 2008), although some of these associations may reflect associations between both obesity and cancer with low levels of physical activity.

Mortality risk is higher at both high and low BMI. Extremely low BMI (BMI $<$ 16 kg/m^2) is associated with malnutrition. For severely or morbidly obese people (BMI $\geq$ 40 kg/m^2) the age-adjusted risk of dying is 1.5 to 2 times as high as for people with a normal-weight BMI (Bray and Gray, 1988). Except for osteoporosis, where high BMI is protective, obesity exacerbates the diseases of old age (Roth et al., 2004a, b). Obesity increases mortality among all men, but the increase is substantially higher for younger men (Drenick et al., 1988). People who are obese in middle age have a higher risk of hospitalization and death from coronary heart disease (Yan et al., 2006).

Mortality and BMI at age 18 years old were associated among women in the United States. Women between the ages of 24 and 44 were enrolled

TABLE 1.3 *Some health issues associated with obesity or showing increased risk in obese people*

Metabolic disease	Cancer
Type 2 diabetes	Kidney cancer
Hypertension	Endometrial cancer
Cardiovascular disease	Postmenopausal breast cancer
Stroke	Esophageal cancer
Hyperlipidemia	Gall bladder cancer
Nonalcoholic fatty liver	Colon cancer
Reproductive disorders	Other
Infertility	Osteoarthritis
Cesarean section	Sleep apnea
Stillbirth	Asthma
Birth defects	Depression
Miscarriage	
Fetal macrosomia	
Preeclampsia	
Maternal death	

in the study and asked to recall their weight at age 18. The women were followed for 12 years. Women who were obese at age 18 had a mortality rate almost three times that of women who had a BMI < 25 kg/m^2 (van Dam et al., 2006; Figure 1.6). Not surprisingly, median BMI of the cohorts classified by their BMI at age 18 was lower than calculated from the baseline measurements at between 24 and 44 years of age; BMI tends to increase with age. The data indicate a general shift to higher BMI in the distribution curve. Again, not surprisingly, the percentage of each cohort that was obese was significantly correlated to their age-18 BMI (Figure 1.7). Very few women with low age-18 BMI were obese. The prevalence of obesity was substantially higher in the higher BMI cohorts, with almost half of women who were overweight at age 18 obese. It is perhaps good news that only 64% of women who were obese at age 18 were obese when older. About one out of three of these women had reduced their BMI with age. However, given that the median BMI of this group increased from 32.6 kg/m^2 to 35.0 kg/m^2 (van Dam et al., 2006), the implication is that those women who were still obese had a substantial gain in their BMI. The data aren't reported, but the implication is that the percent of these women that were very (BMI $\geqslant 35$ kg/m^2) or extremely (BMI $\geqslant 40$ kg/m^2) obese likely had increased.

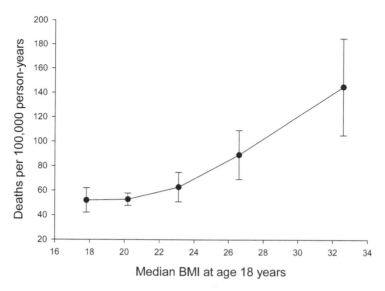

FIGURE 1.6. Deaths among women per 100,000 person-years by median BMI at age 18. Women who were obese at age 18 experienced excess mortality. Data from van Dam et al., 2006.

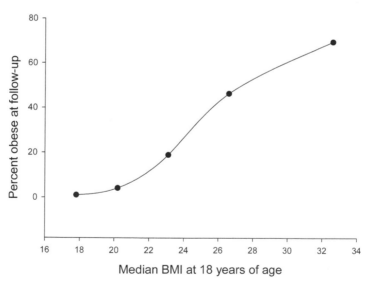

FIGURE 1.7. Percent of women who were obese at follow-up by their median BMI at age 18. Data from van Dam et al., 2006.

Consequences Other Than Health

The increase in the physical size of the population of the United States is having widespread effects beyond health. As we get bigger, so, of necessity, does most of our "stuff." Standard seat width, whether in a stadium, a church pew, or a car, has increased by several inches over the past few decades. Office furniture is changing to accommodate heavier as well as larger workers. Revolving doors have gone from a standard 6 feet in width to 8 feet. The increase in the average airline passenger's weight has had significant implications for fuel use by airlines. It is estimated that the increase in the average American's weight cost the airlines as much as a quarter of a billion dollars in extra fuel costs in the year 2000 (Dannenberg et al., 2004). Automobile gas mileage is also affected by passenger weight.

Society's infrastructure, especially medical infrastructure, is changing, driven by the change in human size. In hospitals, special beds, operating tables, wheelchairs, and other equipment are necessary to accommodate the growing proportion of patients who are obese. As simple an issue as giving intramuscular injections becomes complicated by obesity. Standard hypodermic needles are not effective. They need to be longer to penetrate the fat layer in obese patients in order to reach the muscle (Chan et al., 2006). The task of turning over obese bedridden patients becomes a complicated team effort requiring multiple nurses. Hospitals are faced with a growing population of patients with above-average morbidities that do not fit standard-size gowns, wheelchairs, beds, or MRIs. In fact, standard imaging equipment is not designed to give images of acceptable resolution through the extra layers of fat found in morbidly obese patients. Interpreting x-rays becomes difficult because the thickness of fat creates a haze on the image (Uppot et al., 2007).

Safety can also be an issue. Due to the increased weight of children in the United States, child car safety seats need to be enlarged. It is estimated that as many as a quarter million children are too large to properly fit into current car safety seats (Trifiletti et al., 2006). Children who are too heavy for their safety seats are at greatly increased risk of injury or death in an accident. Adult mortality in automobile accidents is also associated with BMI: Adults with a BMI $\geqslant 35$ kg/m^2 were more likely to die within 30 days of an automobile accident (Mock et al., 2002). The causes of the increased mortality are uncertain and may be related to higher comorbidities and other risks due to obesity (e.g., increased risks due to anesthesia), however, the data indicated that extremely obese people were at higher risk for serious chest injuries from car crashes (Mock et al., 2002). In men there appears to be a U-shaped risk of death with respect to BMI, with both high- and low-BMI individuals being at greater risk (Zhu et al., 2006). There appears to be a higher rate of all types of accidents among people with BMI > 35 kg/m^2 (Xiang et al., 2005).

Most of us never read them, but whenever you enter an elevator somewhere there is a plaque stating the maximum number of people the elevator can safely carry. That number is based on an average-size person. As we get larger the calculations need to change. In 2003 a small commuter jet crashed shortly after takeoff from the airport in Charlotte, North Carolina. A contributing factor to this crash was excess passenger and luggage weight. The safety regulations regarding the number of passengers were based on the average weight of people from 25 years ago. In response to this tragedy the U.S. Transportation Department has updated the safety guidelines for average passenger and baggage weight.

Understanding the Obesity Epidemic

A tremendous amount of research effort has been expended in understanding why some people get fat and others do not. We have learned a great deal about human biology; we know much more about the biology of fat; we have not stopped the increase in human obesity.

We know that BMI and body fat distribution have heritable components (Samaras et al., 1997; Rice et al., 1999; Hsu et al., 2005). Genetics certainly plays a role. There has been a concerted search for genetic factors that underlie causes of obesity, and these efforts have advanced our knowledge of specific, and usually rare, disorders that lead to obesity. So far the search has been less productive regarding general obesity. Less than 5% of obese people have been shown to have identifiable hormonal, physiological, or molecular genetic abnormalities (Speiser et al., 2005). Most human obesity probably reflects complex interactions between genetic, environmental, and social factors often mediated through non-genetically derived changes in metabolism.

The rise in the prevalence of obesity has been too rapid to attribute it to genetic change within the population. This does not mean that there are no genetic components to the increasing incidence of obese people, but does indicate that either a significant proportion of modern obesity is not genetically inheritable or that the underlying genetics making us susceptible to obesity is widespread throughout the population. Both are probably true. Indeed, we argue that genetic polymorphisms that induce susceptibility to obesity will have been more likely to have been selectively favored than selected against through most of our evolutionary history. In many instances a predisposition to accumulate fat may have been selectively invisible, because external circumstances constrained the expression of the phenotype; regardless of genetics the environment may have kept body fat low. The evolutionary perspective predicts that there will be an asymmetry in genetic susceptibility to obesity versus leanness, with evolution being more likely to have favored (or at least not selected against) a susceptibility to accumulate fat. Thus we predict that there will turn out to be many genetic polymorphisms that will be linked with obesity and far fewer that will be linked to a propensity to remaining lean.

Some evidence in support of this hypothesis comes from an examination of growth effects of gene deletions in mouse knockout models. A significant number of mouse knockout models are viable; this is characteristic of an evolved system. There are often redundant metabolic mechanisms. About 34% of viable knockout mouse models examined showed a change in growth reflected in body size compared to wild type mice. From that result a conservative estimate of the number of genes that affect growth and eventual adult size in mice is about 4,000 (Reed et al., 2008). About 9 of every 10 knockout mice that differed from the wild type size phenotype were smaller; to a certain extent this probably reflects that knocking out a gene might very well compromise normal growth. However, it also implies that more genes that favor large size exist compared to those that restrain size.

The anecdotal accounts of extremely obese people in the past often appear to fit a genetic profile. The individuals are said to have been large from an early age, and they continued to gain weight. Some were said to have voracious appetites; others were said to eat normally but still gained weight. There seem to be many paths to obesity. The difference today is that more people have the ability to get there.

Obesity and Evolution

Many authors have hypothesized that aspects of human genetics, physiology, and behavior that are now predisposing individuals to obesity were successful adaptations to conditions in our evolutionary past (e.g., Peters et al., 2002; Chakravarthy and Booth, 2004; O'Keefe and Cordain, 2004; Prentice et al., 2005). The current epidemic of obesity thus results from the interaction of our environment with past adaptations that are now inappropriate given the modern milieu. Our biology and our modern lifestyle no longer match when it comes to weight homeostasis. Obesity leads to pathology, and can even be perceived as pathology, but its causes may be normative, evolved adaptive responses to a past world of high energy expenditure and uncertain and variable food intake.

One obvious and important change from our past is that food acquisition has become disassociated from physical exertion. We evolved as a species that had to work hard for its food (Eaton and Eaton, 2003). Significant energy expenditure was required in order to gain energy intake. We can now have pizza delivered to our door. Acquiring food no longer requires an increase in energy expenditure over maintenance. We are in danger of becoming a species in which a large proportion of our population can and will exist at an energy expenditure level not much different from basal (minimal) metabolic rate and, at the same time, has easy access to highly palatable foods that are easily digestible and have a high energy density. The anatomical, physiological, and metabolic "tools" inherited from our evolutionary past that were successful adaptations to our feeding and foraging strategies of long ago may not be appropriate to our modern feeding and foraging milieu. This mismatch may be responsible, at least in part, for the increasing incidence of obesity under modern conditions. Our species' success at meeting the challenges of our past are now presenting us with a new challenge that our evolved physiology, anatomy, and behavior may be ill equipped to handle.

If that idea is not worrisome enough, there is frightening evidence that the ongoing epidemic of obesity may become self-sustaining. The risks of both childhood and eventual adult obesity appear to be significantly affected by the intrauterine environment, and many of the later associated adult diseases may at least partly result from in utero programming of physiology and metabolism (e.g., Barker, 1991, 1998; Hales and Barker, 2001; Ramsey et al., 2002; Yajnik, 2004). The early work that explored this concept focused on low birth weight babies and proposed the idea of "thrifty" genotypes and phenotypes. The resulting propensity to obesity thus was due to a mismatch between a phenotype adapted to scarcity growing up in a postuterine environment of plenty. However, increased risk of obesity, and more importantly, obesity-associated diseases, appears to follow a U-shaped distribution with birth weight, with both small- and large-for-gestational-age babies at higher risk (e.g., Yajnik, 2004). Thus an intrauterine environment of plenty also appears to predispose to obesity.

Newborn macrosomia (birth weight greater than 4,500g) has significantly increased over the last 20 years (Lu et al., 2001). Both maternal obesity and excessive weight gain during pregnancy are associated with macrosomia (Beall et al., 2004). In a cohort of Britons, maternal BMI, birth weight, and adult weight of the offspring were found to be highly correlated. Maternal BMI was actually a better predictor of the children's future adult weight than was birth weight, suggesting that the association of birth weight with adult weight reflects an effect of maternal BMI on both (Parsons et al., 2001). A mother may pass on her own weight disorders to her offspring in a form of "inheritance of acquired characteristics" (Beall et al., 2004). This leads to the specter of obese mothers passing on to their daughters characteristics that will increase the likelihood that those daughters will be obese and will subsequently have offspring who also are at higher risk of obesity and its associated diseases.

What Causes Obesity?

Obesity is a complex disorder with a deceptively simple ultimate cause: the prolonged consumption of calories in excess of those expended. The epidemic of obesity can be understood as a logical consequence of the fact that it has become progressively easier to consume more calories while expending fewer (Prentice and Jebb, 2004). The factors driving the "super-sizing" of humanity are technological, economic, cultural, behavioral, and psychological, as well as biological. For example, structural societal changes in food economics have resulted in highly palatable, calorie-dense foods (e.g., foods with added sugars and fats) being generally less expensive than their more healthful alternatives such as fresh fruits and vegetables (Drewnowski and Darmon, 2005). In addition, for many people, physical effort has shifted to leisure-time activities and is no longer associated with work. Thus, energy expenditure via physical effort has become an expense of both money and time as opposed to a necessary condition of survival. This results in healthy eating and sufficient exercise becoming more expensive in both money and time than fast food and sedentary leisure activity. A healthy lifestyle is in danger of becoming a luxury item.

Many aspects of food and eating have changed (Table 1.4). Today, a virtually limitless selection of high-fat, high-sugar, and thus high-calorie foods are available. Food is easy to get. It is available at all times of the day and night, with minimal seasonal variation. The proportion of waking time spent in food-related activities can be remarkably small. The costs in energy and time to get food can consist of simply waiting in line. There are no inherent physical dangers associated with food these days. Modern food doesn't fight back and there are no leopards lurking by the supermarket.

Our unhealthy eating habits start young. Research shows that nearly half of U.S. children's vegetable consumption is fried potatoes. Total vegetable intake by U.S. children decreased by 43% from 1977–78 to 2001–2; during the same period consumption of pizza increased by a dramatic 425%. Bread is the greatest source of carbohydrates in children (Subar et al., 1998). Our children are drinking less milk (decreased by 38%) and more soda (increased by 70%) (Isganaitis and Lustig, 2005). Soft drinks are the second leading source of carbohydrates in children (Subar et al., 1998).

Food is not the only issue. People are more sedentary, with many not expending many more calories than the basal metabolic rate. In fact, in 2005 less than half of the adult population in the United States reported getting regular physical activity. But there is some good news. The percent of adults reporting regular leisure-time physical activity has increased

TABLE 1.4 *Comparison of some aspects of food*

Aspect of food	In ancestral environment	In developed-nation environment
Quantity available	Sufficient but not abundant	Superabundant
Temporal availability	Often highly seasonal; episodically rare	Most foods available year-round
High caloric density foods	Rare	Common
Energy expenditure necessary to obtain food	Substantial	Minimal
Time expenditure necessary to obtain food	Substantial	Minimal
Risks inherent in obtaining food	Substantial	Minimal
Function of food	Primarily nutritional with some sociosexual functions	Social often more important than nutritional

since 2001 for all adults (CDC, 2007). However, many people today engage in little physical activity, especially compared with their ancestors.

Of course in the not-too-distant past leisure-time activity was not the important determinant of energy expenditure. People expended considerable energy in the daily activities of life. But now our jobs are more sedentary. We are much more likely to drive than walk, to take the elevator rather than the stairs. The very infrastructure of our living environment often reinforces these decisions. Transportation routes are designed for motor vehicles, not pedestrians or bicyclists (Sallis and Glanz, 2006). In many buildings it is very easy to find the elevators, and very difficult to find the stairs. The built environment of modern humans enables and even encourages low physical activity.

So Why Aren't We All Fat?

As some researchers have pointed out (e.g., Speakman, 2007) the litany of modern-day risk factors that would favor sustained weight gain is so pervasive that the interesting question is why are there so many people that maintain a healthy weight? Not everyone gains weight in the modern environment. That implies that there are factors that influence our susceptibility or vulnerability to obesity. Some of these may be cultural and behavioral. Personal choice can and does play a role as well. But there are genetic, metabolic, and physiologic factors as well. BMI has been shown to have an inheritable component (Samaras et al., 1997; Rice et al., 1999; Hsu et al., 2005). This is not surprising, since height and body build certainly have strong genetic components. But the evidence supports the hypothesis that there are heritable factors that predispose people to gain weight or to be weight stable in the modern world.

We are an extremely variable species, and with over 6 billion people alive in the world today we are expressing our genetic diversity to a greater extent than ever before. Throughout history there have always been obese and thin people. There likely always will be. The modern environment has exposed the extent of our species vulnerability to becoming obese, but also indicates that many humans are resistant.

Summary

A dramatic increase in overweight and obese people in wealthy nations has occurred over the past 25 years. That trend is spreading to developing nations, and in some South Pacific island countries obesity prevalence has already surpassed that in Europe and the United States. Along with this increase in fat on our bodies have come increases in the prevalence in noncommunicable diseases such as cardiovascular disease and diabetes. There is a worldwide obesity epidemic, under a broad definition of the word.

The rapidity of the increase in obesity prevalence implies that it is the result of environmental changes interacting with human biology and that the biology that makes individuals vulnerable to sustained weight gain in the modern environment is widespread in the population. Some individuals appear resistant, however. Thus biological variation does play a role, and understanding the underlying biology of both the vulnerability and resistance to excess fat accumulation is important in understanding the health consequences.

We argue that there are many paths to obesity; genetic variation that in our past was either invisible or possibly even selectively advantageous now results in maladaptive responses. To understand our feeding behavior, and why many of us will overeat given the chance, an understanding of our evolutionary past is critical.

TWO

Our Early Ancestors

···

Human beings (genus *Homo*) have changed considerably since the time when our ancestors were the same as those of the chimpanzee. Of course we retain many features from that time as well. We have more in common with the great apes, especially the chimpanzee and bonobo (genus *Pan*), than we do with other mammals. However, our lineage has been separate from theirs for approximately 6 to 7 million years (Glazko et al., 2005). There are a number of features that are characteristic of our ancestors and distinguish them from the *Pan* lineage; two of the most commonly cited are our larger brains and our bipedal posture.

In 1925 Raymond Dart published a paper in the journal *Nature* describing the skull of a juvenile "man-ape" discovered at Taung in South Africa (Dart, 1925). The Taung child was significant for several reasons. It gave impetus to the hypothesis, favored by Darwin, that humans had evolved originally in Africa. It also indicated that bipedalism preceded brain expansion. The Taung child skull would have connected to the vertebrae in a way that indicated an upright posture, strongly suggesting a bipedal form of locomotion, but the brain would have been the size of a chimpanzee's (Dart, 1925). Dart named the Taung child's species *Australopithecus africanus*.

In the 1950s discoveries by the Leakeys, Louis and Mary, gave further, convincing evidence that our lineage originated in Africa and that bipedalism predated a larger brain. They found fossil skulls dating back almost 2 million years with increased cranial capacity (i.e., larger brains) that were associated with primitive stone tools (Leakey and Roe, 1994). The Taung child does not represent one of our ancestors; australopithecines and early genus *Homo* appeared to be contemporaries.

In the early 1970s, a 3.2-million-year-old relatively complete (40%) fossil skeleton of a female australopithecine was discovered in Hadar, Ethiopia. Known as Lucy (from the Beatles song "Lucy in the Sky with Diamonds") this fossil provided final proof that bipedalism came before a larger brain. Based on Lucy and other fossils of the same species found at Hadar and Laetoli, Tanzania, a new species name was given: *Australopithecus afarensis* (Johanson and White, 1979). This species retained many features from the probable common ancestor of humans and chimpanzees, such as a chimplike brain size, significant sexual dimorphism (males were much larger than females), and long arms and curved finger bones conducive to tree climbing. However, the lower limbs and pelvis indicated that *A. afarensis* was an efficient biped. In addition, fossilized footprints have been found at Laetoli dating to about the same time of two bipedal creatures of the correct size walking side by side, or at least following the same path (Hay and Leakey, 1982).

In the 1990s fossil evidence for an even earlier origin of bipedalism was discovered. The genus *Ardipithecus* (White et al., 1994) is represented by two species that lived from 4.4 to 5.8 million years ago. Other recent discoveries in Africa suggest an even earlier origin. The genus *Orrorin* appears to have been bipedal as well as an agile climber and lived about 6 million years ago (Senut et al., 2001). The genus *Sahelanthropus* may be representative of our earliest ancestor or of the common ancestor of us and the great apes. It lived 6 to 7 million years ago and has a mixture of apelike and *Homo*-like characteristics (Brunet et al., 2002); it is not at all certain that *Sahelanthropus* was bipedal, however.

Interestingly, the fossil evidence and the molecular evidence for the divergence of humans and chimpanzees once again are coming into potential conflict. The molecular evidence, starting with the classical study of Sarich and Wilson (1973), have consistently put the divergence time somewhere between 4 and 6 million years ago. More recent estimates have been in the 5 to 7 million years ago range (e.g., Glazko and Nei, 2003). A provocative study of the great ape (chimpanzee, gorilla, and orangutan) and human genomes concluded that chimpanzee-human speciation occurred about 4 million years ago, although the initial divergence may have been earlier. Evidence from a comparison of the X chromosomes supports the hypothesis that there was an extended period of interbreeding between chimp and human ancestors after the initial divergence circa 6 to 7 million years ago that didn't cease until about 4 million years ago (Hobolth et al., 2007). The evidence isn't definitive, but it certainly supports the idea that our evolutionary history probably is better described as a complex bush rather than a simple tree. If bipedalism really is a defining characteristic of our ancestors, and never existed in the chimpanzee lineage, then the fossils and the molecular genetics are in some conflict.

Regardless of the exact time or manner of the split of our lineage from that of the chimpanzee and bonobo, we know that about 2 million years ago there were a number of bipedal primate species, referred to as hominins by anthropologists (Wood and Richmond, 2000), living in a wide variety of habitats and geographic locations in Africa, but only in Africa. Many of these species had brains about the size of a modern-day chimpanzee. That indicates that they were quite intelligent animals, with a suite of complex and adaptable behaviors with which they solved the challenges of life. That certainly describes modern chimps. These were the australopithecines; they were not our ancestors, but they almost certainly shared a common ancestor with us (Wood and Richmond, 2000). They also lived in the same places and habitats as our ancestors of that time, species in the genus *Homo*. They went extinct. Our ancestors survived, and spread throughout the world, not because of bipedality, but because of a larger brain.

Early Genus *Homo*

We start our investigation with a discussion of what is known about the evolution of the earliest members of genus *Homo* from a few millions of years ago to the emergence of modern *Homo sapiens* within the last 100,000 years. The discussion focuses on "machinery from the past" relevant to appetite, food acquisition, activity and energy expenditure, and energy stores. For example, most people probably are aware that brain size increased in the genus *Homo* from 2 million years ago until about 250,000 years ago. Fewer may be aware that body size also showed a substantial increase (McHenry and Coffing, 2000; Wood and Richmond, 2000). We not only have larger brains than our early ancestors, we are a larger species (Figure 2.1). The increase in body size appears to be an early event in our evolution (Table 2.1).

Which came first, larger brains or larger bodies? At the moment the evidence indicates that there were several species of genus *Homo* living in Africa several million years ago that persisted for a considerable length of time. One species, *Homo habilis,* was physically more similar to the australopithecines, being relatively small and having relatively long arms; *H. habilis* did have a larger brain than the contemporary australopithecines. Recent fossil finds have shown that *H. habilis* apparently survived in Africa until approximately 1 million years ago (Spoor et al., 2007). By that time *H. erectus,* the larger species with more modern limb proportions, had spread out of Africa and was colonizing the world. The evidence favors the interpretation that *H. habilis* was not our ancestor.

Our presumed early ancestors, *H. erectus,* were larger animals than

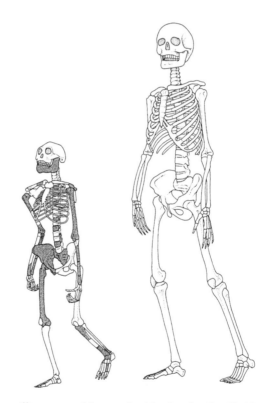

FIGURE 2.1 The 3.2 million-year-old australopithecine fossil called Lucy (*left*) was from a small, totally bipedal creature with a brain approximately the same size as a chimpanzee's. A modern human has a much bigger brain but is also significantly larger. Both skeletons are female. Image courtesy of Milford Wilpoff.

TABLE 2.1 *Mean body mass estimates for adult females and males*

Species	Date range	Body mass females (in kg)	Body mass males (in kg)
Pan troglodytes	Extant	41	49
Australopithecus afarensis	3.9–3.0 Ma	29	45
Australopithecus africanus	3.0–2.4 Ma	30	41
Paranthropus boisei	2.3–1.4 Ma	34	49
Paranthropus robustus	1.9–1.4 Ma	32	40
Homo habilis	1.9–1.6 Ma	32	37
Homo ergaster	1.9–1.7 Ma	52	66
Homo erectus	1.8 Ma–200 Ka	52	66
Homo neanderthalensis	250 Ka–30 Ka	52	70
Homo sapiens	100 Kya–1900	50	65
Homo sapiens	Today (U.S.)	74	86

their australopithecine cousins, with somewhat larger brains. Based on the current fossil record, approximately 1.9 million years ago the species that were likely either our ancestors or at least are representative of that ancestor were taller in stature than australopithecines, had humero-femoral (upper arm to upper leg) ratios more similar to modern humans, and had larger brains than would be predicted for a primate of their body size (McHenry and Coffing, 2000; Wood and Richmond, 2000). Fossil discoveries of relatively small *H. erectus* specimens (e.g., Spoor et al., 2007) have complicated the simple picture; however, our ancestors were, eventually if not from very early on, larger animals than were australopithecines.

Brain size continued to increase, albeit in a nonlinear fashion (Table 2.2). These changes mark the emergence of the genus *Homo* (Wood and Collard, 1999), a transition associated with both morphological changes and a profound change in behavioral ecology. The early members of genus *Homo* had larger brains then australopithecines or extant apes, and this was undoubtedly a key adaptation to their success. However, brain size did not continually increase over time as human brain evolution is sometimes modeled. Although absolute brain size increases from late Pliocene australopithecines to the earliest members of genus *Homo*, measures of brain size relative to body size do not indicate a dramatic change until approximately 500,000 years ago (Table 2.2). A larger absolute brain size can thus be modeled as a part of the overall trend of increased size that characterizes the first members of genus *Homo*, with a selective shift toward dramatically expanded cranial capacity occurring much later in

TABLE 2.2   *Encephalization quotient (EQ) of various extant and fossil species*

Species	Date range	EQ
Pan troglodytes	Extant	2.0
Australopithecus afarensis	3.9–3.0 Mya	2.5
Australopithecus africanus	3.0–2.4 Mya	2.7
Paranthropus boisei	2.3–1.4 Mya	2.7
Paranthropus robustus	1.9–1.4 Mya	3.0
Homo habilis	1.9–1.6 Mya	3.6
Homo ergaster	1.9–1.7 Mya	3.3
Homo erectus	1.8 Mya–200 Kya	3.61
Homo heidelbergensis	700–250 Kya	5.26
Homo neanderthalensis	250–30 Kya	5.5
Homo sapiens	100 Kya–present	5.8

Note: EQ expresses brain size relative to body mass; thus it is a measure of relative brain size.

middle Pleistocene *Homo*. This isn't to say that the earliest members of genus *Homo* did not have larger brains; they did. However, at the beginning, our ancestors were getting larger as well as having somewhat larger brains than would be expected for an ape or australopithecine. Over a long time period brain size remained relatively unchanged until there was a dramatic jump in brain size roughly 500,000 years ago (Table 2.2), as seen in fossils identified as *Homo heidelbergensis* or archaic *Homo sapiens* (Ruff et al., 1997).

Advantages of Larger Body Size

Body size has profound effects on energy requirements, food processing abilities, and the extent of energy that can be stored on the body. The allometry of energy requirements is typically less than one (Kleiber, 1932; Blaxter, 1989; Schmidt-Nielsen, 1994), while the allometry of food processing ability (volume of the digestive tract) and energy stores (adipose tissue and glycogen stores) is typically very close to one (Parra, 1978; Demment and Van Soest, 1985; Schmidt-Nielsen, 1994). Thus, although animal species that are large generally need a higher daily energy intake than smaller species, they can process and store a greater proportion of their daily energy needs. One consequence is that large animals typically can make a living from lower energy density foods than can small animals. Cows, horses, elephants, and rhinos can make a living on difficult-to-digest and low-energy-value foods because they can consume sufficient quantities relative to their requirement. This is the strategy that the gorilla follows.

That isn't to say that all herbivores must be large; voles (often called meadow mice) are mouse-size herbivores. They are quite successful. However, in general they have to be more selective in the plant material they eat compared with elephants. Elephants can survive on plant foods that voles would be physically unable to eat enough of to satisfy energy requirements. The voles would starve to death while full.

But being a large animal doesn't require eating a low energy density diet; tigers are large animals and they are obligate carnivores. Larger animals can go longer without food than can smaller animals because they can store a greater proportion of their energy requirements on their bodies. In other words, "starvation" time increases with body size (Blaxter, 1989; Schmidt-Nielsen, 1994). Thus large animals are more tolerant of variation in food availability. This is seen most dramatically in extremely large animals such as some seals and whales that nurse their offspring while fasting for extended periods (Oftedal, 1993). These mothers are not only capable of surviving off of their body stores for months at a time, but also transfer a significant amount of those body stores to their offspring via milk.

Thus, although increased body size raises total energy requirements, it also brings with it a suite of physiological and behavioral features that allow individuals to meet these new needs. Larger-bodied animals have a larger range of energy flux, the level of energy turnover (Ellison, 2003), in both the positive and negative direction and can thus better accommodate fluctuations in the food supply relative to smaller-bodied animals (Blaxter, 1989; Schmidt-Nielsen, 1994). They can survive on foods that would return too low a rate of energy for a smaller animal. Or they can go longer times between feeding on a high energy density food. And when those high-quality foods are available, being larger enables them to eat proportionately more and to store a greater amount of the excess energy on their bodies.

An advantage to larger body size in genus *Homo* may have been the ability to rely on both low-quality, fallback foods, such as plant underground storage organs (e.g., bulbs, corms, and tubers) (Laden and Wrangham, 2005), and on high-quality but scarce foods (such as animal tissue) when those were available. The ancestral genus *Homo* species of 2 million years ago probably consumed a diet of plant materials (fruits, rhizomes, tubers, nuts, seed pods, tree gums, and so forth) supplemented by occasional animal matter from scavenging and some cooperative hunting. The diet of early *Homo* was likely highly variable, but only a slight increase in the episodic nature of encounters with energy-dense food sources could have provided a significant advantage in food acquisition over australopithecines.

Being large was an advantage in this foraging strategy because it enabled our ancestors not only to subsist on a low-energy plant diet but also to consume large quantities of the rare but valuable animal prey when available, and to store the excess energy via adipose tissue. Thus a surfeit of food could be ingested and the energy used at a later time. The ability to store fat from excess energy intake during sporadic incidences of feasting on a kill, either theirs or some other predator's, was probably a key adaptation that allowed genus *Homo* to move from a grazing- and plant-based foraging strategy to the feast-or-famine strategy of relying on animal prey.

Diet and Adaptation

If you want to understand an organism, the first question to ask is, what is its phylogenetic place in the world? Phylogeny embodies the two fundamental organizing concepts of biology: evolution and genetics. All living things carry the genetic legacy of their ancestors. To understand their present biology you must know who their ancestors were. Arguably, the next most important question to ask after phylogeny is, what does this animal eat? The diet of a species is associated with its morphology, metabolic rate, behavior, social system, and cognitive abilities (Milton, 1988; McNab and Brown, 2002).

Teeth and guts are prime examples; in general, carnivores and herbivores have very different teeth and intestinal tracts. Carnivore teeth are designed for piercing and slicing; herbivore teeth, for crushing and shredding. The masticatory challenges presented by meat are very different from that of leaves. So too are the digestive challenges. Carnivores, in

FIGURE 2.2. Although classified as a carnivore, the giant panda has a diet that consists almost entirely of grass (bamboo). Photo: Jessie Cohen, Smithsonian's National Zoo.

general, have relatively simple intestinal tracts, while herbivores often have large, complex guts. Leaves are harder to digest than meat.

Of course there are always exceptions. The panda eats bamboo (Figure 2.2), a kind of grass, yet has teeth and a simple digestive tract similar to other bears. There are usually many ways to solve dietary problems. In this case phylogeny has constrained the morphology of the panda, but evolution and adaptation have "solved" the problem of being a grass eater in other ways. The panda has a rapid passage rate of digesta through its simple gut (Dierenfeld et al., 1982). This doesn't aid digestion; if anything it reduces the digestibility of the ingested plant material. However, it allows larger quantities of food to be eaten. The panda survives in part by the quantity of food eaten as opposed to the thoroughness of the digestion of that food. This strategy is possible partly because pandas are large animals, and large animals have a greater digestive capacity relative to requirement.

Diet Change in Our Evolutionary History

The diets of our ancestors have changed several times over the course of our evolution. The pregenus *Homo* diet was likely vegetarian, consisting mostly of fruits, flowers, leaves, buds, and other plant tissue. The diet would have had high levels of fiber, tough to chew and to digest. The origins of the genus *Homo* are associated with a change in diet and foraging strategy. The archaeological record suggests an increase in dietary breadth about 2 million years ago, coincident with the transition to *Homo*. The most notable change was an increased consumption of animal tissue (Shipman and Walker, 1989; Milton, 1999a, b; Bunn, 2001; Foley, 2001). This change in diet occurred in association with a stone tool technology. Scavenging was a likely method for accessing animal material, with individuals obtaining meat from carcasses at natural death sites or possibly through direct competition with other predators (Bunn, 1981). Hunting, especially of small animals, was also likely to have been common. Modern chimpanzees, bonobos, and baboons all opportunistically kill and eat animals smaller than themselves (Goodall, 1986; Strum, 1975, 2001; Stanford, 2001). The fossilized remains of small mammals that were consumed, such as monkeys, rodents, or small antelope, likely will be rarer due to the more complete destruction of the smaller bones (Plummer and Stanford, 2000; Pobiner et al., 2007). We probably underestimate the amount of animal prey in early *Homo* diet. Over the course of evolution in genus *Homo*, hunting became progressively more common and efficient. Animal tissue became the dominant food source for many of our ancestors, until the advent of agriculture, when plant foods again became predominant, at least in some cultures.

Certainly a key difference between modern human foragers and our living relatives among extant nonhuman primates is the higher percentage of meat in the diet of humans (between 20 and 50%) and the means by which such meat is acquired (Foley and Lee, 1991; Wrangham et al.,

1999; Bunn, 2001; Foley, 2001). Stiner (1993) describes human predatory behavior as "nearly unique." Relative to nonhuman primates that exhibit predatory behavior (e.g., chimpanzees, bonobos, and baboons), modern humans are more efficient in processing game, and they share a suite of behaviors that are behaviorally similar to members of the order Carnivora, including transport of meat over long distances, caching food, and systematically processing bones for the enclosed soft tissue (Stiner, 2002). Determining where to place fossil hominin species along this spectrum between nonhuman primates and modern humans is a central focus of archaeology and paleoanthropology. Chimpanzee predatory behavior (see chapter 3) is often modeled as the ancestral condition and likely typical of the earliest hominins (Stanford, 2001). There is nothing that chimpanzees do now that australopithecines could not also have done.

Animal carcasses, like fruit, are considered high-quality food items that are scattered on the landscape and can be described as patchy both in space and time. Their nutritional value is tempered by the difficulty and unpredictability in obtaining them. Behavioral strategies that increased both the rate and predictability with which our ancestors were able to obtain animal tissue likely were key factors in our success. In many ways the success of our lineage was due to strategies that increased the nutritional value of the diet.

The diet of early *Homo* probably was not simply an australopithecine diet with more meat, however. It likely included changes in both animal and vegetable components (Leonard and Robertson, 1992). The dietary niche change is best modeled as an improvement in dietary quality, with *Homo* obtaining more calories from plant reproductive and storage parts as well as animal material (Leonard and Robertson, 1992, 1994), and is more accurately characterized as improved omnivory rather than increased carnivory. Increased body size in genus *Homo* may have been an important morphological adaptation in facilitating increased dietary quality through advantages in resource acquisition, energy storage, increased dietary breadth, increased foraging range, and potential increased digestive efficiency (see below).

This new diet was likely lower in fiber and higher in animal tissue and other easily digested material than the diets of australopithecines. However, it was unlikely to have contained large amounts of simple sugars, except seasonally, when certain trees were in fruit, or episodically, if our ancestors occasionally raided beehives for honey. It also was unlikely to contain significant quantities of easily digested starches, such as from processed grains. The glycemic index of this diet was likely quite low. The glycemic index is a measure of the increase in blood glucose concentration following ingestion of a food. Foods containing easily digested and absorbed carbohydrates have a high glycemic index. Difficult to digest carbohydrates, such as fiber, have a lower glycemic index.

It is important to remember where we came from, despite our ability to adjust to radically new circumstances through our technology. We are descended from a long line of frugivore-folivores; our ancestors moved into a novel omnivorous niche. Our nutrient requirements, metabolism, and digestive abilities are heavily influenced by our evolutionary past. The difference between our diet of the past and the diet of today is not just about the increase in energy density of the modern diet; there are many nutrients that are important to our health. Based on a comparison of modern human diets with that of extant nonhuman primates, Milton (1999a) determined that the modern human diet was lower in many micronutrient levels. For example, vitamin C levels of modern human diets are much lower than is typically found in the diet of a wild ape or monkey.

Vitamin C is a key enzyme in metabolism. Vitamin C deficiency leads to significant metabolic pathologies, including the disease scurvy. Scurvy is a classic disease of modern human life; it was likely completely absent from our early ancestors.

Although vitamin C is a required nutrient for all primates, as well as for fruit bats and guinea pigs, most animals do not require a source of vitamin C; rather, they produce this enzyme in their liver. Primates, fruit bats, guinea pigs, and some other vertebrates have lost this ability. The loss was not maladaptive because the natural diet was consistently high in vitamin C. Evolution can't predict need; the scurvy among crews of sailing ships in the 1600s and 1700s was an example of the mismatch paradigm. Humans, via their technological capability, put themselves into an environment where their nutrient needs could not be satisfied with the foods on hand. Of course humans were also able to solve the problem; first through including citrus fruit and juice in the diet of their sailors, and of course now we can give people manufactured vitamin C in a variety of forms.

The Human Digestive Tract

There is nothing intrinsically special about the human digestive tract (Milton, 1987; Milton, 1999b). The stomach is simple and no more capacious than would be expected of an animal our size. The upper intestinal tract (duodenum, jejunum, and ileum) is long and provides considerable absorptive surface area to assimilate fats, amino acids, simple sugars, and minerals, but again, is not out of the ordinary considering our body size. The lower intestinal tract (colon) is sacculated and does provide a significant volume for fermentation of fiber (Milton, 1987). Human beings are capable of digesting hemicellulose and to a lesser extent cellulose via fermentation by symbiotic microorganisms in the colon (Milton and Demment, 1988). Again, this is more a reflection of the fact that we are fairly large animals; there are no obvious specializations other than the colonic sacculations to enhance fermentative capabilities. Our digestive tract fairly well characterizes us as generalized omnivores with some capacity for hindgut fermentation of fiber, implying we are descended from a more herbivorous ancestor.

When compared with that of our closest living relatives, the apes, the human digestive tract has changed in its relative proportions. The human small intestine has a greater volume and the large intestine a smaller volume (Milton and Demment, 1988; Milton, 1999b). The stomach volume is comparable. Overall, the result is that our guts are somewhat smaller than those of chimpanzees after accounting for our larger body size. The more significant difference is the change in gut proportions, with a greater emphasis on the small intestine in humans (Milton and Demment, 1988; Milton, 1999b).

The implication of these relative differences is that apes likely acquire a larger proportion of their energy intake from fermentation of fiber in the hindgut than do humans, while humans may have a higher capacity to absorb fats, simple sugars, and other easily digested food components. Human gut proportions are fairly similar to that of capuchin monkeys (genus *Cebus*; Milton, 1987), moderate-size New World monkeys familiar in days gone by as organ grinder monkeys, and even today used as helper animals for disabled people. Capuchin monkeys feed on a high-quality diet of fruits, high-fat (oil) palm nuts, and animal tissue, both in-vertebrates and small vertebrates (Janson and Terborgh, 1979).

These comparisons imply that for some time human diet has con-tained less fiber, or at least less unprocessed fiber, and a higher proportion of high-quality foods than do modern ape diets (Milton, 1987; Milton, 1999b). It is certainly reasonable to hypothesize that the last common ancestor between chimpanzees and us was more likely to consume a diet similar to modern chimps as compared to human hunter-gatherers or early agriculturists. Thus gut proportions in that common ancestor, and prob-ably for its australopithecine descendants, more likely resembled modern apes than modern humans. At some point in our evolutionary history after the split from the australopithecine lineage, human gut proportions changed. How long ago that change occurred is a matter of debate. It is possible that the shift in gut proportions to favoring larger absorptive surface area in the small intestine at the expense of fermentation volume in the colon was an early adaptation, reflecting the shift to a higher quality diet enabled by the increased brain size that allowed a more opportunistic "high-risk, high-gain" foraging strategy. The greater absorptive surface area of the small intestine would have allowed the rapid assimilation of fats and amino acids from animal tissue and the simple sugars from the breakdown of starch from tubers and other so-called underground stor-age organs of plants (Laden and Wrangham, 2005), especially if they were cooked (Wrangham and Conklin-Brittain, 2003), although cooking is certainly not required for starch digestion (Milton, 1999b). The ability to digest fibrous material would have been reduced, but would not be absent.

On the other hand, the change to a small-intestine-biased gut could have occurred later in our evolutionary history. Indeed, gut proportions in human ancestors may have shifted several times, once during the transi-tion from low-level scavenging to active hunting, and then again during the rise of agriculture, which resulted in a shift back to more plant foods in the diet, albeit with mechanical processing and cooking to reduce the challenge of high-fiber foods.

Indeed cooking has a very significant effect on the digestibility of most foods. When members of genus *Homo* had reliable control of fire and began cooking food, selective pressures to increase the length of the small intestine and decrease the colon probably became stronger. When did cooking arise? Certainly several hundred thousand years ago; there is evidence that perhaps as long as half a million years ago *Homo erectus* had control of fire and cooked food.

Richard Wrangham (2001) suggests that cooking occurred much earlier. Based on his long-term research on chimpanzee ecology and behavior, he proposes that the foods eaten by australopithecines would have been too tough, fibrous, and low energy for early *Homo erectus* to make a living due to the added energetic expense of their larger brain; even adding significant amounts of raw meat to the diet may not have provided sufficient nutrition. He suggests that cooked plant foods were an important staple of the diet of early *Homo;* the increased digestibility of these cooked plant foods allowed the digestive tract to get smaller, freeing up metabolic energy to support the larger brain. The current archeological record does not support this hypothesis, although there is a 1.6-million-year-old site at Koobi Fora in Kenya that might represent the use of fire by early *Homo.* At present the conservative hypothesis is that cooking food arose some time after about 500,000 years ago; coincidently that is also the time for the second major expansion in brain size. Cooking may indeed have been a factor in brain expansion and a decreased gut size, but probably later in time and not at the beginning of the *Homo* lineage.

Gut Kinetics

Anatomical changes to our guts are consistent with an increase in what is called dietary quality, defined as diets with a higher rate of energy extraction. This usually translates into the amount of fiber in the diet; high-fiber diets are called low quality and low-fiber diets are called high quality. This is an oversimplification of the complexity of the diet, gut, metabolism, and behavior interactions, but it can be useful. Our ancestors were successful because they managed to make major dietary strategy shifts, initially from what was probably a high-fiber, plant-based diet to one that progressed to an animal tissue–based diet, and then back to a more plant-based diet with the invention of agriculture, but with cooking and other postingestive processing that increased the digestibility of the plant foods. Our gut morphology undoubtedly shifted with our changing diet.

Interestingly, our gut kinetics does not appear to have changed as

TABLE 2.3 *Comparison of passage rate and digestion of fiber between chimpanzees and humans*

Species/Diet	Mean retention time (in hours)	% of hemicellulose digested	% of cellulose digested
Chimpanzee			
Low-fiber	48.0	76.9	67.5
High-fiber	37.7	62.7	38.4
Human			
Low-fiber	62.4	—	—
High-fiber	41.0	58.0	41.0

Source: Milton and Demment, 1988.

much as our gut proportions. Gut kinetics refers to the rate of passage through the intestinal tract. Despite our having differently proportioned guts from chimpanzees, gut kinetics is remarkably similar between our two species (Milton and Demment, 1988). Our gut kinetics is geared to a slow turnover of gut contents, with relatively long mean retention times (Milton and Demment, 1988; Table 2.3). This digestive strategy is conducive to enhanced digestion and nutrient extraction from food; the long retention time would have been especially beneficial for fiber digestion, in part compensating for the smaller colon volume. However, it does reduce the total amount of food that can be eaten in a given time.

As in the chimpanzee, mean retention time for humans is longest with low-fiber diets and shortest with high-fiber diets. This again is consistent with our being a generalized omnivore with hindgut fermentation capability. We have a respectable ability to digest fiber, especially soluble fibers (such as pectins and gums) and hemicellulose (see Table 2.3). We have some digestive flexibility, with the ability to respond to high-fiber diets by increasing passage rate of digesta. This won't improve digestion of the diet; indeed it will likely lower it. However, it will allow a larger quantity of food to be consumed, by emptying the gut more quickly. This is analogous to the panda strategy for eating bamboo; rapid passage rate with lower digestion but greater total intake. However, we retain food in the digestive tract longer than do pandas, even at high dietary fiber levels (Milton, 1999b).

Why haven't our gut kinetics changed in association with the changes in diet and gut proportions? The evidence from both pandas and humans suggests that gut kinetics may be a commonly conserved attribute (Milton, 1999b). The panda is a carnivore that has become herbivorous. It is reasonable to suppose that pandas might benefit from a slower passage of digesta through the gut to increase digestion time, but instead pandas have retained a rapid rate of passage similar to other bears. Humans have greatly increased the dietary quality of their foods; it isn't clear why a slow rate of passage is still adaptive. We may be stuck with an ancestral pattern of basic gut anatomy and kinetics, though with a shift in gut proportions that would favor digestion of high-quality (low-fiber) foods. On the other hand digestion of fiber probably was an important part of our digestive strategy for most of our evolutionary history.

With regard to the implications of the modern human digestive tract for human diets and susceptibility to weight gain, exactly how long ago these changes in gut proportions occurred is actually not very important (though of great intrinsic interest). Human food technology has progressed at such a rapid rate that modern human diets for most of the world bear little resemblance to the diet to which our digestive tract is adapted. Not completely, of course. We are still generalized omnivores. But the nutrient density, especially caloric density, and digestibility of today's foods is significantly higher than the foods that all but our most recent ancestors ate. In terms of plant foods, although we may have no great ability to digest cellulose, we can certainly very efficiently digest starch.

Digestion of Starch

Indeed, humans appear to be specifically adapted to digest starch, and it starts in our mouths. The oral cavity is the beginning of the alimentary canal, and food processing and digestion begin with chewing and mixing the chewed food with saliva. Amylase is an enzyme that digests starch molecules into simpler sugars. In many animals, including humans, amylase is secreted into saliva, and thus starch digestion begins even before food is swallowed.

There is considerable variation in the amount of amylase secreted into saliva among humans. Interestingly, the number of copies of the amylase gene is also highly variable among people, and the amount of amylase secreted into saliva is strongly correlated with amylase gene copy number

(Perry et al., 2007). This appears to be a trait of our lineage; chimpanzees have only a single diploid copy of the amylase gene, one from each parent. In people the number of amylase genes ranges from 2 to 15, with a person inheriting variable numbers from each parent. For example, one individual had 14 copies of the amylase gene: 10 from one parent and 4 from the other (Perry et al., 2007).

Although food isn't usually retained within the mouth long enough for a large amount of starch digestion to occur, the amylase in the swallowed food continues to be active until it is neutralized by the acid in the stomach. This can result in up to 50% of starch being digested before the food reaches the small intestine. Amylase is also secreted into the small intestine from the pancreas, where most of the rest of the starch in the food is digested. It isn't known if amylase gene number affects amylase secretion in the small intestine in humans, but it is a reasonable hypothesis. This would imply another source of variation in the ability to digest starch among humans, but also implies that starch digestion in humans likely is even more efficient than in chimpanzees.

Humans have become starch digestion specialists. These data also imply variation among people in the amount of starch that reaches the colon for any given starch load; perhaps this is one explanation for variation in the digestive response to starchy foods. Gas, which causes flatulence, is produced by the fermentation of starch (and other polysaccharides such as fiber and oligosaccharides) by symbiotic microflora in the colon; the more efficient starch digestion in the mouth and small intestine, presumably the less flatulence.

When did this change in amylase gene number occur in our lineage? Some authors (e.g., Coursey, 1973; Laden and Wrangham, 2005) have proposed that underground plant parts (e.g., bulbs, corms, and tubers) that serve an energy storage function for the plant were an important food source for early *Homo*. Starch is the plant storage form of glucose (the animal form is glycogen), and these plant parts would be very starchy. An ancient date for gene amylase gene duplications would be evidence in favor of this theory. Based on the divergence in gene sequences among the identified duplicated genes to date the estimate for the divergence at present is on the order of a few hundreds of thousands of years, not one or two million years (Perry et al., 2007). Thus the increased numbers of amylase genes in genus *Homo* does not appear to date back to the earliest members, but at some point in our evolutionary history a series of amylase gene duplications occurred in the *Homo* lineage. This presumably provided an advantage due to an enhanced ability to digest starch; what the starchy foods were is not known, but bulbs, corms, and tubers are candidates, as well as wild grains. The duplication of the amylase gene provided a preadaptation that became further selectively advantageous upon the advent of agriculture, and indeed may have been important for the invention of agriculture. The ability to utilize wild grains more efficiently would have given motivation and selective advantage to behaviors that increased access to grains.

Our Digestive Machinery and Our Current Diet

The staple foods of the modern world provide little digestive difficulties for humans. Because today's foods are typically calorie-dense and readily digestible, our digestive tract is fully capable of processing far more food in a day than we typically need to satisfy energy requirements. This may have not been true of the foods of our distant past. Both higher energy expenditure and at least some portion of the diet consisting of lower quality, high-fiber, difficult-to-digest foods would have meant that the abilities of our ancestors' guts were more in line with their needs. In terms of digestive capability modern humans have significant excess capacity.

We are efficient at starch digestion, and we have a liking for starch. Many of our processed foods are starch based. We have also been able to select for and manufacture starches that are highly digestible. In general we are producing and favoring foods with high glycemic indices. This would be efficient and adaptive if calories were limited; it becomes problematic when calories are abundant.

High glycemic foods favor fat deposition. Partly this is due to defense of homeostasis. By definition, high glycemic foods result in a rapid increase in blood glucose. Insulin is released from pancreatic β cells to increase absorption of glucose into cells where it is oxidized to be used in cellular metabolism or converted into an energy-storage molecule (glycogen or fat). This serves obvious adaptive purposes and also serves to reduce the circulating glucose levels. Glucose is toxic at high enough concentration. One means by which glucose is removed from circulation is to up-regulate lipogenesis (fat production). The energy in glucose is used to produce fat, which is then stored. In a world of episodic food abundance and scarcity, that conversion provides adaptive function; energy from times of abundance can be used during times of scarcity. However, the modern world presents many, if not most, humans with an environment of perpetual food abundance, and the foods are often high glycemic foods. The glycemic response is met with an insulin response, and metabolism becomes shifted to favor energy storage. We store fat more often than we mobilize it, and adipose tissue accumulates.

High glycemic foods also may encourage "grazing," or eating between meals. The rapid and robust insulin response to the rapid rise in blood glucose due to high glycemic foods will quickly reverse the hyperglycemia. Low glycemic index foods result in a shallower rise in both blood glucose and insulin, but both might remain elevated for a longer period of time for an equivalent number of calories. Low glycemic foods are more satiating (Ludwig, 2000; Brand-Miller et al., 2002). A high glycemic index food produces both early hyperglycemia and later hypoglycemia due to the insulin response (Brand-Miller et al., 2002). Appetite is more likely to return quickly after eating a high glycemic index food. The old joke that Chinese food fills you up but you are soon hungry again may relate to the high glycemic index of white rice. Restricting diet to low glycemic index foods has been shown to be successful at promoting weight loss in people, even if total food intake is not restricted (Thomas et al., 2007).

Expensive Tissue Hypothesis

Brain is metabolically "expensive." Not necessarily more so than other organs—liver is also metabolically expensive. But brain accounts for a greater proportion of basal metabolic rate (BMR, the minimal energy expenditure necessary for life) than it does of body mass. The increase in brain size seen in our early ancestors may very well have necessitated an increase in metabolic rate beyond that predicted due to the larger body size (Aiello and Wheeler, 1995). Whether that increase in metabolic rate due to a larger brain also necessitated an increase in total energy expenditure is uncertain—many parameters determine total energy expenditure besides metabolic rate—however it is certainly a respectable hypothesis.

Was the expensive enlarged brain a "cost" to our early ancestors? The answer to that question requires more than a physiological explanation. Metabolic rate certainly is a major portion of an animal's energy budget, and one that is a fixed cost. Total energy expenditure of free-living animals is usually between two and three times resting metabolic rate. During periods of extended energy expenditure (e.g., lactation) that value can go much higher. Lactating mice have been shown to expend energy at rates of seven times BMR or higher (Johnson et al., 2001 a, b, c).

In general, animals with higher metabolic rates expend more total energy. However, other parameters such as activity and thermoregulation have significant effects on total energy expenditure. And perhaps even more important, energy expended is not as important as the balance between energy expended and energy acquired (i.e., food eaten). The key issue is whether any increased energetic cost of a larger brain could be supported by the new foraging strategy it enabled. If it did then the increased energy expenditure was likely moot, at least until a variant of genus *Homo* arose that could do the same things but with less energy expenditure.

Did that happen? Well, the metabolic rate of modern humans is not different from that expected for primates, despite our large brains. We show no signs of this proposed required extra energy expenditure to support a larger brain. Aiello and Wheeler (1995) have proposed that the metabolic increase due to brain has been matched by a metabolic decrease in gut. They argue that the increase in brain size was supported by (and allowed) an increase in food quality in the diet. Their definition of food quality may be expressed as foods that provide a higher rate of energy assimilation and also require less digestive processing. The theory posits that our guts have decreased in size as our brain increased in size. The increased brain size of our ancestors allowed them to gather more easily digestible and energy-dense foods, and to employ external food processing methods (e.g., cooking). Both of these factors allowed for a reduction in our overall gut size, and hence metabolic cost associated with our intestines.

When did this happen? Some scientists have proposed that this shift in energy expenditure from gut to brain occurred early in our divergence from the australopithecines (Martin 1981; Aiello and Wheeler, 1995). But it can also be argued that the energy inefficiency of a large brain was more than matched by the advantages it conveyed over the australopithecine lifestyle. Later in time, however, our ancestors were no longer competing with australopithecines, but rather with other species within genus *Homo*. At that time the ability to do the same with less would have provided a competitive advantage; perhaps it was at this point gut size became reduced in our ancestors, and hence balanced the increased metabolic needs of their large brain.

Human beings have a preference for sweet foods; there is variation among people in sweet preference, and this variation is partly heritable (Keskitalo et al., 2007). People also generally have a liking for fatty foods, though this also is variable. In general, human beings could be characterized as preferring and being motivated to eat foods with a high-caloric density. Our digestive system, on the other hand, is still well suited to processing low-energy density foods, such as those at least moderately high in fiber. We are a mosaic that reflects our evolution.

Summary

Two trends regarding size changes appear to characterize our evolutionary history. Our brains got larger. That is well known. Body size also increased, and that is perhaps less appreciated. A larger body size has both costs and benefits. Absolute energy requirement is likely to increase, but a larger body size also potentially enhances digestion and the ability to store energy relative to requirement. Larger animals can eat more and go longer between eating. Both of these characteristics may have been beneficial to a switch by our early ancestors to a dietary strategy that used rare but high-quality foods such as animal tissue. This may have been the start of our enhanced ability to store energy (fat) on our bodies.

Biological anthropologists (e.g., Martin 1981, 1996; Aiello and Wheeler, 1995) have suggested that the larger brain of our early ancestors also came with an obvious metabolic cost; it took more energy to both grow and maintain. This would encourage a feedback loop in which motivation for acquiring energy-dense foods, possibly especially ones with significant fat, both enabled and was driven by the metabolic needs of a larger brain.

In any case, our digestive tract and metabolism are well suited to ingesting large quantities of low glycemic index foods. We have the physiological flexibility to handle energy-dense, high glycemic index foods, but they were likely rarer in our past. They were a chance to stock up, so to speak, and our metabolism is geared to shift to an energy-storing bias when we feed on high glycemic foods, high-fat foods, or simply large quantities of food. The modern food environment is likely to trigger our energy-storing adaptations.

THREE

The Evolution of Meals

Food and eating are central facets of animals' lives. The search for and ingestion of food occupies a considerable number of the waking moments in most mammals' lives. This is especially true for primates, the mammalian order to which we belong. Primate species in the wild often spend a quarter or more of their awake time in food-related activities (e.g., Janson and Terborgh, 1979; Terborgh, 1983). A diet change in our early ancestors to higher energy density foods was probably key to their success and our eventual existence, but we also changed *the way* we eat as well as *what* we eat. We became less grazers and more meal eaters.

Humans, Food, and Eating

Humans are primates. More specifically, our closest living relatives are the great apes: chimpanzees, gorillas, and orangutans. We share many features of our biology with these apes. Among the great apes, the two chimpanzee species (*Pan troglodytes* and *Pan paniscus*, the common chimpanzee and the bonobo, respectively) are the most closely related to us. All the great apes are frugivore-folivores; they largely eat fruits and leaves. Chimpanzees and bonobos eat meat to a certain extent, but in terms of gut morphology all the great apes resemble hindgut fermenting herbivores, with a simple but capacious stomach, a moderate-size small intestine, and a long and capacious large intestine with a functional cecum.

We certainly evolved from a frugivore-folivore ancestor, but human beings are omnivores. We typically include a significant amount of animal flesh in our diet. Human beings also process much of their food outside of their bodies; grinding, fermenting, and especially cooking foods and thus greatly reducing the masticatory and digestive challenges we face. It is not surprising that our teeth and guts are smaller than those of our ape cousins.

Our eating behavior is different in another fundamental way: People eat meals. Food is brought to a particular place at a particular time and then consumed, usually with other people. And not only do we eat with other people, but it is usually a cooperative event. People pass each other food; people don't steal off each other's plates. We even have "potlucks" where people all bring a food dish to share with everyone else. Eating is social as well as nutritional (Figure 3.1).

Our ancestors also ate meals, probably as far back as several million years ago. There is considerable archeological evidence supporting the idea that members of genus *Homo* regularly ate together. There are ancient hearths, dating back hundreds of thousands of years, which were used for cooking food. These sites were occupied and used by our ancestors or related species for hundreds of years (Jones, 2007). There are assemblages of stone tools and bones with cut marks consistent with butchering dating back at least 2 million years. The number of tools and the quantity of bones are consistent with many individuals acting together.

The basic concept of a meal probably started very early in our evolution, has evolved itself, and has affected our evolution. Meals have shaped our evolutionary history, both as an adaptation and as a selective pressure that we suggest was a key aspect of the adaptive advantages of our larger, more complex brain.

Meals are a fundamental aspect of human feeding biology. However, meals, in the broadest sense, are certainly not unique to humans. Social carnivores can be considered to be meal-feeders, often hunting together and then sharing the kill, but most other primates do not feed in meals. Primates generally gather their own food and eat it as they gather it. This is actually a hallmark of herbivores, both folivores (leaf-eaters) and frugivores (fruit-eaters). Food is eaten as it is obtained, and it is rarely shared. Meal-eating is associated with prey.

In this chapter we examine the possible evolutionary origin and adaptive value of the meal. We explore the significance of meals for our patterns of eating and the reasons why and when we eat. This book is about

FIGURE 3.1. A meal can be about much more than just nutrition. *Luncheon of the Boating Party* by Renoir.

the biology of obesity, so our focus will be on how meals might encourage the overconsumption of food. There are, of course, many other fascinating aspects to the biology, sociology, and politics of meals. To those readers who become sufficiently intrigued with the complexities of meals we recommend the writings of Claude Levi-Straus, Mary Douglas, and Marvin Harris among early social anthropologists.

Technology has greatly increased our ability to learn about the eating patterns of prehistorical peoples. Food archeologists are now using sophisticated molecular techniques to determine what and how people ate long ago. An accessible overview of meals and human feeding behavior from an archeologist's perspective is *Feast: Why Humans Share Food* by Martin Jones (2007).

What Is a Meal?

A meal can be defined as "the food served and eaten in one sitting or a customary time or occasion of eating food." But the concept of "meal" has more meaning than that; a meal is often, even usually, a social occasion. Most of the time people eat meals with other people, not by themselves. People that eat a meal together have a social connection; they could be family, or coworkers, or fellow celebrants at a special event, but there is usually some additional meaning attached to a meal besides simply consuming food. When people eat a meal together they share membership in a group. It could be a casual, tenuous group relationship, such as people meeting for the first time at a conference. It could represent profound, important personal relationships, such as a prewedding dinner where two families are symbolically joined. It could represent romance, politics, or business. And sometimes it is indeed just a time to eat.

We are not trying to mythologize the concept of meals. Meals and eating have nutritional consequences. If eating in meals had not been a successful adaptation to satisfying nutrient requirements for our distant ancestors we might not be here; however, meals are not just about nutrition. Consider the powerful statement that a person makes when they refuse to eat a meal with certain other people, or the social and political strategy of convincing rivals, enemies to sit down together and share a meal. Meals have social significance, and this was probably true from our early evolutionary history. For humans, the act of eating has acquired substantial social-political-sexual significance in addition to its core nutritional function.

Chimpanzees, Meat-Eating, and Meals

Chimpanzees are primarily frugivores; ripe fruits provide the majority of their nutrition in most instances (Goodall, 1986; Stanford, 2001; Gilby and Wrangham, 2007). They usually forage alone, or in mother-offspring dyads. Chimps do not generally share food resources, except within the mother-offspring relationship. The most prominent exception to this is among males when hunting. Food-sharing is almost always meat-sharing, and thus associated with hunting and killing (Teleki 1973; Goodall, 1986; Mitani and Watts, 1999; Stanford, 2001; Gilby, 2006).

Wild chimpanzees include a significant amount of animal tissue in their diets (Teleki, 1973; Goodall, 1986; Stanford et al., 1994; Mitani and Watts, 1999; Watts and Mitani, 2002). They even make and use tools in this endeavor. For example, they use small trimmed sticks and grass blades to fish for termites. But they also hunt and kill mammals smaller than themselves. At several different research sites, chimpanzees are known to have killed hundreds of small- to medium-size mammals (e.g., colobus monkeys, wild pigs, and small antelope) each year (Stanford, 2001). A female chimpanzee has been observed using a sharpened two-foot-long stick, which she poked vigorously into tree holes (Pruetz and Bertolani, 2007), at least once spearing, killing, and then eating a bush baby (*Galago senegalensis*).

Meat still represents only a small portion of the diet; chimpanzees rely on fruit for most of their nutrients. Hunting appears to be somewhat seasonal, occurring more often in the dry season at Gombe, a time of low fruit availability (Stanford, 2001). However, after accounting for other factors such as male group size, which also varies seasonally, energy availability was not associated with hunting frequency (Gilby et al., 2006). Interestingly, hunting is more common during times of relative food abundance at some sites, though this result may be at least partially explained by larger foraging group size during these times (Gilby et al., 2006). However, the Kanyawara chimpanzees of Kibale National Forest appear to engage in hunting during times of fruit abundance (Gilby et al., 2007). Hunting has greater risks associated with it than does foraging for fruit, for example, greater chances of injury and of simply failing to obtain any food. Prey generally tries not to become nutrition. The Kanyawara chimpanzees appear to adopt a more cautious foraging strategy during lean times and a more high-risk, high-gain strategy when there is plenty of fruit available (Gilby et al., 2007).

Meat-eating certainly provides important nutrients; indeed, energy may not be the most important nutritional item that chimpanzees obtain from meat. Protein of course, as well as calcium from bone may be significant. The solitary hunting observed in females and young chimpanzees most certainly is a nutritional act. They are hunting to obtain food, and there does not appear to be any other motivation. But not all hunting and meat-eating appear be a solely nutrition-driven activity in chimpanzees.

Cooperative hunting, mainly done by adult and adolescent males, appears to have social significance in addition to its nutritional significance. Indeed, although hunting success increases with the number of chimpanzees participating in the hunt, the expected reward per individual does not increase, and might even decrease (Stanford, 2001; Gilby et al., 2006). Thus although each additional chimp that joins a group that is hunting colobus monkeys may increase the likelihood the hunt will be successful, that does not necessarily translate into an increased nutritional benefit for every individual chimpanzee participating in the hunt. This fact has been used to suggest that hunting is not strictly cooperative, in the social carnivore meaning of the word. Yet there appears to be substantial motivation among the adult and adolescent males to participate in hunts (Stanford, 2001).

Meat-sharing after the kill has social and political meaning and consequences. The "owner" of the carcass largely determines who receives shares, and how much. Allies are rewarded and rivals snubbed. Sexually receptive females may be allowed a share, and of course copulated with; females with which the owner has a sexual or social history (e.g., his mother) may do well, while most other females get nothing. The owner will likely be harassed to a certain extent; some meat-sharing may be in effect extortion. Or to use an ethologically more descriptive term, the "sharing-under-pressure" hypothesis was supported by the data (Gilby, 2006). The consumption rate of the possessor of the carcass decreased with the number of animals begging or harassing him for a share; beggars were much more likely to leave the vicinity of the carcass (and thus cease harassing the owner) after receiving some meat (Gilby, 2006). Positive associations between individuals often resulted in increased pressure to share; for example, females that were common grooming partners of a male were more successful at obtaining a share of the meat at least partly due to the fact that such females also were more persistent in their begging/harassing (Gilby, 2006).

Another aspect of chimpanzee meat-eating that may differ from that of our ancestors is that it can be a long, drawn-out process. A single colobus monkey carcass may take many hours to be eaten. Perhaps this eating behavior should be labeled an event rather than a meal; by human standards it seems inefficient. The hunting may have been cooperative; the food-sharing afterward is transactional rather than cooperative.

Meals and Brains

There are costs associated with having a large brain. Of course that large brain also gave tremendous adaptive advantages to our ancestors. The genus *Homo* survived and prospered while the australopithecines went extinct. Regardless of its energetic expense, the larger brain of genus *Homo* was a successful adaptation and, by definition, provided more fitness benefits than fitness costs. To understand our evolution we need to understand the adaptive functions of our enhanced brain, from its very beginning.

Tool use was certainly an important aspect to those adaptive advantages. Other animals use tools, but none make and use the variety and complexity of tools that are associated with even early genus *Homo*. Many of the known tools of early humans are associated directly with food, for example, stone tools that were used to cut meat off bone or smash open bones to get at the marrow. Other early suggested tools are sticks for digging up roots and tubers or digging into termite mounds.

Much of the speculation regarding tool use by our early ancestors revolves around food. In fact much of the documented tool use among all animals involves obtaining food. Chimpanzees "fish" for termites using twigs (Goodall, 1986); *Cebus* monkeys open hard-shelled fruits and nuts using rocks (Waga et al., 2006); sea otters use stones to open shellfish (Hall and Schaller, 1964; Figure 3.2); woodpecker finches use cactus spines to extract grubs from wood (Millikan and Bowman, 1967; Tebbich et al., 2002); many animals use tools to obtain food. The behavior isn't unique to us and our ancestors, but tools and food have certainly played a large role in our evolution and our evolutionary success.

The concept of a tool can be quite broad. Ben Beck (1980) defines tool use as "the external employment of an unattached environmental object to alter more efficiently the form, position, or condition of another object, another organism, or the user itself when the user holds or carries the tool during or just prior to use and is responsible for the proper and

FIGURE 3.2. Sea otters crack open crabs and shellfish, such as abalone, by placing a rock on their chest and smashing the shell against it. Photo: © Jane Vargas, 2005.

effective orientation of the tool." The length and complexity of this definition hints at the difficulty of clearly defining what a tool is. This definition requires that a tool be an object that is manipulated physically by the user. This is a useful definition to examine tool use among many different species. After all, most animals modify their environment in some way, but are all such modifications indications of tool use? Is a nest a tool? When an animal rolls in mud to cool off, is the mud a tool?

Scientists who study tool use focus on subtle and sophisticated distinctions. For example, Egyptian vultures crack ostrich eggs by picking up stones in their beak and throwing them at the eggs (Thouless et al., 1989). This is universally agreed to be tool use. However, other birds are known to carry food objects (eggs, shellfish) into the air and drop them on hard objects such as rocks or concrete. This strategy also breaks open the food item allowing the bird to feed, but is this behavior tool use? By many definitions, including Beck's above, the answer is no. There is a subtle and sophisticated distinction between manipulating the object of interest and manipulating another object to affect the object of interest. On the other hand, by the broadest of definitions of tool use, dropping a clam on a concrete barrier until it breaks open and can be eaten can certainly be argued to be a tool-using strategy.

And what about our own ideas of what tools are, restricted to human endeavors. This book was written on a computer using word processing software; those are tools. Does a tool have to be a physical object? We use algorithms, mnemonic devices, and templates to help us solve problems in our work and everyday lives. Are these mental devices "tools"? We refer to them as such frequently. When does a tool grade into becoming a strategy? In the context of this chapter, can a meal be a tool? State dinners are referred to as tools of diplomacy; is that real or semantics?

As fascinating as this topic can be, for our purposes in this book we don't need to answer these questions. Early members of genus *Homo* certainly were tool users, by any definition. They also employed mental and social strategies, just as all animals do to some extent. Both were involved in their feeding strategies; both were necessary for meals to evolve.

Certainly tool-making and tool use were important selective pressures that affected the evolution of our brain, but social and behavioral capabilities enhanced by the larger brain were likely at least as important to the success of the early genus *Homo* as was their ability to change the physical environment. Most primates are highly social animals. Many of their strategies to solve their adaptive challenges are social and behavioral. Humans are no exception. Cooperation has always been a key adaptive strategy for us and our ancestors. Under the broadest definition of tool use it is quite reasonable to propose that the most effective tools humans have are other humans. Power and wealth are usually associated with being able to convince large numbers of other people to work toward your goals. Throughout history this has been the means by which great, as well as terrible, things have been accomplished. The ability to coordinate the actions of many individuals toward a common purpose was a key cognitive ability that allowed humans to spread throughout the world.

The concept of gathering food and bringing it to a communal place where it is shared among the other members of the social network was probably a key event in our evolution. Meals, or at least the earliest rudiments of behavior that formed a cooperative feeding strategy, may have been one of the first behavioral adaptations that separated our lineage from the behavioral and mental adaptive plane of our primate ancestors.

Even in the beginning, meals probably served multiple functions. Nutrition may have been paramount, and certainly solving nutritional needs was necessary. Bringing food to a central place to process, consume, and share presented challenges but also benefits beyond nutrition. The possibilities include defense against predators, scavengers, and competitors. Of course bringing food to a central location might attract these as well. The changes in behavior that are required for meals to become the dominant means by which our ancestors obtained their nutrition also had myriad subtle and complex ramifications for their ecology, antipredator strategies, social structure, reproductive strategies, and so on.

For this strategy to have been successful (and it was) changes in food-related social behavior were likely required. Food-related aggression needed to be reduced; cooperative and sharing behaviors increased. The ability to anticipate and predict the actions of others would have become even more important than it already was. The concept of delayed gratification would become a functional and adaptive strategy. When individuals obtained food they had to consider what would be the more successful strategy: to consume it immediately or to bring it back to the group and share (or options in between)?

The concept of a meal contains many complex and sophisticated parameters that would select for greater intelligence. It requires planning in both space and time. It requires cost/benefit calculations; is it better to eat what I have found or to bring it to the group to share? It is fairly easy to see how this change in feeding behavior opens up considerable avenues of choices that must be made. Choices that depend on sophisticated evaluation of circumstances. Should I eat what I have found and continue foraging, or return to "camp"? How much food have I gathered? What type of food is it? How far from "camp" is it? How long have I been foraging? What other members of my group are out foraging? When are they likely to return to camp? Do I expect them to be more or less successful than I have been? Will they share with me? What do I gain by sharing with them?

Primates have larger brains for their body size than do most other animals. Among nonhuman primates the apes have the largest brain-to-body weight ratio. From a key point in time our ancestors had larger brains than the common ancestor of us and the apes. Brain size increased over our evolution until relatively recently. But not all areas of the brain dramatically increased. The increase was predominately in the cortex.

The social complexity theory of brain evolution posits that species with complex social networks are under selective pressure for increased cognitive ability (Dunbar, 1998). A complex social network doesn't necessarily translate to large social groups or vice versa. Wildebeests live in giant herds, but their social network isn't correspondingly complex. Monkeys and apes, on the other hand, have complex social networks, whether they live in large or small groups (Dunbar, 1998). The social complexity and the increased reliance on behavioral and social strategies to enhance fitness is hypothesized to underlie the larger brains of primates. Even within primates, species that form coalitions in social interactions have more neocortex than do those species that do not. We propose that the additional social complexity of the new feeding strategy of early *Homo*, including at least the rudiments of cooperative, meal-like behavior would have been strong additional selective pressure for greater cortex. Certainly at some point in our evolution the ability to effectively utilize a meal strategy, with all of the cooperative, social, political, and even sexual possibilities inherent in the modern concept, would have been one of many selective pressures enhancing our social cognitive capabilities.

Cooperation and Tolerance

When you carefully consider the matter, eating a meal, in the full, social, modern human sense of the expression, requires a great deal of tolerance. A number of people will all join together to eat, and they will refrain from threatening each other, stealing from each other, denying any individual food, and refrain from overt sexual behavior (ignoring the notion of orgy as practiced in ancient Rome and other places). The individuals involved display a remarkable amount of restraint, compared with other species put into a social feeding context.

We take this for granted, but consider our ancestors of many hundreds of thousands of years ago, perhaps even millions. Food was a valuable, constraining resource. The reflex action when seeing another individual eating a desired food would be to threaten, beg or barter, depending on relative dominance rank and physical attributes. And these actions probably happened, but cooperation and tolerance were also displayed. Reflex behaviors were inhibited.

This is one of the functions of brain cortex; inhibiting and modulating reflexive and emotional reactions. Brain size increased in genus *Homo* over time, but not all parts of the brain expanded. Mainly the increase was in cortex. Brain stem, the circumventricular organs, and the regions of the forebrain often called the limbic system (e.g., hypothalamus, amygdala) are not dramatically different between humans and apes; cortex is.

We propose that the increase in cortex both enabled and was selected for by the behavior of cooperative eating: meals. The corticalization of brain function leads to less reactive behavior, more proactive behavior, and restraint. Consequences are assessed, and delayed gratification and future good will are taken into account. Increased cortex would be both favored and required for the meal feeding strategy to be successfully adopted. Communal eating and food-sharing (meals) were certainly not the only selective pressures that led to the increase in cortex, but we propose that this new feeding behavior had a significant effect on our brain evolution.

Chimps and Bonobos

Among living nonhuman primates the two *Pan* species, the chimpanzee and the bonobo, are our closest living relatives. For example, based on an analysis of monkey, ape, and human Y chromosome breakpoints, a small DNA fragment has been found that has been transposed from chromosome 1 to the Y chromosome, but this transposition is only found in the human, chimpanzee, and bonobo Y chromosomes (Wimmer et al., 2002). We last shared a common ancestor with these apes 4 to 7 million years ago (Glazko and Nei, 2003). Chimps and bonobos are certainly more closely related to each other than either is to us; they diverged from each other a little less than 1 million years ago (Won and Hey, 2005). Thus their evolutionary separation from each other is roughly equivalent to the separation of australopithecines from very early *Homo erectus*.

These two living species make useful models for exploring the ancestral behaviors and the possible extent of behavioral divergence among ancient hominins. Commonalities between us and the two *Pan* species are likely ancestral; the differences between chimps and bonobos are a measure of how different an australopithecine could have been from an early *Homo* even without the difference in brain size.

The differences in social behavior, social structure, and temperament between chimps and bonobos are fascinating. Despite their evolutionary closeness, physical similarities, and even similarities in basic social organization these two species appear to have profound differences in fundamental aspects of their lives, and these differences affect their cooperative abilities.

At the risk of gross oversimplification, chimps make war; bonobos make love. These two facets of these species' biology may be the best known by the public. Of course reality is more complex and equivocal than the dichotomous popular image. Bonobos can be violent; they hunt and kill other mammals (Hohmann and Fruth, 1993) and will physically attack one another (Hohmann and Fruth, 2003; White and Wood, 2007). Chimps are fairly prodigious in their sexual behavior as well. Although the differences between chimps and bonobos may be less than the popular notions, and more in degree than in actual kind, there is a foundation for the above generalization. Male chimps from a community will band together and patrol their territory. Neighboring males they encounter are in danger of being attacked and even killed (Wrangham and Peterson, 1996; Mitani, 2006). Even within the group, chimps commonly use threats, intimidating displays, and physical aggression to gain advantage. Bonobos, in contrast, are known for employing sex in almost all social interactions (de Waal and Lanting, 1997). This is not to say that bonobo life is without conflict. Indeed, conflict often occurs. The key aspect of bonobo behavior is that the resolution of the conflict and reconciliation among the individuals involved generally involves sexual behavior (de Waal and Lanting, 1997).

The relationships between males and females are different as well. The species are similar in that both differ in dispersal pattern from the primate norm. In most primate species females stay in the natal group and form lasting bonds with their mothers and sisters. Males disperse, and must find a way into another group. In both chimps and bonobos, males stay in the community in which they were born. Females leave the community as young adults and must integrate themselves into another community. This has been documented by long-term observations of chimpanzees (e.g., Goodall, 1986) and by fecal DNA analysis of wild bonobos (Gerloff et al., 1999). In a wild bonobo group most adult and subadult males could be DNA-matched to a resident adult female that was likely his mother; this was generally not true of adult and subadult females (Gerloff et al., 1999).

Despite having the same dispersal pattern, male and female social behavior differs between chimps and bonobos. Male chimp behavior fits the predicted behavior based on dispersal pattern. Male chimps form alliances and cooperate with each other. They have known each other for life and are often related to each other. They compete with each other as well, but the amount of cooperation is significant. Female chimps have bonds with males, especially their sons, but adult females generally do not form alliances. Female chimps are generally by themselves or with their offspring. In contrast, and against expectation, in bonobos it is the reverse. Females form alliances, even though they may have known each other for only a short time and are very likely unrelated, while males tend to act independently, even though they may have known each other for all their lives and might even be brothers. Bonobo social organization has been termed female dominated due in large part to these female coalitions, though reality is, as usual, more complex (White and Wood, 2007). Individual males, being larger and stronger, will be able to displace any individual female from food or another desired object. However, woe to him if her friends are around! Of course after chasing him off and putting him in his place they may give him sex. However, in general it appears to be a better strategy for males to defer to females in many circumstances (White and Wood, 2007).

The above is a simplistic, broad-brush description of the differences between these species. Much of our knowledge of bonobo social behavior comes from captive animals. Data from wild animals indicate that bonobos can be more violent and less sexual than the captive portrait. The popular image of the bonobo as a sexy, peace-loving hippie may refer to a mythical animal. But the difference in temperament between chimps and bonobos is real, if possibly exaggerated, and leads to measurable differences in cooperative behavior.

Cooperation and Fairness

Both bonobos and chimps readily learn cooperative tasks in captive settings. These are very smart animals that will readily use tools and other strategies to gain rewards. A simple cooperative task that both species have been trained to accomplish requires two individuals to pull on poles or ropes in order to bring a food bowl within reach (e.g., Melis et al., 2006). Both species are perfectly competent at this task; they differ in the outcomes, however. If two chimps are presented with the task, the frequency of cooperative behavior is lower. A good portion of the time one of the chimps will refuse. This can be explained partly by a common result if they do cooperate; chimps often will not share the reward. The dominant animal (or sometimes whichever one gets there first) will appropriate the food reward, and the partner will get nothing. Tolerant individuals (i.e., those that scored high on a sharing score) were the most successful at this cooperative task, but only if they were paired with a tolerant partner (Melis et al., 2006). In contrast, bonobos always share; both animals in the partnership get something (Hare et al., 2007). Not surprisingly, bonobos nearly always cooperate with each other on the task.

There is another cooperative game, or task, in which chimps have recently been shown to differ from humans. The human version of the game is called ultimatum. A reward is displayed so that both individuals can see it. One individual will get the reward, but only if the other person agrees. The first player proposes a split of the reward; the second player may accept or reject the offer. If the offer is accepted the players get the agreed-upon rewards. If the offer is rejected then the players get nothing.

An economic perspective predicts that the second player will accept any offer; after all, the second player is getting something for free, and rejecting the offer means no reward at all. But humans do not behave like this; if the split is too unequal people will reject the offer. Humans appear to have a concept of fairness that significantly influences their behavior. If an offer is deemed unfair they are willing to make a sacrifice in order to punish greed. Research indicates that there is a genetic component to the behavior; identical twins play the game (i.e., make and accept offers) more similarly than do fraternal twins (Wallace et al., 2007).

Chimps appear to have an economic perspective; they will cooperate with another chimp to retrieve food rewards regardless of the relative amounts as long as they get some reward (Jensen et al., 2007). It would be interesting to repeat this experiment with bonobos to see if they are more like humans; will they refuse to cooperate if the split is too unequal? Or is this notion of fairness or perhaps an evolved tendency to punish "cheaters" something that occurred after the split from the apes? Is this one of the features of increased cortex that was selectively advantageous?

The concept of fairness would seem to be an important aspect of human social eating. Meals are about cooperation and food-sharing. Both chimps and bonobos share. But for chimps it appears mainly transactional; sharing for gain. It isn't the default condition. The behavior of the bonobo provides evidence that the concept of sharing as the norm could have existed from early on.

Predators and Prey

Predatory behavior, in both directions, is believed to have had a profound effect on human evolution: Our ancestors were predators and prey (Hart and Sussman, 2005). A key adaptive advantage of early *Homo* was that they became better predators. They were also prey, however. There were plenty of carnivores in the environment that could kill and eat them. Fossil evidence indicates that australopithecines were preyed upon by big cats and eagles (Hart and Sussman, 2005). Chimps and bonobos, who are of similar size to australopithecines and early *Homo*, are preyed upon by leopards (Zuberbuhler and Jenny, 2002; D'Amour et al., 2006). To avoid being eaten was likely a strong selective pressure in our evolution.

Predation pressure has been suggested to have exerted a limiting influence on body size and fatness in early hominins (Speakman, 2007). Briefly, high predation rates favored a smaller, leaner phenotype. Exactly why this would be true is unclear, though Speakman (2007) relies on data for small mammals that indicate that this generally is the case. Whether the same result from predation pressure would apply to a relatively large animal such as an australopithecine is uncertain. Speakman's argument focuses on BMI more than overall size, and certainly an argument could be made that fatter individuals might make more desirable prey and suffer some deficit in avoiding predation if escape is an important antipredator strategy. Speakman (2007) argues that with the development of improved cooperative antipredator strategies in the genus *Homo,* involving tools (weapons) and eventually fire, there was a relaxation on the upper limit to BMI (body mass index) imposed by predation pressure. External factors still generally constrained individuals from obtaining high BMI, but gene variations that increased the propensity to fatness were no longer actively selected against. As a consequence, members of genus *Homo* became larger and fatter, on average, over time.

Cooperation and Efficiency

It has been suggested that the immediate vicinity of a carcass might well have been a dangerous place for our ancestors. Of course it was nutritionally rewarding, but there was a danger of becoming nutrition for another predator. Antipredator behavior needed to be combined with feeding behavior. Cooperative behavior could have been advantageous, both in detecting and deterring predators and in minimizing the time spent near the carcass, and thus presumably at risk.

There has been much speculation regarding the actual behavior of early *Homo* at a carcass. Was the carcass butchered and eaten in place? Was it butchered in one place and the pieces taken to another location to be eaten? Both behaviors provide advantages and dangers. Eventually the second scenario became the usual event. At some point in our evolution, game and other foods were routinely brought to a camp.

In either case, butchering the carcass quickly would lessen the danger associated with other predators being attracted to the site. There would be an advantage to speed and efficiency that is not evident in chimpanzee meat-sharing. A single red colobus monkey carcass can take a significant part of a day to be eaten by chimps. Presumably our ancestors were faster, at least after meat-eating became the dominant feeding strategy.

Patience is said to be a virtue. Humans are certainly able to delay gratification. Interestingly it turns out that chimps may be as good if not better, at least in certain circumstances. Data from Mark Hauser's laboratory (Rosati et al., 2007) have shown that when offered a choice between one unit of a preferred food (grapes for chimps, raisins or chocolate, depending on their preference, for humans) immediately or three units of the treat two minutes later, chimps were four times more likely to wait for the larger reward. Perhaps that explains why a chimp will hang around for hours, watching another chimp eating a colobus monkey and waiting for a share that may never come. Certainly the accounts of chimps' hunting and meat-eating and sharing behavior seems inefficient to us. Perhaps patience was not always a virtue, or at least not adaptive in all circumstances.

The human strategy to accomplish the quick and efficient butchering of a carcass would be division of labor. At what point in our evolution that cognitive capability arose is unknown. We don't know if all individuals at a kill participated identically in the butchering or whether there were roles to be performed. Did everyone make his own stone flakes to cut the meat off the bone, or were there "experts" who made the tools and others who used the implements? Did the sexes have different roles? In Ethiopia there are tribes where the women still prepare hides using stone tools, which they make. For them, stone tool manufacture is women's work (Weedman, 2005). A million years ago at a carcass did tool-making women create stone flakes that the men used to carve the beast? Did some members of the group act as sentinels, keeping an eye out for predators, ready to warn the group or even to attempt to drive the other carnivores off? Or did everyone rush in and take what he could get?

The fossil and archeological records are likely to be silent about these matters. Sentinel behavior won't leave a record. But it is a behavior practiced by many animals (e.g., meerkats, prairie dogs, red-bellied tamarins). It was certainly in our behavioral toolbox at some point in our evolution. We know that the modern human brain is capable of these behaviors, and thus at some point our ancestors were capable as well. The point is to consider all the complexities of meal behavior and how components of that concept would have provided strong selective pressures regarding our behavior and attitudes toward food and eating.

Consider a modern, formal meal with symbolic significance; there will be people who prepare the food, people who bring it to the table, someone will be responsible for signaling the beginning of the meal, perhaps with a blessing; people will carve and serve the food; and there will undoubtedly be someone who ensures that everyone is given their fair share. If this were a state dinner at the White House there would even be sentinels (secret service agents) keeping an eye on everything. The same individual could be in multiple categories; some individuals may be in only one. In addition to cooperation there is coordination of roles and events.

When our ancestors were capable of such behavioral divisions is not important for this book. What is important is that once the behavior of cooperative feeding arose, and meals became important to our feeding behavior, a whole suite of cooperative and coordinating social behaviors then had potential selective advantage. These behaviors require tolerance, sharing, and even a sense of fairness. They work on trust and delayed gratification. Perform a function now so that you will be rewarded by your group later. They provide a selective pressure for increased cortex. They also begin the intrinsic intertwining of feeding with social behavior.

Summary

Our diet and foraging and feeding behaviors of the past had a great deal to do with our eventual success as a species. A critical difference between our ancestors and the australopithecines was diet—not just *what* was eaten, but also how the food was obtained and how it was eaten. Our ancestors used technology from early on (Leakey and Roe, 1994); stone tools, digging sticks, and fire are some of the tools members of genus *Homo* used to get food and process that food outside of their bodies before eating it. At some point in our evolution, cooperative behaviors became firmly attached to food and feeding. Cooperative hunting, of course, but also the concept of eating together and sharing food became part of our intrinsic feeding biology. These adaptations, among others, enabled genus *Homo* to leave Africa and colonize the world.

To understand how meals contribute to obesity in modern human beings we need to consider the functions of meals in modern societies. To consider eating from the purely nutritional perspective will not fully explain human feeding behavior. Feeding represents nutrition, but also social transaction, politics, sex, and even morality. At some point, feeding becomes dining (see Figure 3.1).

Our feeding behavior was a prime selective pressure for the evolution of our brain. Food and social behavior are intimately linked. Food and

FIGURE 3.3.　Eating contests are very popular. Photo: Jay Kuzara.

feeding perform bonding as well as nutritional functions. Food can be comforting. Food is rewarding in many ways. We have taken eating well beyond a nutritional function. We not only enjoy eating we enjoy watching others eat; especially in ways beyond the social norm, such as the pie-eating contests of county fairs, which have evolved into numerous eating contests, with individuals celebrated for their prodigious feats of ingestion. Modern humans have turned eating into a competitive sport (Figure 3.3)! To understand why we eat and what food represents to us in the modern world we need to go well beyond nutrition and appetite.

Human feeding behavior requires tolerance. We are a competitive species, but we have the ability to restrain that aspect of our temperament in certain contexts. Meals are certainly an example, but there are many others (Figure 3.4). The ability of people to cooperate with each other and coordinate effort is a key aspect of our success.

When did meals first occur? We may never know, but it is reasonable

FIGURE 3.4. Cooperative and coordinated behavior is a key aspect of human beings as depicted in this barn-raising from the early 1900s. Note the low to moderate BMI of the individuals in the picture. Photo: Historical and Genealogical Society of Somerset County, Pa.

to hypothesize that it was an important aspect of the change in feeding-foraging behavior that characterized the change from australopithecines to genus *Homo*. Meal feeding behavior was both enabled by and acted as a selective pressure for increased brain size, especially cortex. Increased cortex would have aided in the planning, tolerance, and cooperative tit-for-tat behaviors that make meals possible. The idea of role division may have also been favored.

The next time you sit down at a restaurant, food court, or other communal eating place, spend a moment considering how unusual this behavior is, how the ability to be so tolerant of others may have arisen in our ancestors, and what role it has played in our evolutionary success.

CHAPTER FOUR

Evolution, Adaptation, and Human Obesity

....

In this chapter we examine what has been called the mismatch paradigm (Gluckman and Hanson, 2006) a central tenet of evolutionary medicine (Williams and Neese, 1991). In general, medicine, both human and veterinary, focuses on the mechanistic aspects of disease: the what and how of pathology. Evolutionary medicine examines the why of disease processes. For example, there are responses to infection that are common to most vertebrates called the acute-phase responses; these include fever, sequestration of iron and zinc, loss of appetite, and increased synthesis and release of acute-phase proteins such as C-reactive protein and fibrinogen (LeGrand and Brown, 2002). These responses can be debilitating, and standard medical practice is to relieve the symptoms, often by reducing some of the acute-phase responses. However, these responses are adaptations to infection; they can in many, but not all cases, enable the infected animal to survive the infection. For example, fever and the sequestration of iron act synergistically to inhibit bacterial infection (Kluger and Rothenberg, 1979). Thus some aspects of pathology caused by infectious disease will be due directly to the disease organism, but others will be related to the host defense response that changes the sick animal's metabolism and physiology in ways that would, on balance, increase the chances of survival, but can lead to short- and long-term ill effects as well.

In the modern medical paradigm, the cause of an infection would likely be treated with an appropriate antibiotic and the host defense responses treated with palliative medications to reduce fever and the feelings of malaise. This of course has many advantages; at the least the infectious agent is killed off while the patient is freed from (or at least buffered from) the debilitating effects of the adaptive host defense response. This enables us to work and lead a normal life even while sick. Of course, evolution works for pathogens as well as hosts; several pathogenic strains have evolved that are resistant to our impressive arsenal of antibiotics. Antibiotic-resistant pathogens are fast becoming a major concern. For many relatively minor infections (e.g., sinus infections) the current standard of medical care is to allow the body's natural defenses to work. The benefit to the individual has to be weighed against the risk of creating resistant strains of bacteria.

Obesity and its associated pathologies certainly differ from an infectious disease, but there are elements in common as well. Obesity is associated with inflammatory aspects of the acute-phase response triggered by cytokines such as interleukins 1 and 6 and tissue necrosis factor-α, all cytokines produced by adipose tissue (see chapter 11). Some aspects of obesity-related pathology appear to be caused by normal, adaptive responses that are, in a sense, out of balance due to the abnormal amount of adipose tissue. Obesity itself is, in many cases, probably due to normal adaptive responses that encourage eating high energy density foods while limiting energy expenditure. Food is rewarding, and there was adaptive advantage in the past to restrict energy expenditure when possible. Indeed, in many ways our modern society is constructed to allow us to eat well while expending little; economic and business decisions reflect our evolved preferences. We have diligently worked to conquer the external constraints that forced our ancestors to expend considerable energy to obtain barely sufficient food; we are now seeing that for many of us the motivations and biological drives to obtain calories appear to exceed those to expend them.

In this chapter we examine the mismatch paradigm and the concepts of homeostasis, allostasis, and allostatic load. Physiological systems are finite; although we can adapt to our surroundings, sustained physiological responses, while necessary, can often exert costs as well. This is termed allostatic load (McEwen, 1998). Organ systems, including brain, can be altered by long-term or dramatic physiological responses (Schulkin, 2003; McEwen, 2007). Pathology can arise from normal physiology if it is extended beyond the normal time frame.

The Mismatch Paradigm

....

Human beings occupy an incredible array of habitats; far more than any other species. We live in environments far removed from our evolutionary past. Our ability to change our environment to adapt it to our biology as opposed to the other way around is unsurpassed. That doesn't mean that we have been liberated from our biology, however. We still carry the past adaptations to the environments under which we evolved. In fact we frequently are faced with a mismatch between our evolved biology and the environments we live in and create for ourselves. Indeed, the mismatch paradigm (Gluckman and Hanson, 2006) is a major component of evolutionary medicine (Williams and Nesse, 1991; Trevathan et al., 1999, 2007). Briefly, our evolutionary past has gifted us (or saddled us, depending on your perspective) with developmental programs that, in conjunction with environmental cues, produce biology and physiology that was appropriate for the expected conditions of our ancestors. In large part the results are still appropriate; humans are pretty well adapted to our world. However, both because we have populated such diverse habitats and, more importantly for this book, because we are so capable of creating new environmental conditions, there are many examples of our biology being out of sync with our environment (see Gluckman and Hanson, 2006). We contend that this is demonstrably true about our feeding biology.

The word *adaptation* has at least two meanings in biology. In evolution, adaptation refers to characteristics of a species that enhanced survival, reproductive success, or both. The use of the past tense is deliberate here. An evolutionary adaptation is a characteristic that addressed a challenge that the organism's ancestors faced. It was adaptive; whether it still is depends on the current circumstances. If the challenge no longer exists, or has changed, then the selected characteristic may no longer be adaptive. Whether it persists or vanishes from future generations depends on many factors: how much variation exists in the population? what are the costs of the characteristic? what are the benefits? is the population growing, declining, or staying stable? Many former adaptations can persist in a population long after the original challenge has ceased to be a selective force.

Evolution tends to favor simple, robust solutions to challenges. That isn't to say that complex, targeted adaptations don't arise; they do. The question is what adaptations are the most likely to persist. Adaptations that provide broad advantages to a related set of challenges are more likely to persist than narrow, "targeted" solutions.

In physiology, the word *adaptation* refers to short-term changes in physiology and metabolism in response to some challenge. Often these responses are in defense of homeostasis; maintaining a constancy of the internal milieu (Bernard, 1865; Cannon, 1935). For example, if you are exercising on a hot day you will rapidly begin to sweat; the evaporative cooling effect will serve to reduce the increase in core body temperature due to your exertion.

Of course the ability to physiologically adapt to circumstances represents an evolutionary adaptation. So evolution has resulted in our bodies having some capability to physiologically adapt to circumstances. Thus we don't have to have a perfect match between the "expected" environment and the actual one. We can adjust to a mismatch. The greater the mismatch, however, the more likely our toolbox of adaptive responses will come up short. And even under circumstances where we can physiologically adapt, our evolved adaptive responses can lead to a lessening of health over time. Physiological adaptation is not without cost.

The Homeostatic Paradigm

The external environment changes in both predictable and unpredictable ways and animals must be able to respond to both challenges (Wingfield, 2004). The internal environment of animals must be buffered from extreme changes in the external environment. Put simply, for an organism to survive, its internal environment must remain within certain constraints. Some of these constraints are fairly broad and some are quite narrow. The science of physiology has been, to a large part, the investigation of these constraints and how they are regulated. A key aspect of physiology for Cannon (1935) was that "the organs and tissues are set in a fluid matrix. . . . So long as this personal, individual sack of salty water, in which each one of us lives and moves and has his being, is protected from change, we are freed from serious peril." For Cannon (1935), homeostasis was "the stable state of the fluid matrix" and homeostatic mechanisms and processes were those that maintained the constancy of the fluid matrix. In order to achieve this necessary condition, animals have evolved myriad adaptations, from cell membranes to complex central nervous systems. These adaptations serve to keep the "internal milieu" (reasonably) constant.

The concept of homeostasis has evolved and matured beyond the "stability of the fluid matrix," but it continues to be perceived as a fundamental explanatory principle of regulatory physiology. The concept of stability, of resistance to change, remains fundamental to homeostasis. Homeostatic systems resist change, and when perturbed, function to return the parameters of the system to within a range of appropriate values, often referred to as a "set point." Restraint and negative feedback are important aspects of homeostatic processes.

Not all physiological processes comfortably fit within the homeostatic paradigm, however. Many scientists (Mrosovsky [1990], Bauman and Currie [1980], Sterling and Eyer [1988], McEwen [1998], Schulkin [1999, 2003], among others) have pointed out weaknesses in the homeostatic perspective. There are examples of physiological regulation that are not strictly homeostatic. Much within the internal milieu is constantly changing and adapting. Many physiological parameters do not remain constant but rather continually adapt to circumstances. This is not an inherent contradiction of homeostasis; Cannon (1935) himself quoted Charles Richet, "We are only stable because we constantly change." But the concept of homeostasis must either be stretched, or other terms and concepts must be added to the lexicon of physiological regulation.

Mrosovsky (1990) proposed the term *rheostasis* to describe circumstances where physiological set points are changed and then maintained/defended at the new level. Bauman and Currie (1980) proposed the term *homeorhesis* to describe alterations of physiology to meet demand states (e.g., reproduction). Moore-Ede (1986) suggested that circadian rhythms of physiological parameters could be incorporated into homeostasis by labeling them "predictive" homeostasis as opposed to the "reactive" homeostasis, which meets acute, unpredictable challenges. Indeed, the central coordination of physiology, anticipatory physiological responses, and the interplay of physiology and behavior appear to be given short shrift by the classic homeostatic paradigm. Sterling and Eyer (1988) proposed the concept of allostasis to account for regulatory systems that appeared to fall outside of the classic concept of homeostatic processes. For example, regulatory systems in which there are varying set points or no obvious set points at all (e.g., fear) or where the behavioral and physiological responses are anticipatory and do not simply reflect feedback from a monitored parameter.

Stability is perhaps a misleading word when considering evolved physiological adaptations. Physiological systems serve the survival and reproductive capabilities of the organism (fitness). Of course stability in some parameters is a necessity, but a truly stable organism, in the strictest (and unreasonable) sense, would end up extinct. Organisms have to be able to change and react to challenges. Perhaps a better term than *stability* is *viability*, defined as the capability of success or ongoing effectiveness. In an evolutionary context, this means the ability to pass on genetic material. Regulatory physiology functions to enable an organism to achieve and maintain viability, and that requires an organism to change its physiological state as appropriate to the season, its age, or in response to acute need or challenge. There must be physiological processes that are not homeostatic, and that oppose, at least temporarily, stability. The term *allostasis* has been proposed for such processes (Sterling and Eyer 1988; Schulkin 2003). Briefly, allostasis is the physiology of change (Sterling and Eyer, 1988; Schulkin, 2003), defending viability by changing state, as opposed to homeostasis, which is defending viability by resisting change (Schulkin, 2003; Power, 2004).

Homeostasis and allostasis can be considered as complementary components of regulatory physiology. Homeostatic processes maintain/regulate physiology around a set point, and allostatic processes change the state of the animal, including changing or abandoning physiological set points. Homeostatic processes are associated with negative restraint and resistance to perturbations; allostatic processes are associated with positive induction, perturbing the system, and changing the animal's state. We would modify Sterling and Eyer's (1988) original definition of *allostasis* from "achieving stability through change" to "achieving viability through change," and define *homeostasis* as "achieving viability through resistance to change."

Is the concept of allostasis necessary for understanding regulatory physiology? Homeostasis can be predictive (Moore-Ede, 1986) as well as reactive. The conception of allostasis outlined above is very similar to

Mrosovsky's (1990) "rheostasis." It is broader in that Mrosovsky restricted rheostasis to changes in a regulated level that is then defended, and allostasis does not require a new level that is defended or stable. Many circadian phenomena are considered both anticipatory and homeostatic, at least under a broad conception of homeostasis (Moore-Ede, 1986). The distinction between homeostasis and allostasis can often be difficult to make.

An important aspect of allostatic regulation is the concept of centrally coordinated physiology. Much of homeostatic regulation centers on local feedback (usually negative). In contrast, anticipatory, feed-forward systems are a consistent feature of allostatic mechanisms. Many of the examples of allostatic regulation (see Schulkin, 2003) involve the induction of neuropeptides by steroid hormones, and the concept that the hormones that regulate peripheral physiology in response to a challenge are also involved in changing central states of the brain, and thus induce behaviors that aid the animal to meet the challenge. Peripheral physiology and function is represented in the brain, and often coded and regulated by the same information molecules that regulate function in the periphery. The brain and the periphery are linked via numerous information molecules (see chapter 7). The concept that behavior and physiology act together to preserve viability is central to the concept of allostatic regulation.

Allostatic Load

McEwen (1998) extended the concept of allostasis to regulatory systems that were vulnerable to physiological overload, with the resulting development of disease. A relatively new concept in regulatory physiology as it pertains to health, disease, and pathology is allostatic load (McEwen, 1998, 2000, 2005). The concept of allostatic load arises from the fact that many physiological adaptations are short-term solutions. They have a cost that can be borne over a limited time period, but will begin to lessen health if they remain continuously activated. Consider our sweating example. If the environment remains too hot and you continue to exercise, eventually sweating will lead to dehydration and sodium depletion, a potentially serious and even life threatening condition. Sweating solves the problem of reducing the rise in body temperature over the short term, but other responses (e.g., drinking water, retreating to a cooler place) eventually will have to be activated.

The congruence between the mismatch paradigm and allostatic load is fairly straightforward: The more an organism's biology and physiology is a mismatch for its environment the greater the effort/cost it will incur in its attempts to adapt physiologically, and the more likely that the physiological adaptations will be insufficient and even possibly inappropriate. The greater the mismatch the greater the allostatic load. The adaptive responses may be turned on, and then never turned off, or regulated inappropriately. Normal, adaptive (in the evolutionary sense) responses become maladaptive.

In the case of human obesity, we did not evolve under conditions of high food availability and little required exertion. We evolved as a species that had to work hard for its food. The modern world presents its own challenges, but for much of humanity (but not all, unfortunately) struggling to obtain enough food is no longer an issue. But food was often a limiting resource in our past, and we appear to carry formerly adaptive responses that lead to overeating and an accumulation of excess adipose tissue. Adipose tissue is not just a passive store of energy in the form of fat; it is an active player in metabolism and physiology (see chapter 11). It is, in effect, an endocrine organ that interacts with other end-organ systems. We propose that when adipose tissue mass becomes oversized relative to the rest of the body, then conditions of allostatic load can accumulate and lead to pathology. This allostatic load is caused by both normal as well as abnormal adipose tissue function.

Machinery from Our Past

As a species, we are biologically adapted to our past. In this way we are no different from every other extant species on Earth. We differ in that we are constantly changing the world in which we live. Our technological capabilities enable us to construct circumstances that provide challenges that in our past were at best extremely rare. Even the food we eat differs greatly from our not-so-distant past (Eaton and Konner, 1985). It is no surprise that our inherited biological responses are not always sufficient or even appropriate.

The concept of homeostasis is applied to body weight by many researchers. There is abundant and credible evidence that animals attempt to remain weight-stable under many circumstances. More recently a lipostatic theory has come to be favored among researchers studying human obesity. The discovery of the peptide hormone leptin (Zhang et al., 1994), produced by adipose tissue (the main fat cells of the body) and generally secreted in direct proportion to the amount of adipose tissue, has provided a possible proximate mechanism for the intellectually appealing theory that our food intake and energy expenditure are regulated (at least to a certain extent) by the amount of fat on our body. Adipose tissue is now thought of as both a significant energy storage organ and an endocrine organ important in regulating metabolism (Kershaw and Flier, 2004).

The roles adipose tissue and numerous information molecules like leptin play in regulating metabolism and feeding behavior are discussed in more detail in later chapters; the focus of this chapter is on possible evolved tendencies and characteristics of our species that might predispose us to violate "lipostasis," that is, adaptations that served to increase fitness in the past, possibly even by aiding in maintaining total body energy (fatness), but that in the modern world result in a nonlipostatic creature.

Is Being Lazy Adaptive?

We argue in this book that our ancestors generally had to exert more physical effort and expend more energy to live than we typically do today. But that doesn't mean that there was no advantage to being lazy. Doing nothing conserves energy.

But *lazy* is not really the right word; animals in the wild act according to evolved behavioral strategies that have, on balance, produced the highest fitness. And anyone who has observed wild animals for any length of time knows that doing nothing appears to be a successful strategy in many circumstances. Resting is a common behavior that many animals spend a good part of their day employing; for example, colobus monkeys spend more than half of their waking time resting (Figure 4.1). Performing calisthenics, expending energy for the sole purpose of exercising one's muscles, is largely a human endeavor. Of course many animals will expend energy in physical play, both social and solitary. An argument can be made

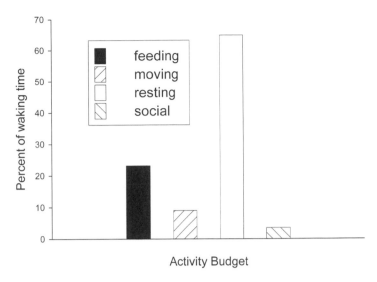

FIGURE 4.1.   Doing nothing is quite common in nature. Colobus monkeys spend most of their waking time resting, and they sleep through the night. The chart shows daily activity for six troops of *Colobus vellerosus* in Ghana. Data from Wong and Sicotte, 2007.

that there is adaptive function to enjoying physical effort, but effort has costs as well. Good arguments can be made that animals should generally expend as little as possible to accomplish their fitness objectives.

Human beings vary in how energetically they choose to live their lives. Some people are always doing something; others are quite content to kick back whenever the opportunity arises. It is interesting to consider that both strategies would have had adaptive advantages in our past.

The Paleolithic Diet

Our distant ancestors didn't eat the foods we eat today. The foods of our past differed in form, digestibility, and nutrient content, among other characteristics (Eaton and Konner, 1985). All of these aspects of our ancestral foods are important for understanding the evolution of our feeding adaptations. The differences in nutrient content may be especially key factors; although animals eat food, what they require are nutrients. Nutrient deficiencies will have significant effects on physiology and metabolism, and thus likely on behavior. We are defining deficiency in a very broad way, not in a medical sense in which deficiency results in measurable pathology. We have physiological adaptations that enable us to accommodate low intakes of most nutrients, at least over short to medium time frames. However, those adaptations change our physiology and metabolism. They can affect feeding behavior.

Again, consider our example of sweating under the hot sun. After a sufficient amount of time you would become thirsty and motivated to drink. In addition, your reactions to sodium will change. Sodium has a salty taste. After several hours of sweating that salty taste likely will be perceived as both less intense and more pleasant. You might find yourself craving salty foods in addition to something to drink (Schulkin, 1991; Fitzsimmons, 1998).

Calcium is an important nutrient that was likely found in higher concentration in the diet of our ancestors compared with today (Eaton and Nelson, 1991). With the advent of agriculture there was a significant increase in seed products (grains) in the human diet. Seeds tend to be high in phosphorus and low in calcium. The modern human diet is likely much lower in total calcium and higher in phosphorus. Human calcium regulatory physiology may be evolutionarily adapted to the opposite: high dietary calcium intake and a high calcium/phosphorus intake ratio (Eaton and Nelson, 1991). Does this possible change from our ancestral condition have any relevance to weight gain and obesity? Well, some interesting data suggest that it is possible.

A number of studies have found that habitual calcium intake is inversely related to BMI (body mass index), weight gain, and total body fat (reviewed in Heaney et al., 2002). High calcium intake is associated with being lean; low calcium intake is associated with greater fat mass. In transgeneic mice subjected to calorie restriction, low calcium diets decreased fat loss while high calcium diets accelerated fat loss (Shi et al., 2001). The mechanisms for the interaction between dietary calcium and body fat are not yet understood, but in addition to calcium itself the calciotropic hormones (vitamin D and parathyroid hormone or PTH) appear to play a role (Zemel, 2002). A side effect of low calcium intake is an up-regulation of the calciotropic hormone system, which appears to affect adipose tissue metabolism (Sun and Zemel, 2004, 2007; Morris and Zemel, 2005) (see chapter 11).

What are other important characteristics of our past diet that might affect our current feeding behavior? We argue that nutrients and food types that were either rare or ubiquitous in the diets of our ancestors will have the greatest probability of molding our feeding behavior and preferences.

What Is Rare Becomes Valuable

Value often relates to scarcity. This concept operates in human economics: An item that is rare will usually be more valuable than a comparable item that is common. An analogous principle operates in evolution. There are resources in the environment that serve as constraints, because they are scarce or difficult to obtain. They require effort. If obtaining them would enhance survival or reproduction, then selective pressures will likely favor individuals who have higher motivation to obtain them. These individuals would be more willing to expend the effort. If the resource is a food or food type, then perhaps the taste of the food will become preferred.

For example, marmosets are squirrel-size New World monkeys that feed extensively on gums from trees (Coimbra-Filho and Mittermeier, 1977; Figure 4.2). This is a fairly unusual dietary strategy, but a successful one for marmosets. Not surprisingly, captive marmosets have a fondness for gum arabic solutions, and these are often offered to them in zoos as enrichment (McGrew et al., 1986; Kelly, 1993). One of us (MLP) has fed pygmy marmosets gum arabic solutions from a syringe and had the 100g animals frantically grasp the syringe in an attempt to keep it from being withdrawn. To humans, the solution tastes rather bland. Other related small New World monkeys that are not gum feeders in the wild showed little enthusiasm for the solution, and either ignored or actively avoided the syringe. All animals were used to being fed sweet-tasting solutions from a syringe as part of training to enhance medicine delivery. The marmosets found the gum solutions "tasty" and were motivated to feed on them; the other animals did not.

In the wild, marmosets obtain gum by gouging holes in the bark of trees to stimulate gum flow (Coimbra-Filho and Mittermeier, 1977). They spend considerable time in an exposed position working on these gum sites. They expend considerable energy and expose themselves to the risk

FIGURE 4.2.   The common marmoset (*Callithrix jacchus*) is a squirrel-size monkey native to Brazil. Photo: Michael Jarcho.

of predation. They are highly motivated. Their taste biology is one aspect of the coordination between their motivated behavior and their dietary strategy.

Consider another example. About 20 years ago one of us (MLP) was asked by the animal care staff at the National Zoo to review the diet fed to the colony of kangaroo rats. During routine physical examinations, several animals had suffered fractured legs. This was not due to rough handling by the veterinary staff; rather, the animals' bones appeared to be weak. The veterinarians and the animal care staff were concerned that diet might be contributing to poor bone mineralization.

The diet consisted primarily of a rodent chow, several types of seeds (kangaroo rats are highly granivorous in the wild), and iceberg lettuce, which was a preferred food. A calculation of nutrient content based on the proportions of these diet components indicated that the offered diet was well balanced and healthy. However, the total quantity of food being offered was significantly more than the animals would eat, and probably even could eat. Thus the animals were selecting from among these components. For the next week actual food intake was monitored. All food given was weighed, and all uneaten food was collected. The results were enlightening.

The rodent chow was rarely consumed. The seeds were eaten readily, with some individual preferences among animals for certain seeds. The iceberg lettuce was always completely consumed. The nutrient composition of the eaten diet was deficient in calcium.

The fix was simple. The amount of seeds offered was greatly reduced, and iceberg lettuce was restricted to the occasional treat. The animals responded by increasing their consumption of the rodent chow, and fractured legs were a thing of the past.

Why did granivorous kangaroo rats prefer iceberg lettuce even over seeds? Well, kangaroo rats are desert animals. They feed on seeds, insects, and leaves, when they can get them, which is only in the short periods of plant growth around seasonal rains. Juicy, water filled leaves, like iceberg lettuce, are rare, but potentially very seasonally important to kangaroo rats. It would appear that they are highly motivated to eat them. The captive kangaroo rats had responded as their biology directed, but under the conditions in which they evolved, external circumstances always limited their access to this kind of food. We had inadvertently placed our captive animals under conditions of a perpetual spring, and their biological adaptation was now inappropriate.

Captive animals have been removed from the environments under which they evolved. This isn't always a bad thing; real life and nature can be hard, even cruel. Captive animals benefit from many aspects of captivity, but they are also exposed to evolutionarily novel circumstances. The chance for mismatches between their evolved systems and adaptations and their captive environment can be high. In this way captive animals are analogous to human beings in the modern environment. We confront evolutionarily novel circumstances daily. Many have no adverse effects, and indeed many are beneficial. Our modern environment has many advantages over the past. But we carry preferences and tendencies from our past that evolved to solve challenges that are greatly reduced or even turned upside down in today's world. Our food preferences are a good example. Modern human capabilities basically turn the concept of what was rare and therefore valuable on its head. Our food economic system seeks to produce foods that people like; taste preferences that may have arisen to give motivation to make great efforts to obtain rare or risky foods now provide motivation for food producers to make those kinds of foods ubiquitous. We remain highly motivated to eat them, despite their prevalence and how easy it is to get them. There is now, in effect, a mismatch between the drive to obtain foods that were important in our past because they were difficult to obtain and the modern ease of obtaining those kinds of foods. Not surprisingly many of us overindulge.

What foods were rare and valuable in our past environment? What foods would our ancestors been very motivated to exert themselves and even put themselves at risk to obtain? A good argument can be made that foods high in fat and simple sugars were rare and valuable for our ancestors. Certainly these are the kinds of foods that we show considerable preference for today. What kinds of foods in the past would have had these characteristics?

Honey

Honey is a calorically dense, high-simple-sugar food. It has a relatively high fructose concentration (often more than 10%). Fructose is perceived by people to be very sweet, sweeter than glucose or sucrose (Hanover and White, 1993). People have been eating honey for as long as we have history, and undoubtedly for much longer. Beehives containing honey would have been fairly common in the environments of our ancestors, including the African environment of our earliest ancestors. If our early ancestors were able to make stone tools, digging sticks, carrying devices, and all of the other technological devices that have been proposed to be in our earliest toolboxes then they would have been able to find beehives and extract the honey. Once they controlled fire the task would have been much easier; smoking bees out of their hive is a relatively safe way to obtain honey.

In Africa there is a bird called the greater honeyguide (*Indicator indicator*) that obtained its name due to its reputation for leading mammals to beehives. The greater honeyguide does not eat honey; it feeds on the bees' eggs, larvae, and pupae, and on the beeswax itself. Honeyguides are one of the few birds that can digest wax; the other birds with this ability are generally marine birds that feed on aquatic fauna that produce wax esters (Place, 1992). The greater honeyguide is believed to attract and lead ratels (honey badgers), baboons, and humans to beehives, and then scavenge the leftover hive after the mammal has taken the honey. There are no scientifically documented cases of honeyguides leading ratels or baboons to hives; the evidence is strictly from secondhand anecdotal reports, mainly from indigenous people. However, the behavior has been observed and recorded for humans (Friedman, 1955). The greater honeyguide will follow men in motor vehicles (Friedmann, 1955), is attracted to the sound of chopping wood (Friedmann, 1955), and will respond to a human whistle (Dean and McDonald, 1981). The birds will make distinctive calls, fly toward a hive, and then stop and repeat the calls (Short and Horne, 2002).

This fascinating connection between bird and human may be one of the earliest symbiotic associations between humans and another vertebrate species. It could well predate our expansion out of Africa. It may be vanishing, however; in more developed areas, where the people generally buy sugar from the store and rarely search for wild honey, the behavior is disappearing. It soon may be restricted only to the wild areas of Africa (Friedman, 1955; Dean and MacDonald, 1981).

Honey may have been an important food for much of our evolution. Exactly how long ago our ancestors began to routinely eat honey will be difficult to determine; honey-eating is unlikely to leave definitive fossil or archeological evidence. It is reasonable to hypothesize, however, that liking the taste of honey was a motivating factor for our ancestors to devise means to obtain it, and the benefits accrued by individuals that made the effort may have reinforced our sweet tooth.

Of course honey was not the only sweet-tasting food available to our ancestors. Many ripe wild fruits are also sweet, though rarely as sweet as the domesticated fruits we have produced. Most sweet foods would have been relatively rare (e.g., seasonal), difficult to obtain, or both. We propose that one aspect of our preference for sweet-tasting foods was to motivate our ancestors to make the effort to obtain them; it now motivates us to overindulge in eating them. High-sugar foods were beneficial when external conditions limited their availability. When they are plentiful and cheap they effectively swamp our behavioral and physiological systems.

Highly palatable foods may change our appetite regulation system (Erlanson-Albertsson, 2005). Palatable food can activate the reward circuits in the brain and tends to stimulate eating. Foods high in sugar and fat are particularly attractive to humans; from an evolutionary point of view such foods would have been highly rewarding in terms of providing energy. They also appear to activate our internal reward system. There are parallels between the overconsumption of desired foods and drug addiction (e.g., Berridge, 1996; Nesse and Berridge, 1997).

Fat

Dietary fat has been linked with the increase in brain size during the early evolution of genus *Homo* (Leonard and Robertson, 1994; Aeillo and Wheeler, 1995; Cordain et al., 2001). Fat is proposed to be both an advantage accrued by the change in foraging strategy presumably enabled by increased brain size, and also as a necessary component to being able to metabolically support a larger brain. Evolutionary models for a possible mechanism, or "prime releaser" (Robson, 2004), for size increases argue either directly (Leonard and Robertson, 1992, 1994; Leonard et al., 2003) or indirectly (Martin, 1983, 1996; Foley and Lee, 1991; Aiello and Wheeler, 1995; Aiello et al., 2001; Cordain et al., 2001) for the critical role of increased dietary quality in supporting the higher energetic demands of a larger brain and body born by *Homo* relative to australopithecines. An increase in dietary fat is one of the proposed means by which early *Homo* benefited from the change in dietary strategy.

An ecological perspective suggests a possible selective pressure for increased brain size as a result of increased dietary breadth in *Homo* relative to australopithecines. It is hypothesized that the task of locating and obtaining significant amounts of animal tissue required more cognitive judgments than a grazing strategy. It suggests a possible feedback loop wherein higher cognitive abilities (assumed to be associated with increased relative brain size) would allow individuals to increase their foraging returns and their dietary quality, which would in turn select for higher cognitive abilities for resource acquisition (Aiello and Wheeler, 1995). The data from archeological sites of early *Homo* indicate a relationship between encephalization (increased brain size) and increased abilities to obtain high-quality food items, such as prime-age ungulates (Stiner, 2002).

Fat was probably rare and valuable in the diets of our ancestors. We have evolved a liking for fat. We may even have specific fat taste sensors on our tongues (see chapter 9). Animal fat may have been necessary in the evolution of our larger brain, both to fuel the costs of its growth and maintenance and as a selective pressure; eating more fat may have required more brain and allowed more brain.

Brain and Fatty Acids

Brain is a high-fat organ; perhaps as much as a third lipid (fat). There are certain functional aspects of brain morphology and structure that are composed of lipids. Certain fatty acids are considered essential for proper brain growth and development. Thus there are reasons beyond energy that an increase in dietary fat may have been beneficial or even necessary to support the growth of a larger brain (Decsi and Koletzko, 1994).

Long-chain polyunsaturated fatty acids (LCPUFA) are polyunsaturated fatty acids with 18 or more carbons. Certain LCPUFA (e.g., docosahexaenoic acid and arachidonic acid) may play an important role in regulation of fatty acid metabolism and gene expression in the brain (Kothapalli et al., 2006; 2007). Brain growth in mammals is associated with increased incorporation of LCPUFA, primarily in the cortex (Farquharson et al., 1992). The human fetus obtains the majority of the LCPUFA from placental transfer (Brenna, 2002); after birth, milk presumably becomes the prime source, as conversion of precursors to docosahexaenoic acid and arachidonic acid is relatively inefficient (Brenna, 2002).

A source of dietary LCPUFA may have been critical to our ancestors during both pregnancy and lactation. Animal tissue, especially brains and marrow, would have provided significant amounts of LCPUFA. The absolute amounts needed are not large, and brain growth is spread out over many years; this was probably true for our early ancestors as well. It is possible that LCPUFA was not a constraining resource in our evolution, but the hypothesis is well worth considering. These fatty acids have been shown to influence brain growth and development in baboons and in humans. Breast milk replacer formulas have now added the appropriate LCPUFA. The conflicting advice to women regarding fish consumption during pregnancy and lactation arises from the positive effects of LCPU-FAs that are found in high concentrations in fish as opposed to the negative effects of mercury, which unfortunately is also now found in high concentrations in fish (U.S. FDA, 2007).

Martin (1981, 1983) suggested that human milk may have changed to support the increased need for LCPUFA due to our larger brain. Most of human brain growth occurs after birth and is thus supported by lactation. This was likely the case for our early ancestors as well. Pelvic girths of species in genus *Homo* are not adequate for infants to be born with brains much bigger than 250 to 300cc, approximately the size of an adult chimpanzee brain, and only one-fourth the final adult human brain volume. Did the milk of early *Homo* have to change in order to support the growth of our larger brains?

Human milk does contain the essential LCPUFA and their precursors, though not at particularly high concentrations. Human milk concentrations of LCPUFA are much higher than what is typically found in cow milk, however (German and Dillard, 2006). That is why LCPUFA must be added to cow milk–based infant formula. But what would a comparison of milks from early *Homo* and australopithecines have found? We can never test this directly, but we can compare the milks of modern humans with those of other primates.

Lauren Milligan (Milligan, 2008) examined milk from 14 species of monkeys and apes, with samples from both wild and captive species. Her sample included milk from our closest relatives the chimpanzee (*Pan troglodytes*) and the bonobo (*Pan paniscus*), as well as mountain (*Gorilla beringei*) and lowland gorillas (*Gorilla gorilla*; Figure 4.3), and the orangutan (*Pongo pygmaeus*). Ape milk differed from monkey milk; it was less variable, but on average was lower in fat and hence energy (Milligan, 2008). There were some fatty acid composition differences, but for LCPUFA most differences among species appeared to relate to diet and not phylogeny (Milligan et al., 2008). More importantly, ape milk did not appear to be appreciably different from human milk in fatty acid composition or gross amounts of fat, protein, and lactose. On the basic nutritional constituents, human milk is not distinguishable from that of our ape relatives.

FIGURE 4.3.   A lowland gorilla nurses her infant. Gorilla milk is similar to human milk in most nutrients. Photo: Jessie Cohen, Smithsonian's National Zoo.

Still, fatty acid composition of milk is affected by dietary fatty acid intake (Milligan et al., 2008). The fatty acid composition of the milk will reflect both present and past intake, as fatty acids stored in adipose tissue are mobilized to meet lactation's demands. It is quite plausible that the metabolic needs of their infants' growing brains acted as a selective pressure to enhance our female ancestors' motivation for eating fat and ability to store fat in adipose stores (scc chapter 12).

Summary

Human beings create their own habitats; not completely, of course, but to a remarkable degree. We create our living spaces, our food, our social institutions; we control the temperature and humidity where we work and live; we can determine when it is light and when it is dark. What we have created reflects to a certain extent, we argue to a great extent, our evolved preferences and characteristics. But that does not mean that we create environments that are in complete synchrony with our biology. We have created environments so different from those we evolved under that it is no wonder that we have environmentally induced diseases. Obesity is but one example.

Our evolved biology frequently comes in conflict with our environment; sometimes because we have chosen to live in extreme places (high altitude, extreme cold or heat, extreme aridity) and sometimes because we have created aspects of our environment that are novel. We can adapt to these circumstances, but the more there is a mismatch between the environment and our biology, the more likely that we may suffer some health cost. The mismatch paradigm and the concept of allostatic load capture this idea. The greater the mismatch the more likely our physiological responses will be unsuccessful over the long term. Physiological adaptations to circumstances generally have metabolic costs; we are finite organisms, and metabolically costly adjustments to our environment can eventually lead to breakdown and pathology.

We have evolved preferences concerning how we like to live. We have temperature ranges we find comfortable; we like sweet-tasting foods and also fatty foods; we are capable of remarkable exertion, but that doesn't mean we will choose to exert ourselves if we have a choice. All of these preferences make adaptive sense when our past is considered. However, we have built a world where we can live in comfortable temperatures most of the time, regardless of the weather. We can have as much sweet and fatty food as we like. And we don't have to exert ourselves physically if we don't want to. We have created an obesogenic environment. Our behavior has created a potential problem for our physiology. For example, we can indulge in preferred foods to an extent far greater than our physiology was ever likely to have been exposed to in the past. We evolved on the savannahs of Africa; we now live in Candyland.

CHAPTER FIVE

Evolution, Adaptation, and the Perils of Modern Life

....

The preceding chapters examine some aspects of our evolutionary history as they pertain to our biology. Now we explore the ways that the modern environment, interacting with our evolved biology, might make us vulnerable to sustained weight gain leading to obesity.

Modern foods and the ways we eat have changed dramatically over the last 50 to 60 years, let alone from what our prehistoric ancestors experienced. Even the meat we eat differs from that of our hunter ancestors; meat from grain-fed cattle is generally higher in fat and has a different fatty acid profile compared to wild African ruminants (Cordain et al., 2002). Of course our eating habits are only part of the puzzle. Physical activity in modern, well-off countries differs both quantitatively and qualitatively from our past. We also examine the structural form of our living habitat, the so-called built environment, to see how it contributes to decreased physical activity. All of these components, along with familial traits—genetic, cultural, and socioeconomic—contribute to our modern obesogenic environment (Figure 5.1).

We examine the changes in developing countries due to modern economic and scientific progress. The obesity epidemic that is in full swing in economically advantaged countries is just beginning in the developing world (Prentice, 2005). This gives us a chance to examine the features of demographic, dietary, and cultural change that might underlie our species' vulnerability to obesity.

An ironic aspect of human obesity is that it is frequently associated with malnutrition, not only in the same population but also in the same individual! Obesity and poverty are not uncommon today; indeed obesity is becoming more prevalent among the poor compared to the rich in many

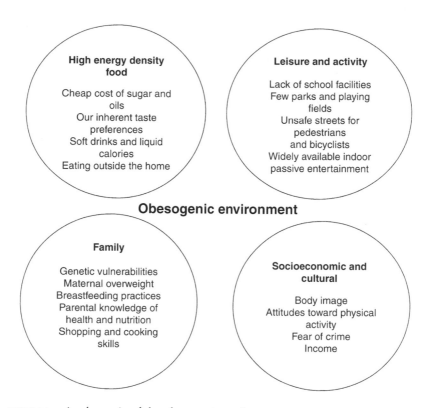

Obesogenic environment

High energy density food

Cheap cost of sugar and oils
Our inherent taste preferences
Soft drinks and liquid calories
Eating outside the home

Leisure and activity

Lack of school facilities
Few parks and playing fields
Unsafe streets for pedestrians and bicyclists
Widely available indoor passive entertainment

Family

Genetic vulnerabilities
Maternal overweight
Breastfeeding practices
Parental knowledge of health and nutrition
Shopping and cooking skills

Socioeconomic and cultural

Body image
Attitudes toward physical activity
Fear of crime
Income

FIGURE 5.1.   A schematic of the obesogenic environment.

countries (Brown and Condit-Bentley, 1998). Unfortunately, being calorie sufficient does not necessarily translate to satisfying nutrient requirements. We examine the evidence of nutrient deficiencies linked with obesity.

Finally, we turn to some intriguing ideas about how obesity may be spreading through the human race. Is obesity actually contagious in some manner? There are both social and biological methods by which obesity could spread by contact.

The factors in our modern world that affect our diet, activity, body image, and eventually our motivation to behave in ways that make us vulnerable or resistant to gaining excess fat cannot be all explained by biology. They all impact our biology, and are influenced by our biology, but there are nonbiological factors to the obesity epidemic. We try, in this chapter and elsewhere in the book, to acknowledge them, and we do not discount their importance. However, we are biologists, and this book is about human biology and its relation to human obesity. To the readers who believe we are giving sort shrift to nonbiological factors driving the increase in human obesity we can only say that we can write with authority only about that which we know.

Modern Food

Food is a logical place to start. Although a vulnerability to obesity comes from many factors, at base it arises from eating more food than is necessary to maintain weight. The extra food energy is stored on the body as fat. This is normal adaptive biology. The question is, why are so many people these days eating more than they expend on a consistent basis?

The foods of today are generally high in one or more of the following food types: meat, starch, simple sugars, and fat. They tend to be low in fiber and other difficult to digest materials. Modern foods are easily digestible and have a high energy density. They are the kinds of foods that would have been both rare and desired in our past. The types of foods our ancestors would have been motivated to expend considerable effort and risk to obtain. Of course nowadays we can have them delivered. Unfortunately that doesn't appear to have reduced our taste for them.

Why have modern foods become so energy-dense? Some of the answer is an understandable market response. Companies are delivering foods we like, and many of our taste preferences were set long ago when the desire for high-calorie foods, and thus being motivated to exert the effort and chance the risks to obtain them, was adaptive. External factors limited our access to high-calorie foods in the past. The modern economy and technology have made them readily available. Our motivation and desire for calorie-dense foods no longer serves the past adaptive purpose, but it still exists.

There are also some more recent historical factors to consider. In the early 1900s the concern in England and the United States was about undernutrition, especially for the working classes. A lack of calories was cited as a reason for poor work performance. There was concern that the diet of the working man did not contain enough fat and thus didn't provide enough calories to sustain the physical labor that most work required. This represented a change in attitude in Britain. Previously the prevailing economic wisdom was that wages should be set at subsistence levels (Oddy, 1970). Hunger was thought to be motivating. According to Townsend (1786, quoted in Oddy, 1970) hunger was "the most natural motive to industry and labour, it calls forth the most powerful exertions." Estimated mean energy intake by working people was either stable or actually declined in England from the mid-1800s to the end of the century, and then rose in the early 1900s. The level of energy intake (2,000 to 2,300 calories) would be considered barely adequate today, especially considering the higher levels of physical activity required back then. Of course average height was lower then, indicating that a significant proportion of the population was underfed and not able to reach their growth potential.

With this change in economic and business philosophy—regarding worker health and welfare as a positive factor in productivity—producing inexpensive foods with high-caloric density was put forth as an economic issue; it would be good for industry. Well-fed workers would be more productive.

Modern data show that the concern was likely well founded. A survey of workers in Bangladesh in the early 1990s examined the proportion of men who were unable to work during the previous month due to illness (the study specifically excluded accidents). There was a strong effect of BMI (body mass index) on the result (Figure 5.2). More than 40% of men with a BMI less than 16 kg/m² and about 35% of men with a BMI of 16 to 17 kg/m² had been unable to work. The proportion decreased significantly for BMI between 17 and 20 kg/m² and was stable at about 12%. Men with a BMI greater than 20 kg/m² had the lowest rate of being unable to work due to illness at less than 5% (Pryer, 1993). A higher BMI appeared to be protective against illness, and it increased annual productivity. Of course none of the surveyed workmen in Bangladesh had a BMI greater than 25 kg/m². The evidence from developed countries is that high BMI is associated with greater absenteeism at work (Bungum et al., 2003) and at school (Geier et al., 2007) due to health. Thus both low and high BMI are associated with health risks.

Undernutrition and malnutrition were serious concerns in many countries throughout the twentieth century and still are a concern in some countries. The causes for malnutrition have changed. The main cause of malnutrition in the past was lack of food. Malnutrition now is often the

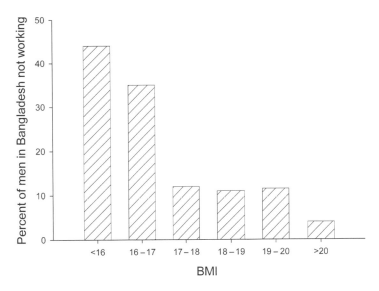

FIGURE 5.2.   Men in Bangladesh not working in the previous month due to illness. Accidents were excluded. Low BMI is associated with missing work. Data from Pryer, 1993.

result of political instability and violence. That leads to lack of food for certain people in certain areas, but not because sufficient food doesn't exist; rather, it cannot be delivered to those who need it. The concern about worldwide calorie malnutrition was and still is both about economic production and also about morals. People in the richer nations were motivated to help reduce starvation in the poorer countries. The impetus to produce more calories at a cheaper cost was a worldwide priority.

We succeeded in producing cheap, high-calorie foods. Our market systems combined with our technology have solved the problems of distributing that food around the world (at least in the presence of political and social stability). Unfortunately, our industrial output no longer relies on brawn and impressive, sustained physical effort. We have produced a solution to a problem that, in many countries, no longer exists, and contributed to a new health problem.

Ironically, the economics of food appears to be changing and may be reversing the cheap food "problem." A number of circumstances, including climate change, the increase in meat consumption that diverts grain to food animals instead of people, and the diversion of grains and other foods to biofuels have resulted in a worldwide increase in food prices. Malnutrition and starvation may again become predominant among the world's poorest people, even as obesity increases.

Liquid Calories

We also have developed a relatively new (evolutionarily speaking) category of calorie source: drinks, alcoholic and high-sugar. Alcohol has been ingested by humans for all of known history and much of our prehistory; alcohol consumption likely predates agriculture, as hunter-gatherers in some areas are known to be able to produce fermented beverages from fruits (e.g., coconut). Alcoholic beverages certainly provide calories, but that is not usually the purpose of their ingestion. High-sugar drinks, basically sweetened flavored water, are an even more recent concoction. There is concern that these drinks, having been virtually absent for most of our evolution, may be a poorly physiologically monitored source of calories.

People of all ages apparently like to drink sodas, and the market system has responded. The number of different flavors and brands of soda on the market are amazing. In supermarkets an entire aisle is usually devoted to sodas alone. The size of the individual units being sold has also increased. The original contour bottles of Coca-Cola were 6.5 ounces. Now the individual can or bottle of soda holds roughly 2 to 3 times that amount. When soda was introduced at restaurants, McDonald's offered only 7 ounces, a size that is not even on the menu today. McDonald's now offers 12-, 16-, 21-, 32-, and 42-ounce soft drinks. The biggest soda size recorded was a whopping 64 ounces at 7-Eleven, and these sodas are rarely shared. In addition, sodas at restaurants often come with free refills; this is not true of juice or milk. Sodas represent some of the cheapest calories available in our modern environment. Not surprisingly, individuals who drink large amounts of soda have increased odds of being overweight (e.g., Schulze et al., 2004; Fowler et al., 2005).

And it isn't just sodas anymore. There are sweetened fruit punches; sweetened, flavored sports waters; and sweetened, flavored coffees and teas. There are whole companies geared to sell calorie-laden liquids to people. The coffee outlets alone are ubiquitous, and coffee drinks now come with the equivalent calories (from added sugar and fat) of a small breakfast.

How well does our body assess these liquid calories? Are liquid foods as satiating as solid foods? Do we account for our double latte espresso in our regulation of food intake? The amount of calories from a single 16-ounce soda daily would result in a 20-pound weight gain in a year if the caloric intake was not compensated for by increased activity or a decrease in other food intake. And it appears that people do not completely compensate for the calories in sugar-sweetened drinks. Women who increased their consumption of sugar-sweetened drinks from less than one per week to at least one per day increased their average daily calorie intake by an average of 358 kilocalories. Women who reduced sugar-sweetened drink consumption from one or more per day to at most one per week decreased average calorie intake by almost the identical amount (319 kilocalories/day; Schulze et al., 2004). Women who maintained a consistently high level of soda consumption gained on average a kilogram per year over the eight-year study and had almost twice the relative risk of developing type 2 diabetes (Schulze et al., 2004).

But of course we have diet sodas with zero calories so they won't contribute to weight gain. Apparently this is not true. In a recent study, people who drank diet sodas were significantly more likely to become obese (Fowler et al., 2005). Of course causality is tricky here; perhaps gaining weight leads to drinking diet sodas. That probably is true. What can be said is that drinking diet sodas does not appear to be a successful strategy to combat weight gain.

And there is evidence that artificial sweeteners can actually increase food intake. Rats fed artificial sweeteners had higher intakes of rat chow and thus higher total energy intake (Tordoff and Friedman, 1989). There are plausible mechanisms that would cause a sweet taste to increase appetite. The cephalic-phase insulin response (discussed in chapter 9) can be triggered by noncaloric sweet-tasting substances (Powley and Berthoud, 1985; Tordoff and Freidman, 1989). The increased insulin secretion will result in a transient decrease in blood glucose concentration if the diet soda was ingested by itself and would moderate the rise in blood glucose concentration due to any food ingested with the diet soda. It also would gear metabolism to storage and away from oxidation, in anticipation of a surge of metabolic fuel entering the bloodstream (Tordoff and Friedman, 1989). So drinking a diet soda before your meal arrives could lead to higher food consumption and to enhanced fat storage.

Fructose

Many of the drinks mentioned above are sweetened by sucrose or high-fructose corn syrup. Sucrose is a disaccharide; that is, it is composed of two sugar molecules (glucose and fructose) bound together. Fructose is perceived as very sweet by human beings, and indeed by many other mammals; sweeter than sucrose (e.g., Hanover and White, 1993). High-fructose foods were rare and probably desirable in the past environment (e.g., honey and many ripe fruits are moderately high in fructose). The use of fructose in foods undoubtedly increases their palatability and our liking for them.

Fructose has metabolic effects that may contribute to its being an obesogenic food. Hepatic metabolism of fructose is conducive for de novo fat synthesis (Bray et al., 2004; Havel, 2005) and high fructose consumption is associated with hypertriglyceridemia (Lê and Tappy, 2006). Fructose consumption does not stimulate insulin secretion (fructose uptake into cells is via an insulin-independent mechanism using GLUT5 rather than GLUT4) and does not facilitate leptin secretion (Havel, 2005). Consumption of a high-fructose meal by humans led to lower secretion of insulin and leptin, and a lesser suppression of ghrelin (Teff et al., 2004). This endocrine pattern would be expected to produce less satiation and thus a lower reduction in appetite, possibly resulting in increased caloric intake.

High Glycemic Foods

The glycemic response to food is considered an important indicator both of the potential effects of the food in question and of the state of the glucose regulation system of the individual consuming the food. The response depends on the characteristics of the food and of the person eating the food. The glycemic response is generally measured as the incremental area under the blood glucose response curve (IAUC). From this measure several relative measures of glycemic response can be calculated and used to compare foods: most notably glycemic index, glycemic load, and glycemic impact (Monro and Shaw, 2008).

High glycemic foods are thought to have significant impact on long-term health and the propensity to weight gain; however, the evidence is uncertain as to what extent. Low glycemic foods do appear to reduce subsequent energy intake (Flint et al., 2006). The evidence supports the hypothesis that low glycemic response foods are beneficial for people with impaired glucose regulation, though the effect is not particularly large; it is not clear that these foods have any beneficial effects for people with unimpaired glucose regulation (Howlett and Ashwell, 2008).

More Than a Source of Calories?

Studies have shown that food is more than a source of metabolic fuel and raw materials for tissue synthesis. Food components can act as signaling molecules. Food can have epigenetic effects. Dietary supplementation of pregnant rats with folic acid, vitamin B12, choline, and betaine changed coat color in the offspring via a flipping of a transposon in the *agouti* gene, probably through DNA methylation (Waterland and Jirtle, 2003).

Fatty acids can act as signaling molecules as well. Long-chain fatty acids amplify pancreatic β cell secretion of insulin via cell-surface receptors (Poitout, 2003). Both the amount of calories and the metabolic fuel in which calories are delivered can program physiology in young, growing animals. A high-carbohydrate diet fed to rodents for 2 days immediately after birth (instead of normal high-fat milk) results in adults that are hyperinsulemic with a heightened sensitivity to glucose (Patel and Srinivasan, 2002).

Numerous events occur in fetal and early postpartum life that will affect the adult's vulnerability to disease. Our modern environment and its interaction with our biology can affect not only adults but also their offspring, even before they are born. Maternal diet and food-related behaviors can affect the unborn child. Glucose itself can act as a mutagen and has been implicated in the etiology of certain birth defects in the children of diabetic mothers (Lee et al., 1995). Early programming of physiology that has lifelong consequences is considered in more detail in chapter 13.

Eating Out

We haven't just changed what we eat and how much we eat, but also how we eat. With our busy work lives, and possibly more discretionary income than in the past, much more of our eating is done away from the home. Between the late 1970s and the mid-1990s, the percent of calories contributed to our diets by eating outside of the home increased from 18% to 34%. Much of this food is higher fat, lower nutrient (particularly fiber and calcium) food (Bowman et al., 2004; Bowman and Vinyard, 2004). Even breakfast, the meal most likely to be eaten at home is now often eaten at a restaurant or purchased and then eaten on the go. More than 25% of McDonald's U.S. business is accounted for by their breakfast menu. And again, we have separated effort from eating. People frequently do not even prepare the foods they eat anymore. There is often almost no effort associated with eating.

What is consistent with our past is the association of food with sociality. The advertisements for restaurants stress not only good (and plentiful) food but also a social atmosphere. Eating is what you do with friends and lovers. Eating together is associated with bonding. Eating has social rewards; at least that is the message of many restaurant advertising campaigns. And it probably does really reflect our biological heritage.

We evolved as a species for which communal eating and cooperation in food gathering and consumption was a key adaptation. Food and eating became intrinsically linked with our social behavior and our social identity. Our group was who we ate with, and individuals' value within the group may very likely have been strongly affected by their effect on the group's food supply. Our culture is laden with references that support that idea. The bread winner; bringing home the bacon; such expressions convey the concept that providing food for the group is valued. Of course these reflect modern culture and values, but their origins may be very ancient.

Portion Size

In the United States it seems that everything about eating has been getting bigger. The sizes of food packages; the portions served in restaurants; even the plates we eat off of are significantly larger now. A bagel 20 years ago was about 3 inches in diameter; now bagels are usually 6 inches in diameter, which means they contain more than twice as many calories per bagel. According to the U.S. Department of Health and Human Services, many foods today contain double or more the calories compared to how they were routinely prepared 20 years ago (e.g., cheeseburgers, muffins, chocolate chip cookies, even coffee drinks).

Portion size influences total energy intake. Increasing portion size by 50% resulted in higher daily energy intakes that were sustained for an 11-day period (Rolls et al., 2007). Reducing portion size leads to overall reductions in energy intake (Rolls et al., 2006).

There are cultural differences in portion size and in eating behavior. For example, the social and cultural behaviors regarding food and meals differ between France and the United States. Portion size of meals in restaurants is smaller in France, and people in France, on average, spend more time eating a meal than do people in the United States. Thus they consume fewer calories per meal even though they spend more time eating (Rozin, 2005). In the United States all-you-can-eat buffets are common and popular, especially for workday lunches where there is limited time to eat.

Physical Activity

We evolved to be a physically tough, hard-working species. Human beings are quite capable of daily energy expenditures of 3,000 calories (that is 3,000 kilocalories to animal nutritionists) or more indefinitely. In fact in the past that probably was very close to the expenditures of people engaged in hard physical labor. That would be about twice the resting metabolic rate (see chapter 6), which is a reasonable energy expenditure for a free-living animal. But that doesn't mean that we are driven to expend that much energy daily. One of the asymmetries that likely contribute to overweight and obesity is that motivation to indulge in food can be greater than motivation to indulge in physical activity.

In the past a majority of our waking hours would have likely been spent in physical effort. Many of us now spend the majority of our waking hours sitting in a chair. Physical activity has become optional for many people. It is something that is done during nonwork hours, as recreation or for health.

NHANES performed a study measuring not only the number of activities but also their intensity and found that less than 5% of individuals meet the recommended 30 minutes of moderate physical activity 5 days per week. Self-report of physical activity was significantly less accurate than direct measures, with individual reports indicating that 30% of individuals believed they met current exercise guidelines—another example of how powerful and pervasive self-deception is. Many people think they are more active than actual measurement reveals.

Perhaps that reflects a bias within our cognitive system stemming from our past. Maybe effort is judged to cost more than the gain from equal calories of food. Or maybe it is a lack of experience to judge by.

People certainly are capable of rather amazing feats of consistent, intense hard work. The Erie canal was dug by men and mules using shovels and wheelbarrows (the men not the mules). Loggers in West Virginia in the 1800s would hike several hours to a work site, work a 10- to 12-hour day, and then hike several hours back home to eat dinner and sleep. It was very humbling to find that an enjoyable but strenuous backpack trip was in the past covered twice daily by a man carrying an ax and his lunch on his way to a day's worth of hard physical labor. What is unusual used to be routine.

Going hand in hand with this sedentary lifestyle is an increase in the amount of time spent watching television, with 52.3% of the population watching 3 or more hours per day in 2001–2002 (NHANES, C. Tabak unpublished analyses). In children, trends are similar. The U.S. Centers for Disease Control and Prevention (CDC) Youth Risk Behavioral Surveillance System in 2003 found that over 35% of children watched 3 or more hours of TV each day, and over 21% used computers for the same length of time. Children who watched 4 or more hours of television per day had greater body fat and a higher BMI than those who watched less than 2 hours per day (Andersen et al., 1998). Only 36% of children met physical activity recommendations, and less than 33% have a daily physical education class.

A lack of physical activity can contribute to weight gain. There is convincing evidence that physical activity can ameliorate the health effects of obesity (reviewed in LaMonte and Blair, 2006). Cardiovascular fitness has independent effects on health and well-being.

The Built Environment

The built environment consists of the man-made features that provide the setting for human activity. It ranges from the largest scale of endeavor (e.g., railroads, interstates) to our personal spaces. It contributes in obvious and not so obvious ways to our lifestyle and to our vulnerability to obesity (Brownson et al., 2001; Gordon-Larsen et al., 2006).

Humans live in an incredible array of habitats, under conditions ranging from extreme heat to extreme cold. We originally evolved in Africa, however, and it is likely that heat loss was more important than heat conservation. We are well adapted to losing heat; our relative hairlessness and our sweat glands that are distributed across almost our entire body enable us to transfer heat to the environment. Because of our built environment we are able to live far from the equator, in environments that seasonally get extremely cold. We now regulate our temperature more through technology than through metabolism and physiological adaptation. Energy expenditure due to thermoregulation is quite minimal in today's world, at least in developed nations with climate-controlled buildings.

Our modern built environment does not encourage routine exertion. For example, our transportation routes are designed for mechanized transport: cars and buses, not pedestrians and bicyclists. This is a change from land-use patterns before the early twentieth century that were of necessity designed for convenient pedestrian travel for common activities such as shopping and going to school (Sallis and Glanz, 2006). With the increase in automobile ownership and the resulting rise of suburbs, land-use policies were put in place that were favorable to automobile travel but often unfavorable to pedestrian travel. These include zoning codes that separated residential from commercial and industrial areas. Jobs and stores were now no longer within easy walking distance, and most travel required the use of an automobile (Sallis and Glanz, 2006).

That the built environment influences our behavior, our choices, and hence our vulnerability to obesity has intuitive appeal. Research has demonstrated that aspects of this theory are supported. For example, adults who live in communities deemed "walkable" are more physically active and less likely to be overweight (Sallis and Glanz, 2006). Of course the direction of causality is a concern; perhaps people that value physical activity are simply more likely to choose to live in walkable communities. Still, the evidence that the built environment influences physical activity is reasonably well supported. It is certainly the case that aspects of the built environment can significantly discourage walking and other activity.

Because of the changes in where we live, how we travel, and how we work, physical activity has become mostly recreational. This is especially true for adolescents and children. Very few work on farms these days or have physically demanding chores. Most do not even walk to school (Zhu and Lee, 2008). Even in urban schools, where many children live near to the school, safety concerns due to crime, traffic, and poor street and sidewalk conditions reduce the proportion of students who walk (Zhu and Lee, 2008).

The availability of recreational facilities has a significant affect on the levels of physical activity in children and adolescents. For example, access to a safe park was positively associated with physical activity in urban adolescents (Babey et al., 2008). Unfortunately, recreational facilities that encourage physical activity (e.g., schools, swimming pools, parks, playing fields, basketball courts) are unequally distributed within the United States. The relative odds of adolescents in a community being overweight decreased almost linearly with the number of physical activity facilities in the community (Gordon-Larsen et al., 2006; Figure 5.3). The good news is that even one such facility was associated with a decease in adolescent overweight prevalence; the bad news is that many poor neighborhoods, where adolescent overweight and obesity rates are highest, are the least likely to have such facilities (Gordon-Larsen et al., 2006). The neighborhood environment also affects adults' likelihood of encouraging children to use available recreational facilities. The presence of litter, graffiti, and other markers of neighborhood disorder increases safety concerns and

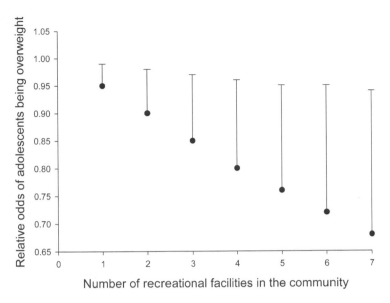

FIGURE 5.3. The existence of even one recreational facility in a U.S. community significantly decreases the relative risk of obesity in adolescents. Data from Gordon-Larsen et al., 2006.

reduces adults' willingness to encourage children to use local parks (Miles, 2008).

Land-use patterns and the built environment may also affect food choices. The relative ease of access to fast-food restaurants and convenience markets versus grocery stores has been shown to be associated with obesity prevalence (reviewed in Papas et al., 2007). Low-income neighborhoods generally have less access to good quality produce and other healthy foods and fewer healthy menu options in their local restaurants (reviewed in Sallis and Glanz, 2006). The evidence for the effects of the built environment on dietary choices is significantly less than for effects on physical activity, however. Indeed, most low-income consumers were found to have access to healthy food alternatives, however at significantly greater cost than the less healthy standard foods (Jetter and Cassady, 2005).

A final consequence of our modern lifestyle, technology, and material culture is the chemical environment. We produce a lot of novel materials that have become ubiquitous in our life: plastics, silicon chips, even nano-particles. The manufacturing process uses many chemicals that are usually either absent or in very low concentration in the environment (e.g., heavy metals, various solvents, polychlorinated biphenols, bisphenol A). The levels of these chemicals in the environment have increased; thus chemical exposure and the obesity epidemic are associated.

Is there any causal link? The evidence is certainly not definitive, but there are data to support chemical exposure as one of many factors that might cause weight gain in vulnerable individuals (Heindel, 2003). In many toxicity tests of chemicals, doses given to laboratory animals well below the toxicity limit (and therefore deemed safe) have resulted in consistent weight gain (reviewed in Baillei-Hamilton, 2002). There are many chemicals being released into the environment that have estrogenic or estrogen-blocking characteristics; for example bisphenol A (Heindel, 2003; Crews and McLachlan, 2006). These potentially endocrine-disrupting chemicals have been linked to human health and disease (reviewed in Crews and McLachlan, 2006).

Sleep

We evolved as a species that likely slept through most of the night. All but one species of anthropoid primate (the monkeys, apes, and ourselves) are habitually active during daylight and sleep at night. The owl monkey of Central and South America is the only nocturnal monkey. Because of technology, many humans are now able to shift their life schedule to be less dependent upon the natural light cycle. In many professions there are work shifts throughout the 24-hour day; some occupations (e.g., bartenders, night security guards) are predominantly performed at night.

Night shift workers are subject to a number of potentially disruptive factors that affect their lives. These can be external, such as having to accommodate the majority of the world that largely functions on a daylight-centered schedule; they also are internal, in that their sleep and eating patterns are now out of sync with the evolved norms in circadian patterns of metabolism and endocrine function. For many people, maintaining a reversed circadian activity schedule results in a lessening of health over the long term (Boulus and Rossenwasser, 2004). There is significant variation among people's ability to adapt to a time-shifted circadian pattern. Some people are largely incapable of doing so for any extensive period of time. Sleep disturbances are perhaps the most common disorder for night shift workers (Boulus and Rossenwasser, 2004). Sleep disorders have been linked to many diseases, including cancer (Spiegel and Sephton, 2003). Night shift work may become listed as a carcinogen because of the documented increased risk for certain cancers such as breast and prostate cancer (Straif et al., 2007).

Sleep disruptions and sleep loss have metabolic consequences that may contribute to the development of obesity and type 2 diabetes if they persist. Sleep-disordered breathing (e.g., sleep apnea) is associated with diabetes, largely in association with obesity (Resnick et al., 2003). Sleep disturbances (Nilsson et al., 2004) and short sleep duration (Mallon et al., 2005) are associated with diabetes in men; both short and long sleep duration are associated with diabetes in women (Mallon et al., 2005; Ayas et al., 2003).

In the United States the mean number of hours of continuous sleep has decreased. In 1960 the modal sleep duration for adults was between 8 and 9 hours (Kripke et al., 1979); in 1995 that value had dropped to 7 hours (Gallup Organization, 1995). Recently, almost 1 in 3 adults between 30 and 64 years of age reported routinely sleeping fewer than 6 hours per night (National Center for Health Statistics, 2005). Sleep loss and short duration of sleep have increased in parallel with the increases in obesity and diabetes (Spiegel et al., 2005; Knutson et al., 2007)

Short sleep is associated with several metabolic conditions that would predispose to weight gain and poor glucose control. Indeed, sleep deprivation is associated with increased insulin resistance; in addition, pancreatic beta cells in sleep-deprived people do not respond to insulin resistance with increased insulin secretion. Sleep loss altered pancreatic β cell secretion in healthy young men (Schmid et al., 2007). It decreased basal glucagon secretion and enhanced secretion in response to hypoglycemia. Sleep loss also increased feelings of hunger (Schmid et al., 2007). Sleeping less than 6 hours per night doubles the risk of developing type 2 diabetes in men (Yaggi et al., 2006).

Sleep deprivation leads to hyperphagia in rats (Rechtschaffen et al., 1983). Hypocretins, also called orexins, provide a molecular basis for linking sleep and feeding; these molecules have potent affects on both food intake and awakening (Sutcliff and de Lecea, 2000). Circulating catecholamines are reduced during sleep; disrupted sleep is associated with an increase in nocturnal circulating catecholamines (Irwin et al., 1999). Circulating levels of the appetite-suppressing peptide leptin have a circadian pattern, being highest during the middle of the normal sleep period. Subjects restricted to 4 hours of sleep per night for 6 nights had significantly lower mean and peak levels of circulating leptin (Spiegel et al., 2004). Short sleep also increased circulating levels of ghrelin (Taheri et al., 2004), a gut-brain peptide associated with increased appetite (see chapter 7).

Sleep disruption also affects weight retention after pregnancy. Short sleep (less than 5 hours per day) was significantly associated with post-partum weight retention. At 6 months postpartum, women who slept 5 or fewer hours per day on average were likely to have retained more than 5 kg of their pregnancy weight gain one year after giving birth (Gunderson et al., 2008).

Night eating syndrome is characterized by evening hyperphagia and frequent awakening during night sleep to eat small amounts of food (Stunkard et al., 1955). The night eating syndrome is associated with depressed mood and obesity (Stunkard et al., 1955; Birketvedt et al., 1999; O'Reardon et al., 2004). Women who displayed the night eating syndrome ate significantly more food at night than did controls, but interestingly did not have greater daily food intake. In the early morning hours they had lower levels of circulating ghrelin and higher circulating insulin, probably caused by their habitual awakening to eat. They also displayed greater depressive symptoms than did control women (Alison et al., 2005).

Chronic sleep loss and sleep disturbances are common in today's world. Many sleep disturbances are associated with obesity and diabetes. Indeed many sleep disturbances are exacerbated by obesity, possibly providing a feedback mechanism by which disrupted sleep favors weight gain, which further disrupts sleep patterns. Our changed sleep patterns may be another factor in our vulnerability to obesity.

The Nutrition Transition

The obesity epidemic is in full swing in the rich nations such as the United States, Great Britain, and most of Europe. It has hit certain geographic regions even harder, such as the South Pacific island nations. It is begin-

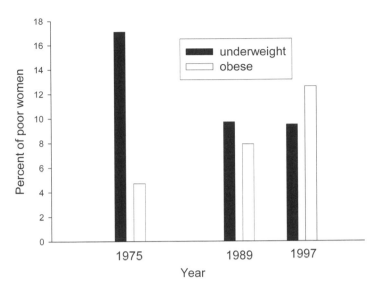

FIGURE 5.4. In the past, poor women in Brazil were more likely to be under-weight than obese, but that is no longer true. Data from Monteiro et al., 2004.

ning to develop in many Asian countries and certain African nations (Prentice, 2005). It has begun to affect the indigenous peoples of Amazonia (Lourenço et al., 2008). The start of the obesity epidemic can be seen in formerly poor nations that are becoming economically better off, at least in parts of the country. Being poor in the poorest nations of the world is associated with being underweight and malnourished. Being poor in a developing country is associated with an increased risk of obesity (Hossain et al., 2007).

Obesity is becoming a disease of the poor. For example, in Brazil poor women were much more likely to be underweight than obese in 1975. By 1997 this had reversed (Monteiro et al., 2004; Figure 5.4). The proportion of obese poor women has steadily increased, until in 1997 it exceeded the proportion of rich women who were obese (Monteiro et al., 2004; Figure 5.5).

The pattern of lifestyle changes that accompany the increase in obesity can be seen in developing nations. The changes associated with increased rates of obesity and its health consequences are demographic, occupational, lifestyle, as well as nutritional (Popkin, 2002). A suite of transitions characterize the change from a poor, economically disadvantaged society to a wealthier society. There is a demographic transition, in which high fertility and mortality rates give way to lower fertility and mortality. This change is accompanied by, and likely partially caused by, a decrease in the prevalence of infectious diseases associated with hunger and poor sanitation. Life span increases. The population structure shifts to include a higher proportion of older people.

The population also begins to shift from being predominantly rural to being predominantly urban (Figure 5.6). The cities have more employment options. Associated with this shift is a decrease in physical activity associated with work. Urban jobs are much more likely to be sedentary.

Diet begins to change, incorporating more foods with higher fat and refined carbohydrate, the so-called Western diet (Popkin, 2001; Prentice, 2005). Drewnowski (2000) said: "Wealth is associated with better diets." However, a further reading of his paper clarifies that the more accurate statement is that wealth is associated with higher-calorie diets. As average income levels rise in a country the diet begins to change, generally becoming higher in fats and simple sugars and lower in fiber. This change certainly delivers more calories per capita; it does not ensure better total

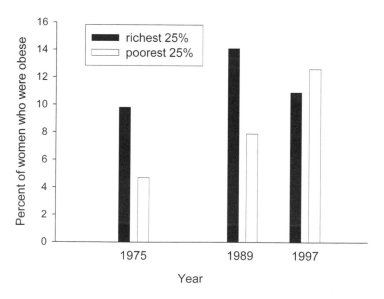

FIGURE 5.5. Poor women in Brazil are now as likely as rich women to be obese. Data from Monteiro et al., 2004.

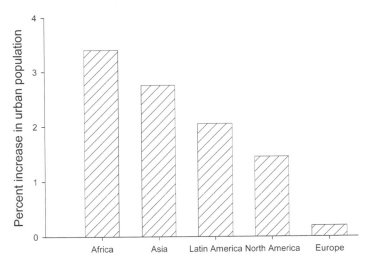

FIGURE 5.6. From 1995 to 2005 urban growth was faster in developing countries. UN, World Urbanization Prospects: The 2005 Revision, Table A.6, accessed at www.un.org/esa/population/publications/WUP2005/2005WUP_FS1.pdf.

nutrition or good health. In the U.S., low socioeconomic status is associated with low intakes of nutritious foods (Bowman, 2007).

The cost of fat and sugar has steadily declined (Drewnowski, 2000; Drewnowski, 2007). Technological advances, as well as agricultural subsidies in developed nations, have greatly decreased the cost of vegetable oils. Refined sugar is also remarkably cheap these days. A dollar's worth of either sugar or vegetable oil in the United States would provide a person with enough calories for about 2 to 4 days.

In conjunction with these changes in diet and lifestyle, average adiposity increases. The positive aspect of this change is that the prevalence of extreme thinness rates declines; unfortunately, however, obesity rates begin to climb, along with increases in the prevalence of the associated diseases. In 2000 there were an estimated 171 million people with type 2 diabetes in the world; that number is expected to more than double by 2030 (Wild et al., 2004; Hossain et al., 2007).

An unfortunate irony is that obesity is associated with poor nutrition, not just within the same country or community, but also within the same individuals. It may seem counterintuitive, but on reflection it makes sense. An excess of calories does not guarantee adequate amounts of all other nutrients. This was far less likely to be the case in our past; especially during preagricultural times when animal tissue was a dominant food. The foods in our past diets probably contained what we needed; the problem was getting enough. Satisfy your energy requirement and you would have eaten enough food containing enough other nutrients to satisfy requirements. The increased diversity of our foods and the production of foods with low nutritional content other than energy (e.g., sodas) have complicated that simple equation, and increased the chances of malnutrition.

This is a familiar complication of feeding wild animals in captivity. It is very difficult to re-create the natural diets of most wild animals with commercially available foods. Luckily animals need nutrients, not specific foods, at least for most species. For a few dietary specialists (koalas and eucalyptus leaves, for example) diet needs to be fairly food specific as well as nutritionally replete. However, most of the time a diet of commercial foods can be formulated and will be successful. Over- and undernutrition are always a concern, however. The kangaroo rats that deliberately selected a diet low in calcium was a case of a likely adaptive preference for high-water foods by a desert animal, which led to poor nutrition when the preferred food (iceberg lettuce) was offered every day (see chapter 4). A calorie-sufficient diet wasn't a nutritionally adequate diet.

There have been cases in the past where carnivores in zoos were fed predominantly raw meat. Makes sense; the animals were obligate carnivores, so why wouldn't pieces of beef or chicken be a good diet? In one case a young owl was fed just such a diet. The animal developed severe bone disease. A diet of meat, without the bone that it is usually attached to, was calcium deficient. Meat alone is not nutritionally complete. After all, carnivores, such as owls and tigers, don't eat meat. They eat other animals, and all that is in that animal. Owl pellets are still used today in science classes because of the bones of small animals (mostly rodents) that can be found in them. And although you can still find and identify the bones after they have passed through the owl's digestive tract, a good amount of the calcium in those bones was digested and absorbed by the owl.

What we have inadvertently done to captive animals in the past we appear to be doing to ourselves. Both children and adults who obtain a large percentage of their calories from fast foods consumed more calories but fewer micronutrients (Bowman et al., 2004; Bowman and Vinyard, 2004). There are some disturbing data on obesity in children and their nutritional status. For example, in the United States obese children are more likely to exhibit iron deficiency (Brotanek et al., 2007). Iron deficiency anemia is associated with significant behavioral and cognitive delays. Using data from the NHANES IV survey from 1999–2002, Brotanek and colleagues found that iron deficiency was present in 8% of toddlers (1- to 3-year-olds). The prevalence of iron deficiency ranged from 5.2 to 20.3% in different subgroups. The data indicated a protective effect of attending day care (only 5.2% iron deficient) and a greater risk among overweight/obese toddlers (20.3% iron deficient). Toddlers of Hispanic ethnicity had higher rates of iron deficiency (12.1%) than white (6.2%) or African American (5.9%) toddlers; however that result may have been due to the lower rates of day care attendance and higher rates of obesity among Hispanic toddlers (Brotanek et al., 2007). These data are in agreement with previous studies that have shown iron deficiency to be associated with obesity in children and adolescents (Nead et al., 2004). The causes of iron deficiency and its association with overweight in children are not fully understood, but prolonged bottle-feeding may lead to high intakes of milk and juice (Brotanek et al., 2005, 2007). This type of liquid diet can lead to excessive weight gain, and milk and especially juice are usually poor sources of iron. Thus more-than-sufficient calories are matched with insufficient iron, and anemic, obese toddlers are created.

Another nutritional concern related to obesity is folate status. Obese women generally have lower serum folate status than do normal-weight women (Mojtabai, 2004). This may partly explain the higher rate of neural tube defects in the babies of obese women (see chapter 12). For example, even after folic acid fortification of flour in Canada, maternal obesity still conferred an increased risk of neural tube defects; indeed, the risk was even more pronounced than before mandatory folic acid fortification of flour (Ray et al., 2005).

Is Obesity Contagious?

Evidence indicates that friendships and social companions have a great influence on the propensity to gain weight (Christakis and Fowler, 2007). The social distance between people was a significant predictor of mutual weight gain. Geographic distance was unrelated. For example, weight gain by a near neighbor had no statistical affect on weight gain of the subject. Weight gain by a close friend of the same sex, however, did significantly increase the probability the subject would gain weight, regardless of geographic separation. This connection seems to transcend the intuitive connections between social partners and behavior. For example it isn't that friends necessarily share exercise habits and attitudes toward physical activity, though that certainly happens. The evidence is consistent with people taking the norms of appearance from their network of friends. The authors hypothesize that weight gain by a significant friend may increase the acceptance of weight gain by the subject (Christakis and Fowler, 2007).

Other intriguing ideas have been proposed that could indicate that a vulnerability to weight gain may truly have an infectious component. There is an association between obesity and the kind of gut flora a person has. Some symbiotic intestinal microorganisms are more efficient at fermenting substrates and providing energy for themselves and the host (reviewed in Dibaise et al., 2008). Obese people were more likely to have gut flora dominated by these kinds of gut microorganisms (Ley et al., 2006; Turnbaugh et al., 2006). In addition, lipopolysaccharide from gram negative bacteria interacting with a high-fat diet may be one triggering mechanism for the chronic inflammation associated with obesity and the metabolic syndrome (Dibaise et al., 2008).

Our food choices certainly affect our gut microbe populations. A fascinating hypothesis posits that the reverse may be true; our gut microbes may influence our food preferences. Metabolic products in both plasma and urine associated with gut microbial activity differed between chocolate-craving and chocolate-indifferent men, implying differences in gut microbial metabolism possibly derived from differences in microbial population (Rezzi et al., 2008). These differences were seen even when no chocolate was consumed.

There has also been a suggestion that viruses can affect weight gain and might be linked with obesity as well (Vasilakopoulou and le Roux, 2007). Several viruses have been linked to obesity in animal models (e.g., rats, dogs, and chickens). Human adenovirus 36 (Ad-36) causes significant fat gain in chickens, mice rats, rhesus monkeys, and common marmosets (Dhurandhar et al., 2000, 2002). A higher proportion of obese humans were seropositive for Ad-36 antibodies than were nonobese humans (30% versus 11%; Atkinson et al., 2005).

Summary

The modern environment differs in significant ways from that of our ancestors. These differences have consequences, both good and bad. We have changed our environment to make life easier; less exertion is required; and calories are easy to come by. The physical structures we design and build generally serve to reduce energy expenditure. We moderate our environmental temperature so we don't get too hot or too cold. Mechanical devices (elevators, automobiles) move us from place to place. A desire to reduce calorie malnutrition among people has led to the development of inexpensive, calorie-dense foods. Our market system has taken this to an extreme in many cases, responding to people's preferences for high-sugar and high-fat foods. These preferences are likely formerly adaptive behaviors that served to motivate our ancestors to acquire the rare and potentially hazardous-to-obtain energy-dense foods of nature. The technological capability of humans has resulted in making foods with these formerly rare characteristics ubiquitous. It is not surprising that many people get fat in this new environment; it is perhaps more surprising that many people remain lean.

CHAPTER SIX

Energy, Metabolism, and the Thermodynamics of Life

Obesity, at its core, results from a sustained period of positive energy balance. More energy is ingested than is expended in the processes of life. The excess energy is stored on the body, primarily as fat. The increase in adipose tissue is central to the metabolic cascades that will lead to a lessening of health. The simplistic and generally unrealistic answer is to reverse the process; expend more calories than are consumed. This is very easy to say; extremely hard to accomplish.

In this chapter we explore the concept of energy as it pertains to biological systems. Energy is a powerful, subtle, and sometimes confusing concept. It is central to modern science. The term *energy* was first used in the modern sense by Robert Young of England in 1807; he substituted it for the Latin term *vis viva* used by Newton and Liebnitz to refer to what we now call kinetic energy. The principles of thermodynamics require the modern concept of energy. Indeed the first law of thermodynamics is that energy is conserved; that energy is neither created nor destroyed. Energy and the laws of thermodynamics are central to understanding metabolism, the collection of biochemical processes involved in living organisms.

What is energy? First of all energy is not a physical thing; rather it is a physical quantity that can be calculated for a system. This fact may have played a substantial role in the slow acceptance of the modern concept of energy. In the late 1700s, Antoine Lavoisier, considered by many to be the father of modern chemistry, made a number of key research findings that are central to our understanding of chemistry and metabolism. He showed that mass is conserved in chemical reactions; in other words, in a chemical reaction the total mass of the products is always equal to the mass of the reactants. He also showed that what we would now call heat energy is conserved. From this came the caloric theory of heat. The caloric theory of heat proposed that heat was an indestructible fluid that flowed between objects, from hot to cold. Heat was a thing. Just as mass was conserved, caloric was conserved. This was certainly a start toward the modern conception of energy, but it lacked the fundamental property that energy can be transformed among many manifestations without loss.

Energy is a powerful concept that can also be abstract and confusing. What is elegant and pleasing to some can be mind-numbing and sleep inducing to others. Potential energy, kinetic energy, the ability to do work, the energy in photons, the energy in electron orbitals, the energy in chemical bonds, the energy in electric and magnetic fields, the mechanical energy in a spring, all of these are different conceptions of what is ultimately the same quantity. And all of the above forms of energy are used by living things.

But energy is not a thing; it is a scalar quantity that can be calculated for a system. If the system is closed, such that nothing enters or leaves, then that calculated quantity called energy will remain constant, no matter how much other characteristics of that system change. The physics Nobel laureate Richard Feynman (1964) perhaps expressed it best: "There is a fact, or if you wish, a law, governing natural phenomena that are known to date. There is no known exception to this law—it is exact so far [as] we know. The law is called conservation of energy; it states that there is a certain quantity, which we call energy that does not change in manifold changes that nature undergoes. That is a most abstract idea, because it is a mathematical principle; it says that there is a numerical quantity, which does not change when something happens. It is not a description of a mechanism, or anything concrete; it is just a strange fact that we can calculate some number, and when we finish watching nature go through her tricks and calculate the number again, it is the same."

Hermann von Helmholtz may have been the first scientist to apply the principles of conservation of energy to physiology. He argued forcefully that physiology should be founded on the principles of physics and chemistry, and he rejected the notion of vital forces somehow separate from the nonliving world (Helmholtz, 1847). He showed that the conservation of kinetic energy was a mathematical consequence of the assumption that work cannot be produced from nothing; he went on to demonstrate that in circumstances where it appeared that energy was lost, in reality it was converted to heat energy.

The conservation of mass and energy are fundamental to our understanding of the biochemical processes that living organisms employ, and that we refer to as metabolism. Of course we now know that mass is another form of energy. Einstein's famous equation ($E = MC^2$) tied together two fundamental entities in physics, energy and mass, and linked them with a third, the speed of light (Einstein, 1905). Mass and energy are thus interchangeable, with both exhibiting aspects that previously had been thought to relate only to one or the other. For example, light can be bent by gravity, the fundamental attraction between masses. Mass can be converted to photons, and nuclear weapons made.

In physics, mass/energy is a fundamental unit. In biology, mass and energy remain separate, though we realize they are forms of the same intrinsic quantity. However, as far as we know, no living things use nuclear decay directly in their metabolism. In the metabolic pathways of life both mass and energy are conserved, and one is not transformed into the other. There are important differences between the physics and biological concepts of mass as well. In much of physics, what is important is the quantity of mass; its composition does not matter. The force needed to accelerate 1 gram of mass by 9.8 m/s² (the acceleration due to gravity at the earth's surface) is the same regardless of whether it is a gram of lead or a gram of feathers (in vacuum, of course). In other words, a gram of lead weighs the same as a gram of feathers. They both contain the same amount of total energy as well, given by Einstein's equation.

In biology, the amount of biological energy in a gram of matter depends on its composition. A gram of feathers does indeed represent an amount of biological energy, while a gram of lead has none. More to the point of this book, a gram of adipose tissue represents a significantly greater amount of biological energy than a gram of muscle, or skin, or bone.

Energy and Metabolism

Living organisms can be viewed as biological systems that cycle energy through themselves. Those energy cycles can be studied at the level of cells, organisms, or whole ecosystems. For the purposes of this book, which is to examine human obesity, analysis at the level of the ecosystem is not particularly relevant, though perhaps an analogy could be made with the economics and sociology of food in society. And, indeed, an understanding of our past ecology is important in understanding our current biology. However, this chapter focuses on energy and energy metabolism from the organism as the top level, down to cellular metabolism.

Metabolism, at its core, is the means by which biological entities cycle energy through different forms to produce the necessary molecules and perform the necessary functions of life. Metabolism is the link between the raw materials that come into the body through eating, drinking, and breathing and the functions of living. Metabolism is the totality of chemical processes that function to enable life. Molecules are broken down into their components releasing energy (catabolism), or built up from precursors using energy (anabolism). Spontaneous reactions generally release energy; other reactions require energy. A key aspect of metabolism is the coupling of energy-releasing reactions with energy-requiring ones, generally via enzymes.

In 1838 Germain Henri Hess published results of his investigations showing that the heat released by a chemical reaction only depended on the starting and end points; it didn't mater how many intermediate steps there were. This is fundamentally how life uses chemical energy. Metabolism is, in its most simplistic form, linking multiple steps in chemical reactions to produce necessary molecules or the energy to drive other reactions.

Many metabolic reactions are coupled to energy-storing and energy-recovering reactions. Very little heat will be generated. Energy is released by oxidation of ingested food and transferred eventually to energy-rich phosphate bonds in a variety of phosphorylated molecules (e.g., adenosine triphosphate [ATP]). The chemical energy in the bonds of the food is cycled through metabolic pathways to become chemical energy stored in the bonds of these phosphates. Evolution has acted to produce efficient metabolic pathways that lose very little energy during the process of transforming food into energy for life processes.

But wait, you might say, we were just told that energy is conserved. It is never lost. That is true: The total energy of the universe is conserved and is never lost, but energy can enter and leave any particular part of the universe. There are very few, if any truly closed systems, and life is certainly not a closed system. Energy enters and leaves a living organism in a variety of ways. A key aspect of this chapter is examining the components of energy intake and expenditure, what their consequences are for weight gain and obesity, and how scientists measure and study them. But before we discuss the various aspects of energy intake and expenditure another fundamental concept inherent in the energy metabolism of living things needs to be discussed. The amount of energy in a molecule is not the same as the amount of available energy, that is, the amount of energy that can be used in metabolism. A fundamental implication of the laws of thermodynamics is that there is a limit on the proportion of energy released in chemical reactions that can be used to perform work, that is, be metabolically available.

Thermodynamics of Life

Life exists within the constraints of the laws of thermodynamics. Indeed, living systems are biological "machines" that have evolved to use thermodynamic properties to exist and reproduce. The first law of thermodynamics is that within a closed system energy is conserved. The second law of thermodynamics states that the total entropy of any closed system must increase over time. A unique feature of this law of physics is that it gives direction in time to processes.

But what is entropy? Entropy may be as confusing, and powerful, a concept as energy. Like energy, entropy is not a thing; rather it is a measurable quantity of a system. Unlike energy, entropy is not conserved; indeed the total entropy of the universe continually increases. There are many functional definitions of entropy. The statistical definition of entropy is a measure of the number of possible microstates of a system; it is in effect a measure of the uncertainty about a system. Another way to think of it would be that entropy measures our ignorance of the system. Perhaps more relevant to our examination of metabolism, it is a measure of the ability of a system to undergo spontaneous change. A system in equilibrium has maximal entropy; we know little about the history of the system except for its conserved quantities (e.g., mass and energy) and that it will not change unless it receives input from (or loss to) the outside world. A system with low entropy is likely not in equilibrium, and we should be able to predict the direction of spontaneous changes to the system. Low-entropy systems have a low probability of stasis and a high probability of change; high-entropy systems are the reverse.

A misconception is that living organisms somehow violate the second law of thermodynamics because they represent an increase in order. Actually, living systems are based on and use the principles of the second law. Certain biological processes proceed spontaneously; these increase entropy. Many other important biological processes decrease the entropy of the system; these take energy. Metabolism links entropy-reducing reactions with entropy-increasing processes. The local entropy of the organism may decrease, but the total entropy of the organism and its environment always increases.

A good example of the use of entropy within biological systems is a high concentration of ions within a cell, for instance, calcium ions in intracellular fluid. The probability that the intracellular concentration would remain higher than the extracellular concentration naturally is low; this is a low-entropy condition. The concentration gradient is maintained by the low permeability of the cell membrane to the calcium ions. If the cell membrane permeability is increased (by opening voltage-gated channels in the cell membrane or other mechanisms), calcium ions will spontaneously flow from the intracelleular fluid to the extracellular fluid. Entropy has increased. To reverse the process, and thus return to the low-entropy state, energy is required to pump the calcium ions back into the cell. Of course the reverse situation, with lower concentration in the cell relative to extracellular fluid, simply works in the opposite fashion. Calcium ions would spontaneously flow into the cell if the voltage-gated channels are open, and it would take work to pump them back out. Skeletal muscle contractions are a nice example of both of these circumstances (Table 6.1).

In both these cases the spontaneous action (calcium ions flowing from high concentration to low concentration) increases the entropy of the system (defined as the intracellular space and the surrounding extracellular space). To decrease the entropy of that subsystem (re-create the ion gradient) energy is required. A substantial proportion of our energy metabolism (discussed in more detail below) consists of the energy necessary to pump various ions against a gradient (e.g., into or out of cells).

The most common forms of energy important to animals are the energy in chemical bonds; mechanical energy; energy inherent in various

TABLE 6.1   *Skeletal muscle contractions*

1. An action potential originating in the central nervous system reaches an alpha motor neuron, which then transmits an action potential down its own axon.

2. The action potential opens voltage-dependent calcium channels on the axon; calcium ions enter the cell from the extracellular fluid due to the concentration gradient.

3. Calcium ions cause vesicles containing the neurotransmitter acetylcholine to fuse with the plasma membrane, releasing acetylcholine into the synaptic cleft between the motor neuron terminal and the motor end plate of the skeletal muscle fiber.

4. The acetylcholine diffuses across the synapse and binds to and activates nicotinic acetylcholine receptor on the motor end plate. Activation of the nicotinic receptor opens its intrinsic sodium/potassium channel, causing sodium ions to enter the cell and potassium ions to exit, due to the relatively high-sodium/low-potassium concentration of the extracellular fluid. Because the channel is more permeable to sodium, a net of positive ions enter the cells and the muscle fiber membrane becomes more positively charged, triggering an action potential.

5. The action potential depolarizes the inner portion of the muscle fiber, which activates voltage-dependent calcium channels that are in close proximity to calcium-release channels (ryanodine receptors) in the adjacent sarcoplasmic reticulum, causing the sarcoplasmic reticulum to release calcium.

6. The calcium binds to the troponin C present on the actin-containing thin filaments of the myofibrils. The troponin then allosterically modulates the tropomyosin. Normally the tropomyosin sterically obstructs binding sites for myosin on the thin filament. Once calcium binds to the troponin C and causes an allosteric change in the troponin protein, troponin T allows tropomyosin to move, unblocking the binding sites.

7. Myosin (which has ADP and inorganic phosphate bound to its nucleotide binding pocket and is in a ready state) binds to the newly uncovered binding sites on the thin filament (binding to the thin filament is very tightly coupled to the release of inorganic phosphate). Myosin is now bound to actin in the strong binding state. The release of ADP and inorganic phosphate are tightly coupled to the power stroke (actin acts as a cofactor in the release of inorganic phosphate, expediting the release). This will pull the Z-bands toward each other, thus shortening the sarcomere and the I-band.

8. ATP binds myosin, allowing it to release actin and be in the weak binding state. (A lack of ATP makes this step impossible, resulting in the rigor state characteristic of rigor mortis.) The myosin then hydrolyzes the ATP and uses the energy to move into the "cocked back" conformation.

9. Steps 7 and 8 repeat as long as ATP is available and calcium ions are present.

10. Calcium ions are actively pumped back into the sarcoplasmic reticulum. The active pumping of calcium ions into the sarcoplasmic reticulum creates a lower concentration in the fluid around the myofibrils. This causes the removal of calcium ions from the troponin. Thus the tropomyosin-troponin complex again covers the binding sites on the actin filaments and contraction ceases.

gradients, especially electrical; and heat. Heat is often termed a waste product because it cannot be used to produce tissue. Heat is an inevitable by-product of biological systems, in concordance with the second law of thermodynamics; it is energy expenditure, defined as energy lost to the organism. However, it is quite useful as well. Many metabolic processes are most efficient within a certain range of temperatures. Mammalian physiology has quite a narrow range of temperatures (compared with the environment) within which the animal remains viable. In general the environment is cooler than the lower limit of mammalian body temperature. Thus heat will flow from the animal to the environment, and energy is lost (Blaxter, 1989; Schmidt-Nielsen, 1994). This is actually very important and adaptive; for many animals the greater danger is that metabolism and activity will cause body temperature to rise too much. Heat needs to be lost to the environment. To maintain body temperature, animals must regulate heat flow; sometimes by increasing heat loss and other times generating heat to replace what was lost.

Getting Rid of Energy

Heat is the main component of energy expenditure. Antoine Lavoisier and Pierre-Simon Laplace constructed the first direct calorimeter, a chamber surrounded by ice into which they placed a guinea pig; the heat produced by the guinea pig caused the ice to melt. By measuring the amount of melted ice they were able to calculate the heat production of the guinea pig. By comparing the results of this experiment with a guinea pig to that from using a flame, Lavoisier was able to demonstrate that animal metabolism is in effect slow combustion. This concept was embodied by Kleiber (1932) in *The Fire of Life*, his book on animal metabolism.

Heat is often considered a waste product. It can be a useful one in a cold environment. It can be deadly in a hot environment, however. Metabolism creates heat. If that heat is not lost to the environment then the animal's body temperature will rise. Mammals have a fairly narrow range of temperatures within which their metabolism will correctly function. Mammals have evolved many and varied anatomical, physiological, and behavioral strategies to conserve or lose heat: for example, blubber layers in marine mammals to conserve heat; extremely large ears with extensive blood supply to lose heat in elephants.

Heat transfer to the environment can be a limiting factor in metabolism. In a very illuminating set of experiments Kruk and colleagues (2003) studied lactating mice at different environmental temperatures. Compared with standard laboratory temperature (21°C [70°F]), at high temperature (30°C [86°F]) both food intake and milk production were decreased. Total energy expenditure decreased. The females weaned fewer pups and a smaller total litter weight. At cold temperature (8°C [46°F]) females increased food intake and milk output and weaned normal litters of pups. What is interesting about these results is that at 21°C other manipulations to raise female energy expenditure, such as giving her extra pups, were not able to increase either food intake or milk production (Johnson et al., 2001a). Thus the increased expenditure at 8°C cannot be explained simply as an up-regulation of metabolism in response to need. The lactating mice at 8°C did up-regulate their metabolism to meet a need, but the mice were unable to up-regulate metabolism in response to other manipulations at higher temperature (e.g., 21°C). The hypothesis of Kruk and colleagues (2003) is that metabolism is constrained by the ability to lose heat. At warmer temperatures an up-regulation of metabolism beyond the observed asymptotic levels would lead to overheating. At 8°C the increased flow of heat to the environment due to the steeper temperature gradient enabled the females to raise their metabolism and successfully cope with the additional energy requirement.

This illustrates an important point. Many factors constrain metabolism. Some are internal to the animal; some are external; and some are interactions between internal and external factors. Animals are finite, so there has to be an internal upper limit on energy metabolism and thus energy expenditure. There are also minimal energy requirements for life that animals cannot go below, at least for any considerable amount of time. However, external circumstances often limit metabolism to well below the theoretical maximum, as well as forcing animals to expend well above their theoretical minimum. Humans, with our technology and societal infrastructure, have removed many of the external constraints on the lower end of energy expenditure. Our thermoregulatory and activity energy requirements can be very low.

Eating and Entropy

Eating consists of bringing low-entropy materials into the body (produced by other living organisms), breaking them down to release energy (and increase entropy), and harnessing that energy to drive entropy-decreasing processes that are necessary for the organization of biological material that defines life. Living organisms are biochemical machines that cycle energy through themselves to decrease entropy locally while increasing it globally. Raw materials and energy enter the system; metabolism harnesses their energy through processes governed by the laws of thermodynamics; and the end products are increased chemical organization within the organism, physical waste products that are excreted, and heat.

Our bodies acquire the energy needed for life from the oxidation of food. Complex energy-rich compounds are oxidized to simpler energy-poor compounds, and the energy released is used in various ways. For the energy released by the oxidation of food to be used directly in metabolism it must be stored by transferring it to other molecules. Phosphate esters are an important class of energy-transducing molecules. These molecules react with water to release a significant amount of energy. However, even though hydrolysis of these molecules is thermodynamically favored, many of them are quite stable in water. This allows them to serve as chemical intermediaries to provide the energy to drive thermodynamically unfavored, but biologically important, reactions.

Thus energy cycles through metabolic pathways, carried by coenzymes such as adenosine triphosphate (ATP). Catabolic reactions produce ATP, which anabolic reactions then use as an energy source. Catabolic reactions, which are generally spontaneous, release energy and increase entropy; anabolic reactions decrease entropy and require energy input. Enzymes enable thermodynamically unfavorable reactions to occur by coupling them with thermodynamically favored reactions. Metabolism, in effect, is an enormous array of linked chemical reactions.

A common theme running through this book is that of ancient information molecules that evolution has adapted and co-opted to perform multiple and diverse functions in different end-organ systems and in diverse metabolic pathways. There is an intriguing link between the base information molecules for the machinery of life and the primary phosphate esters that regulate energy flow through the organism. Ribonucleic acid (RNA) is the bridge between the genetic code (deoxyribonucleic acid [DNA]) and the functional molecules necessary for life. Very simplistically, the information encoded in the DNA is translated to RNA, which then directs the assemblage of the amino acids to form the functional peptides. The base molecules that form RNA (adenine, guanine, cytosine, and uracil) are attached to a ribose sugar molecule and a phosphate group and linked together to form a linear polymer. These same base molecules attached to a ribose sugar molecule and then phosphorylated to the triphosphate form are the primary energy-regulating and energy-transferring molecules of life: adenosine triphosphate (ATP), guanosine triphosphate (GTP), cytosine triphosphate (CTP), and uracil triphosphate (UTP). The major workhorse of metabolism is ATP, but GTP is involved in protein synthesis, CTP in lipid synthesis, and UTP in carbohydrate synthesis. Thus the information molecules that form the information-transducing system are also the base of the energy-transducing system. Evolution has produced multiple systems using the same key building blocks.

Energy Expenditure

The basic components of energy expenditure are basal metabolism, thermoregulation, thermic effect of food (previously specific dynamic action), activity, reproduction, growth, and change of body composition (Kleiber, 1932; Brody, 1945; Blaxter, 1989). All of these components are variable to a certain extent, and under some degree of regulatory control. The sum of all these components equals total energy expenditure. A change in one of these components may result in a change in total energy expenditure, but it might be balanced by a change in one of the other components.

The various components of energy expenditure can be categorized into three classes: unavoidable and necessary, necessary but reducible, and optional (Figure 6.1). The prime example of the first category is basal metabolic rate (BMR). Activity is an example of a necessary but reducible energy expenditure. Reproduction is an evolutionary imperative, but on short time scales it is optional. Animals have reproductive strategies that determine when or if they put energy into reproduction.

Basal metabolic rate (BMR) is the minimal energy expenditure neces-

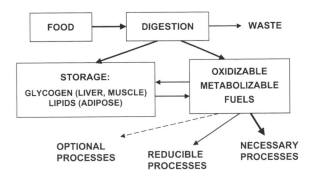

FIGURE 6.1.   Components of energy expenditure classified as necessary, reducible, and optional. Modified from Wade and Jones, 2004.

sary for life (McNab and Brown, 2002). It is a measure of energy expenditure under well-defined circumstances that should minimize reducible energy expenditures and eliminate most if not all optional expenditures. Variation in BMR is associated with a number of biological parameters. It is higher in endotherms (so-called warm-blooded animals: birds and mammals) than in ectotherms (fish, amphibians, and reptiles). For the purposes of this book we restrict our discussion to mammals.

A significant proportion of BMR is accounted for by basic cellular processes, such as moving ions and molecules across cellular membranes (Schmidt-Nielsen, 1994). In general, BMR accounts for about half of total energy expenditure in captive animals (e.g., Power, 1991) and between a half and a third of total energy expenditure in wild animals. Recommended caloric intakes for modern humans in industrialized societies, such as the United States, are only about 1.5 times BMR or less. For example, BMR for a 60-kg woman would be about 1,500 kcal/day; if she were to keep to a recommended diet of 2,000 kcal/day and maintain weight, her energy expenditure would represent 1.33 times BMR. Her BMR would account for three-quarters of her energy expenditure. This is another indication that our energy expenditure in the modern environment may be low relative to our evolutionary history.

Phylogeny, diet, and other adaptations, such as to the particular environment of the animal (e.g., desert, marine, and so forth), can and do affect BMR (McNab and Brown, 2002), but the primary explanatory parameter for mammalian BMR is body mass (Kleiber, 1932; Brody,

1945; Figure 6.2). Over almost the entire range of mammalian body sizes (the relationship changes at very small body mass) and across all mammalian orders, BMR is proportional to body mass raised to approximately the 0.75 power. The exact exponent is often debated (ranging mainly between 0.72 and 0.76), but matters little for the purposes of this book. The important point is that the exponent is significantly less than 1. Another important point to remember is that an allometry of approximately 0.75 describes the interspecific allometry of BMR; within a species the exponent is not necessarily 0.75 and in fact is often less than that value. For example, estimates for the intraspecific allometry of BMR in golden lion tamarins (small New World monkeys) ranged from 0.439 to 0.609 (Thompson et al., 1994). Thus required minimal energy expenditure in humans may increase even more gradually with mass than the rate predicted by the Kleiber equation.

Human BMR is unremarkable (Figure 6.3). Despite our large brains, technology, social adaptations, and everything else that supposedly separates us from other mammals, at the basal level our energy metabolism looks no different from other primates our size, or indeed other mammals our size. The regression line for primate active-phase BMR is little differ-

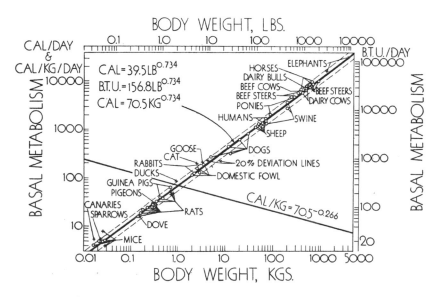

FIGURE 6.2.   The famous mouse-to-elephant curve for BMR (basal metabolic rate) from Brody, 1945.

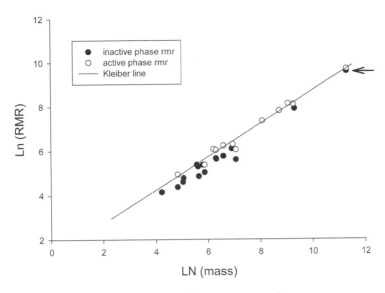

FIGURE 6.3. The plot of the logarithm of BMR (measured as oxygen consumption in ml O_2/hr) versus the logarithm of body mass in grams for primates. Open circles represent measurements taken during the active phase and filled circles represent measurements taken during the inactive phase. Human BMR is indicated by the arrow. The line is from the Kleiber (1932) equation 3.48 × body mass$^{.75}$.

ent from the original Kleiber curve (Kleiber, 1932). Of course a careful examination of the primate metabolic rate data indicate that the time of day a measurement was made has a significant effect. The pattern of metabolic rate across body mass for primates during their active time period (when they are habitually awake) differs from the allometric pattern for when they are asleep. In general inactive (sleeping) measurements are below the Kleiber line while active-phase measurements are on or above the line. The difference is more pronounced for smaller animals.

Humans, like other primates, spend a good portion of their lives sleeping. Metabolic rate in people while asleep is approximately 10% lower than when measured under the defined conditions for BMR. It has been demonstrated for many animals that metabolic rate during the time period when an animal is habitually inactive (asleep) can be significantly lower than when measured during the habitually awake and active period (Aschof and Pohl, 1970). This is certainly true of small primates (Thompson et al., 1994; Power et al., 2003). The smallest monkeys, New World primates in the subfamily Callitrichinae, reduce body temperature by several degrees at night and lower their metabolic rates by 25 to 40% (Power et al., 2003). This result is not restricted to primates; rodents less than 1 kg in body mass routinely lower their metabolic rate by an average of 25% during sleep, in this instance generally during daylight hours (Kenagy and Vleck, 1982).

Metabolic rate depends on many factors: nutritional status, environmental temperature, activity, even time of day. Metabolic rate is regulated; it is not a fixed quantity. Animals regulate their energy expenditure to a significant degree. The concept of BMR has utility as a measure of minimal energy expenditure, especially the cost of sleeping for animals like us that spend a significant part of their lives asleep. There are also upper limits on total metabolism; it appears that these are usually about 5 to 7 times greater than BMR. Thus BMR can give you a good estimate of the minimal energy expenditure necessary for life, and decent rough estimates of the normal (2 to 3 times BMR) and maximal (5 to 7 times BMR) rates of metabolism, depending on external constraints on heat loss.

Total Energy Expenditure

Total energy expenditure is the sum of all the components of energy use. Often the heuristic model used is a simple bar graph with the components of energy expenditure stacked on top of each other. This model can be useful but also slightly misleading. Energy is fungible to a certain extent, and changes in one component of energy expenditure may or may not result in a change in total energy expenditure because other components may be adjusted to compensate.

Animals regulate their energy expenditure. An increase in thermoregulatory or reproductive costs may be matched by a decrease in physical activity. For example, lactating mice decrease physical activity (Speakman et al., 2001). Thus part of the energetic cost of lactation is borne by a behavioral change. This could have fitness consequences; not being as physically active might produce an opportunity cost, that is, advantages gained by being active are forgone. But the energetic cost of lactation does not simply add onto other costs to raise the total energy expenditure;

the reduced cost of lower activity absorbs, in effect, some of the cost of lactation.

In another example, mice were fed 80% of their normal food intake for 50 days. Compared to mice fed ad libitum during that same time period, the food-restricted mice were indeed lower in body weight. However, the difference in total body energy content between restricted and control mice only explained 2.2% of the difference in energy intake between the two groups. A lower resting metabolic rate (22.3%) and especially decreased activity (75.5%) enabled the restricted mice to compensate almost completely for the reduced energy intake (Hambly and Speakman, 2005).

In the mouse lactation example, reduced activity might actually be required as opposed to being a choice, as at peak lactation the lactating mice were expending energy at about 7 times their BMR (Johnson et al., 2001a, b). They may have been close to the absolute metabolic limit; at least under the temperature regime at which they were kept. However, at lower environmental temperatures lactating mice were able to increase metabolism above this limit. The reduction in activity may have been more to limit an increase in body temperature due to high metabolic expenditure and thus large production of endogenous heat. Reduced activity would have served to reduce the heat generated by muscle.

Certainly energy expenditure must be constrained between upper and lower bounds; there must be a minimum necessary for life and a maximum imposed by the combination of internal and external constraints. Within those two extremes there appears to be a great deal of flexibility, however. Animals have multiple strategies to compensate for increased energy expenditure or decreased energy intake.

Expensive Tissue Hypothesis Revisited

We discuss the expensive tissue hypothesis for human brain evolution in chapter 2. Basically, brain is very metabolically active tissue. Human brains account for proportionately more energy expenditure than do the brains of other animals simply because our brains are larger in relation to the rest of the body. Some authors have hypothesized that early genus *Homo* may have been faced with an additional component of energy expenditure due to that larger brain (e.g., Leonard et al., 2003). They predict that the BMR of an early *Homo* would have been above the Kleiber line due to the larger brain. Modern human BMR is not unusual (see Figure 6.3). The hypothesis is that the improvement in diet quality eaten by genus *Homo,* first from the foraging shift to include more meat in the diet and eventually due to cooking and other external food processing, enabled our guts to get smaller, reducing our energy cost in one compartment (gut) as we increased it in another (brain).

This intriguing hypothesis relies heavily on the additive theory of energy expenditure. It is inherently structural and anatomical in its concept of energy. It doesn't reflect a regulatory perspective about metabolism and energy expenditure. For example, it assumes not only that the increase in brain size would increase metabolic rate (reasonable, but not certain) but also that the increase in metabolic rate automatically translates into an increase in total energy expenditure. But the components of energy expenditure are not fixed (except, to a large extent, BMR). There is no a priori way to predict how the increased metabolic cost of a larger brain might or might not have affected energy spent in activity, thermoregulation, or other components of energy expenditure. Perhaps the metabolic cost of a larger brain did increase total energy expenditure, but perhaps the larger brain enabled energetic savings in thermoregulation, or perhaps there were savings in other components of the energy budget.

For example, resting metabolic rate measured by oxygen consumption and estimated daily energy expenditure using doubly labeled water in short-tailed field voles at different sites were positively correlated, even after accounting for the positive association between body mass and these measures of energy expenditure. However, the association was only significant when examined across sites. Within a site there was no association between an individual's metabolic rate and daily energy expenditure. In other words, at sites with, on average, higher metabolic rates the animals also had, on average, higher daily energy expenditure, but within a site animals with above average metabolic rates did not necessarily have higher daily energy expenditure (Speakman et al., 2003). The extrinsic effects of between-site differences had a greater effect on energy metabolism and expenditure than did intrinsic effects due to individual differences.

That is not to say that the expensive tissue hypothesis doesn't have merit. It does. But as is usually the case in biology, reality was undoubtedly much more complex and messy than elegant theory would lead us to believe. We really don't know how the larger brain affected average energy expenditure in our earliest ancestors. We do know that whatever energetic challenge it presented, our ancestors were able to meet it, whether by reduced expenditure in other compartments or simply by increased caloric consumption, perhaps via improved foraging strategies enabled by that large brain.

Does the expensive tissue hypothesis have implications for our food preferences and thus our modern-day vulnerability to obesity? The general thrust of the hypothesis implies that a motivation toward obtaining and eating easily digestible, energy-dense foods was an important aspect of being able to support the cost of a larger brain. And this certainly does appear to match what we know of our evolutionary history. As we improved our new foraging strategy and became a more efficient and deadly predator, brain continued to get larger, until a few hundreds of thousands of years ago. We probably became a more efficient exploiter of other foods as well: wild grains, honey, and so forth. The expensive tissue hypothesis gives an additional evolutionary reason for why we would prefer energy-dense foods. Our guts are well adapted to such foods, and our brains may have needed them.

Energy Intake

Energy expenditure must eventually be met by energy intake; food must be eaten. All organic nutrients contain energy that can be released through metabolism. The major categories of food components that usually are accounted for when determining the caloric value of food are fat, carbohydrate, and protein. Of course other food components also have metabolizable energy. For example, although we usually don't drink alcohol to satisfy our energy requirement, alcohol certainly provides usable energy to us when we ingest it; the metabolism of alcohol into the final excretion products produces ATP molecules. However, the vast majority of our calories come from fat, carbohydrate, and protein, and our energy metabolism is geared to using these three different substrates.

There are different ways of expressing the amount of energy in food, and they have different biological and metabolic meanings. The simplest measure is called gross energy (GE). This is the combustible energy in a food. In other words, gross energy is the total amount of heat energy that would be released if the food was completely oxidized by combustion. It represents the maximal biological energy that an organism could obtain from a food. Protein in particular is not completely oxidized. There is still available energy in urea.

Digestible energy (DE) is the gross energy from food minus the energy lost in feces. The digestible energy of a food depends not only on the food itself but also on the animal doing the ingesting. For example the DE of hay would be very different for horses and people. There is an additional further loss of energy in urine. This is mostly from incompletely oxidized protein and usually only accounts for a few percent of GE (Blaxter, 1989). The net amount of energy obtained from a food after accounting for fecal and urinary losses is termed its metabolizable energy (ME). These are the values listed in calorie tables.

Simple carbohydrates (predominantly glucose) and fatty acids are the primary metabolic energy substrates we use. Protein (amino acids) is not a major source of energy except during starvation when we catabolize our own tissue. The general estimates for the ME of fat, protein, and carbohydrate are 9 kcal/g, 4 kcal/g, and 4 kcal/g, respectively (Maynard et al., 1979; Blaxter 1989). These are approximations; the specific ME depends on the specific fat, protein, or carbohydrate oxidized.

Is ME always the most accurate and appropriate measure of energy intake? Not necessarily. It depends on how much tissue deposition occurs. For example, in a growing, nursing mammal a significant proportion of the milk protein (and often milk fat) is directly deposited into tissue and is not metabolized. The energy value of that material is its GE, not its ME. The actual metabolic energy value of the milk can certainly be calculated if balance trials (measuring intake and output) or other methods to measure tissue deposition are conducted on the infants.

Energy Balance

Obesity has a deceptively simple cause: sustained food intake beyond what is needed to match energy expenditure. This concept is usually expressed by nutritionists as being in positive energy balance (Maynard et al., 1979; Blaxter, 1989). The concept of energy balance has intuitively straightforward implications. Negative energy balance usually results in lost weight; positive energy balance usually results in gained weight. If weight is neither lost nor gained then the energy balance is likely at zero, but not necessarily. The simple concept of energy balance is not quite so simple when it is examined in greater detail. Energy balance is not directly related to weight change; it is directly related to changes in total body energy. Total body energy depends not just on the total mass of the body but also on the composition of that mass.

Of course obesity is not actually about weight gain either; it is about excess gain of *fat*. In general we speak of positive energy balance leading to weight gain; in particular, from a nutritionist's or physiologist's perspective we would speak of it in terms of an increase in total body energy. In practical and medical terms the key issue is the increase in fat stored in adipose tissue. The medical concerns over the increase in humanity's average weight are not about us becoming more muscled, more hydrated, or having denser bones, all things that would increase weight. No, the medical and health issues revolve around the extra weight being predominantly fat. And fat is intimately tied to energy. It is the most efficient means of storing energy on the body. Thus energy balance and fat/adipose tissue are linked. Positive energy balance leads to an increase in energy stores, which generally translates into an increase in adipose tissue.

Balance Trials

The concept of nutrient balance is an important tool in nutrition. The first documented metabolic balance trials were conducted by Santorio Sanctorius (1561–1636), one of a circle of learned men in Italy that included Galileo. Santorio Sanctorius was a physician and professor in Padua, Italy. He invented a thermometer and a device to measure pulse, though there is some controversy over whether these were his independent inventions or were made in collaboration with Galileo and others. What is known is that he was the first to apply a numerical scale to the "thermoscope." Indeed, his genius is perhaps best understood as his commitment to describing natural phenomena in terms of numbers, that is, measurement, instead of their "essence" in the Aristotelian sense. Santorio argued that the fundamental properties of things were mathematical ones; properties that could be measured. He also argued that in investigations of nature the most important evidence to consider was that of the senses, followed by reason, and only then was authority to be relied upon.

Many consider Santorio the father of metabolic balance trials. Over a period of 30 years he weighed himself, everything he ate and drank, and all his waste excretions. To account for the fact that his excretions were so much smaller than his intake, he proposed a theory of insensible perspiration. Although his theory has little merit (though given the knowledge base of his day it was reasonable), Santorio is rightfully credited for his empirical methodology. The idea of rigorous, empirical measurement of intake and excretions to understand the inner workings of the body remains an important tenet of nutritional and metabolic research.

Any nutrient can be examined in a balance equation. For example, calcium balance is very important in terms of bone health. If a person is in sustained negative calcium balance they will lose bone mineral; eventually the loss will be irreversible and bone strength will be permanently compromised. That person will be vulnerable to bone fractures (Power et al., 1999).

The idea of nutrient balance is slightly different for different categories of nutrients. Calcium is a mineral, and mineral balance is expressed as intake minus excretion. Calcium intake, from ingested foods, fluids, and supplements, is the positive input. Calcium excretion, from feces, urine, and other body fluids is the negative input. The difference is the calcium balance.

A balance equation for any element could be so constructed. However, most nutrients are not elements, but rather biological entities of more complexity. For many nutrients it is not just the amount excreted that matters but also the quantity metabolized. In some cases the two methods can be used in a complementary fashion. Nitrogen balance, intake minus excretion, can be used as a proxy for protein balance (intake minus excretion and metabolization).

Energy represents an extreme in terms of the definition of a nutrient. Energy isn't a thing, the way a calcium atom is. But it is no less real. It can be measured; it can be tracked as it moves through metabolic pathways. It cycles through the organism; although it can never be destroyed, it will eventually be lost to the system, generally as heat, but also as other excretions.

Metabolism is the dynamic process whose outcome is measured in balance trials. A simplistic idea of nutrient balance is unrealistic, however. It conjures up the notion of taking a nutrient from one place and putting it in another; like moving money between bank accounts. But biology is rarely static; most nutrients don't wait around in a "store" until the organism needs to withdraw it to pay for some life process. Metabolism is dynamic; it consists of multiple metabolic pathways that are continuously in operation, shuttling nutrients, hormones, and other molecules into and out of different forms (metabolites) and different pools within the body.

For example, bone represents, among its many functions, a substantial store of minerals, especially calcium and phosphorus. Bone, like all living tissue, is not static. It is constantly being remodeled. Bone remodeling serves to repair microdamage and to allow bone to respond and adapt to mechanical stress. Bone remodeling also aids in maintaining extracellular fluid calcium homeostasis (Power et al., 1999).

At any given time some fraction of the bone surface contains resorption cavities, created by osteoclasts. This "missing" bone, from which calcium has been released into the extracellular fluid, is termed the remodeling space (Heaney, 1994). Osteoblasts will repair the resorption cavities; in general osteoclast and osteoblast activity are in equilibrium. If the bone remodeling rate increases, the remodeling space increases and total bone mineral content decreases. If the bone remodeling rate decreases, the opposite happens. Mineral (mostly calcium and phosphorus) will have a net flux into or out of bone. The rate of bone remodeling is regulated by hormones and by calcium intake (Power et al., 1999).

Similarly, energy is constantly being transformed among its various forms within the body. The immediately usable energy of metabolism is in the form of the phosphorylated molecules (ATP, GTP, CTP, and UTP). One step removed are the oxidizable fuels such as glucose and fatty acids. And finally there are the storage forms of energy, primarily glycogen and fat.

Energy Stores

Animals don't just need energy; they need oxidizable fuel. If an animal is in negative energy balance then a net amount of metabolizable substrate will be mobilized from the various storage depots; if the animal is in positive balance the net flow will be into the storage depots. The main fuels are glucose and other simple sugars and fatty acids. Energy substrate is important in biology. Different metabolic fuels have different advantages and disadvantages, and regulating the metabolic fuel availability can be as important as energy regulation itself.

Energy is stored in the body in multiple ways. The two main ways of storing readily accessible energy are glycogen, the storage form of glucose, stored mostly in liver and muscle tissue, and lipid or fat, stored in adipose tissue. Protein is an energy store of last resort. Under conditions of extreme energy expenditure or sustained starvation, protein from muscle and organs will be metabolized. This is a successful short-term adaptation, but life-threatening if it continues for an extended time. The metabolically preferred sources of stored energy are glycogen in liver and muscle tissue and fatty acids released from adipose tissue and to a lesser extent muscle.

Fat has significant advantages as an energy storage medium. It contains approximately twice the amount of metabolizable energy per gram dry weight than does either carbohydrate or protein. Furthermore, it is stored on the body in association with very little water. In contrast, a gram of glycogen is likely stored with anywhere from 3 to as much as 5 grams of water (Schmidt-Nielsen, 1994). Therefore, in terms of the energy value per unit weight stored on the body, fat can be as much as 10 times more efficient than storing glycogen (Table 6.2). A single kilocalorie of glycogen will have a mass of about 1 gram; a kilocalorie of fat has a mass of 0.11 gram (Schmidt-Nielsen, 1994). This has some obvious consequences in terms of the energetic and other costs of locomotion. For example, consider migrating birds, for which it has been shown that up to 40 to 50% of their body weight is fat (Schmidt-Nielsen, 1994). This is an extremely weight-efficient way for a flying animal to store energy. If they were to store the same quantity of energy as glycogen, they likely could not get off the ground.

TABLE 6.2 *Approximate metabolizable energy values for carbohydrate, protein, and lipid*

	Kcal per gram dry wt	Kcal per gram wet wt	Kcal per liter of O_2
Carbohydrate	4.2	1.0*	5.0
Protein	4.3	1.2*	4.5
Lipid (fat)	9.4	9.1	4.7

*Exact value depends on the amount of water associated with the carbohydrate or protein.

Although for humans the consequences of excess weight are not as extreme as for a flying animal, being heavier still costs more in terms of locomotion. Therefore you would predict that there is an optimal upper limit on the amount of glycogen that would be stored on the body, and that beyond that energy would be stored as fat, which would add considerably less additional weight. Of course the main organs of glycogen storage (liver and muscle) are relatively fixed in size, while adipose tissue can become larger and more extensive, another advantage of fat as an energy storing medium.

Why then should energy be stored as anything but fat? Glycogen has at least two advantages over fat as an energy store. First, glycogen can be used in anaerobic metabolism (Schmidt-Nielsen, 1994). Thus it can be used in high-exertion activities in which energy expenditure exceeds the aerobic capacity. The second advantage is that it is easily converted to glucose, which is the preferred energy substrate for the brain, the placenta, and the fetus. During pregnancy, mammalian glucose metabolism takes on an especially important role.

The Organs of Energy Storage

Liver and adipose tissue are the two most important energy storage and energy metabolism organs. Liver is more involved in glucose metabolism, while adipose tissue is obviously the main storage organ for fat. The two organs act on different time scales. The liver is more associated with short-term energy metabolism and energy needs. Adipose tissue is important for and responds to longer-term aspects of energy metabolism and energy balance. Both organs play important roles in regulating food intake. The roles may be complementary or possibly opposing, depending on the immediate circumstances.

Energy stores are not static. Physiology and metabolism do not work that way. There are many metabolic processes continually acting that mobilize and store glycogen in muscle and liver, and fatty acids in adipose tissue. Nutritional homeostasis should not be thought of as a static process; rather it is a dynamic equilibrium, with nutrients being continually cycled into and out of stores. Also, adipose tissue is not merely a static store of energy. Adipose tissue consists of active endocrine cells that produce potent information molecules that act to regulate energy metabolism and appetite, among other physiological functions (Kershaw and Flier, 2004). The liver is also an important regulator of metabolism (Friedman and Stricker, 1976). Energy stores are active players in the energy balance equation.

Energy Stores versus Energy Requirement

The allometry of energy requirement scales to less than linear. The allometry of energy stores is linear, or possibly even greater than 1. These are key biological principles, based on empirical measurements (Schmidt-Nielsen, 1994). This is a major advantage of large body size. Large animals can carry proportionately more of their energy requirement on their bodies. They can go longer between eating, and when they find a large quantity of food they are more capable of ingesting excess amounts to be stored for later use.

Consider elephants and voles (sometimes called meadow mice). Both are herbivores; both are hindgut fermentors. A significant portion of energy needs are supplied by the fermentation of plant cell wall constituents for both species. They are also, of course, dramatically different in size. Elephants are measured in tons; voles in grams. An elephant is roughly 150,000 times larger than a vole, yet an elephant's basal metabolic rate is only about 7,600 times that of a vole. In another example, if we compare the BMRs of a .7-kg golden lion tamarin (a small Brazilian monkey), a 70-kg human being, and a 7,000-kg elephant, we see that the mass ratios are the same for the comparisons: A human is 100 times bigger than a golden lion tamarin, and an elephant is 100 times bigger than a human. The BMR ratios are only about 32, however: Human BMR is 32 times more than golden lion tamarin BMR and 32 times less than elephant BMR. There are significant advantages to being big.

Throughout our evolutionary history, human beings, like most other mammalian species, went through both predictable and unpredictable phases of positive energy balance (laying down stores) and negative energy balance (mobilizing stores). Indeed the circadian rhythm of activity also represents a circadian rhythm of energy balance. We are a species that sleeps at night; at least we did so in the past. During the sleep period our bodies are in negative energy balance. During the active phase (daytime for us and most anthropoid primates; nighttime for most rodents) food is being ingested generally in excess of immediate need, and thus energy stores are being replenished. We are large animals, which allows our cycles of positive and negative energy balance to be more extreme than can be tolerated by a small mammal. We can go a longer time in negative energy balance; we can also process and store a much greater amount of energy during the replenishing phase.

For example, a golden lion tamarin needs to eat about 114 kcal/day of food in order to meet basic maintenance requirements in captivity (Power, 1991). This is about double its BMR requirement (Thompson et al., 1994). A human, about 100 times bigger than a golden lion tamarin, requires only about 2,500 kcal/day of food to meet requirements; that is only about 22 times the energy requirement of the golden lion tamarin. Small monkeys such as golden lion tamarins can probably have about 10% of their body mass as fat in captivity; often it will be less (Power et al., 2001). That corresponds to about 5 days of its energy requirement; for a human a 10% fat mass (which would be a very lean person) corresponds to almost a month of energy requirement. Our large body size requires more food in total but allows a greater portion to be stored. We are buffered from famine, but unfortunately apparently not from food excess.

Summary

Energy metabolism is a key component in the obesity picture. Our large body size gives us metabolic advantages in terms of being able to store a sizable proportion of our energy requirement on our bodies. The most weight-efficient form of energy storage is fat. The evolutionary advantages to our ancestors from having the ability to store fat on their bodies can be derived from the basics of energy metabolism, energy expenditure, and energy intake. The presumed increased metabolic cost of the larger brain that characterizes our lineage, and is the definitive adaptive advantage of our ancestors, may have placed additional pressure on energy storing and conserving mechanisms.

CHAPTER SEVEN

Information Molecules and the Peptide Revolution

....

In the last chapter we reviewed the concepts of energy, metabolism, and energy balance: the difference between energy intake and energy expenditure. Metabolism and energy expenditure are regulated. There are mechanisms to regulate energy intake as well. In the next two chapters we examine some of the molecules and pathways that are involved in the regulation of food intake, or appetite.

In this chapter we travel back in time, in some cases possibly all the way back to the common ancestor of vertebrates and invertebrates, to examine the evolution and function of what we are calling information molecules, also referred to as signaling molecules or regulatory molecules. Most of the examples in this book will be recognized by the reader as steroid and peptide hormones. These potent molecules are widespread among living things, and thus their origin dates back to very early in the history of life on earth. Their original functions may never be fully known. What has become increasingly apparent is that evolution has resulted in most, if not all of these information molecules having varied functions that differ among tissues and by context with other regulatory processes.

The Evolutionary Perspective
....

A fundamental concept in evolution is that all species share a common ancestor. For any two species, if you trace back their descent far enough you will arrive at a species that is a common ancestor. Phylogenies can be constructed that estimate the relationships among species and estimate the length of time since the different lineages shared a common ancestor.

These phylogenies are based on anatomy, fossils, and molecules, from proteins to DNA. An implication of this concept is that the molecules involved in the processes of life also share common ancestor molecules. In other words, the metabolic and signaling pathways of two related species are likely to have similarities based on having derived from the same ancestral pathways. The more distantly related the species the less likely the molecules and pathways will be similar. This forms the underlying principle for the molecular evolution studies that have been used to estimate the time of divergence of the human lineage from the great apes (approximately 5 to 7 million years ago), and recently to compare bone collagen proteins recovered from 65-million-year-old Tyrannosaurus fossils with bone collagen from modern animal species. Of those tested, the fearsome dinosaur is most closely related to the modern chicken (Schweitzer et al., 2007; Figure 7.1).

In this chapter we are, at least in part, exploring the descent of molecules. Evolutionary processes, such as gene duplication, mutation, selection, and sexual reproduction result in divergence over time for both molecules and metabolic pathways. However, certain molecules and certain pathways appear to be remarkably conserved. These molecules and pathways are likely involved in critical life processes, and that constrains their variation.

Molecules found among different species that appear, due to DNA, amino acid, and structural similarities, to have derived from the same ancestral molecule are called orthologs. If you went far enough back in time you would find ancestors of these species in which the molecules were identical. Because, after the divergence of the two species from their common ancestor, the evolutionary histories are now independent, differences between the two orthologs will accumulate over time. The orthologs may or may not perform the same function(s). The more closely related the organisms, the more likely that function and structure are similar, but function can diverge, especially as most of these molecules have multiple functions that can vary with tissue, age of the organism, and external circumstances.

Gene duplications are a major source of the genetic variation on which evolution acts. In other words, occasionally a segment of DNA containing a functional gene becomes duplicated in the genome. Then there are two (or more) identical genes initially. The amylase gene in humans is

FIGURE 7.1.   Modern chickens and the *Tyrannosaurus rex* seem to share a common ancestor. Most paleontologists believe birds evolved from a lineage of bipedal dinosaurs. Photo of chicken: Jessie Cohen, Smithsonian's National Zoo.

a good example of gene duplication (see chapter 2). The genes can have independent evolutionary paths after the duplication event; in fact that is the expectation, though selective pressures may act similarly on both genes.

In the case of the amylase gene duplications in humans, the duplicated genes retained their original function. More copies of the amylase gene just means more amylase excreted into saliva (Perry et al., 2007) and possibly the digestive tract. The duplicated genes still have accumulated "silent" mutations that do not change function. It is even possible that some of the mutations have some effects on amylase secretion, though that has not been examined to our knowledge. The accumulation of differences among the multiple copies of the gene allows an estimation of the time since the original duplication events.

The evolutionary importance of gene duplication is that it creates the possibility for function in one of the genes to diverge from the original without the organism losing the original function. Information molecules coded by different genes that are similar due to common ancestry with a gene that was duplicated are called paralogs. For example, the genes coding for myoglobin and hemoglobin are thought to be ancient paralogs, descended from a common gene that was duplicated early in the history of life. The pancreatic polypeptide-fold family of peptides also is most likely made up of paralogous genes.

Information Molecules

There are many examples of ancient regulatory molecules that, over time, became adapted and co-opted to serve multiple, diverse functions. These regulatory molecules serve as "information molecules," transmitting information among organ systems and coordinating the responses of peripheral organs and the central nervous system to external and internal challenges to an organism's viability. For obesity, some of the important information molecules that will be discussed in this book are the peptides leptin, insulin, cholecystokinin (CCK), and corticotropin-releasing hormone (CRH); the steroids cortisol, estrogen and testosterone; and of course the receptors for these hormones.

Evolution is limited in that it can only work on existing variation. However, once genes for a potent regulatory molecule and receptor system exist, the potential for differentiation of function, regulation, and mode of action exists as well. An evolutionary perspective predicts that these molecules will have multiple and diverse functions and that their regulation, function, and mode of action will vary among different taxa, among different tissues within taxa, and even with developmental stage. Every ligand probably can bind to multiple receptors; every receptor probably binds multiple ligands. Variation in function can come from changes in either the ligand or the receptor. Whenever scientists discover a function for a molecule, that function is likely merely one of many.

The Peptide Revolution

The peptide revolution that has been in progress since the mid-1970s continues to accelerate; new peptides and new functions for previously known peptides continue to be found. As an instructive example consider corticotropin-releasing hormone, which was first identified in extracts from sheep hypothalamus (Vale et al., 1981) and is probably best known for its initiating role in the hypothalamic-pituitary-adrenal (HPA) axis (Dallman et al., 1995). Twenty-five years later we know that CRH is part of a hormone family consisting of four known ligands (CRH, urocortin, urocortin II, and urocortin III); two known receptors, each with multiple splice variants; and a binding protein, CRH-BP. These molecules serve not only as neurotransmitters and neuromodulators but also as autocrine, paracrine, and endocrine hormones in peripheral tissues, and they are found throughout the body (see Power and Schulkin, 2006 for a review).

The CRH hormone family is an excellent example of the principles behind the evolution of molecules and signaling pathways. The CRH molecule itself is remarkably conserved within mammals. Primate, rodent, carnivore, and equid CRH are identical; bovine CRH differs from the others in several amino acids (Seasholtz et al., 2002). This probably reflects the rapid adaptive radiation of the ruminants during the late Eocene and early Miocene, when grasses began to dominant many environments.

The CRH molecule is ubiquitous in vertebrates, with orthologs being found in birds, amphibians (Stenzel-Poore et al., 1992), and fish (Okawara et al., 1988), and it plays multiple roles in regulating and coordinating the metabolic responses to external and internal challenges to an organism's viability (Denver, 1999; Seasholtz et al., 2002). Residues 9 through 21 are identical among all CRH orthologs so far sequenced, and fish CRH shows 75% or greater homology with mammalian CRH (Seasholtz et al., 2002). Thus CRH is an example of an information molecule whose structure and function appear to be remarkably conserved over lineages separated by 500 million years or more.

At the same time this potent molecule has given rise to three other paralogous peptide hormones in mammals, the urocortins, most probably through gene duplications and subsequent independent evolutionary change within lineages. In amphibians a peptide found in skin (sauvagine; Montecuccchi and Henshen, 1981) derives from a common molecular ancestor to the urocortins. In fish the peptide urotensin I also stems from this ancestor molecule common to all vertebrates (Seasholtz et al., 2002). Since sauvagine and urotensin I are more closely related to the urocortins than to mammalian CRH, the implication is that urocortin, sauvagine, and urotensin I are orthologs of each other, and paralogs of CRH. Thus the original separation of the urocortins from CRH occurred before the separation of mammals from the other vertebrates (Seasholtz et al., 2002).

A comparison of the CRH receptors supports this hypothesis (Figure 7.2). The type 1 receptor (CRH-R1) is generally associated with CRH while the type 2 receptor (CRH-R2) appears to be primarily a high-affinity receptor for the urocortins (Lewis et al., 2001). Both CRH receptor types are found in fish, amphibians, and birds, as well as in mammals (Arai et al., 2001). Thus an ancestral CRH hormone family existed in the common ancestor of all extant vertebrates. A third receptor type, more closely related to CRH-R1, has been identified in catfish (Arai et al., 2001). This receptor (CRH-R3) is sufficiently different from catfish CRH-R1 that the following possibilities cannot be distinguished: CRH-R3 is unique to fish; CRH-R3 was present in the common ancestor of fish and tetrapods but was lost in the tetrapod radiation; or there is a third as yet undiscovered CRH receptor type in mammals, birds, and amphibians.

The CRH binding protein, CRH-BP, is also found in these lineages and appears remarkably conserved (Huising et al., 2004). All vertebrate CRH-BP so far sequenced has 10 cysteine residues that form 5 consecutive disulfide bonds. Interestingly, CRH-BP has been found in insects, including honeybees, the malaria mosquito, and fruit flies (Huising and Flik, 2005). The insect CRH-BP shares 23% to 29% amino acid identity with vertebrate CRH-BP and 8 of the cysteine residues are conserved (Huising and Flik, 2005)

The existence of a honeybee CRH-BP ortholog is supporting evidence that an ancestral form of the vertebrate CRH signaling system was in existence before the split between the insect and vertebrate lineages. Insect diuretic hormone I has been proposed to derive from a common ancestor

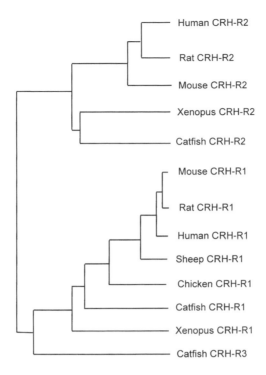

FIGURE 7.2. A phylogenetic tree of CRH receptors. Adapted from Aria et al., 2001.

with CRH, based on similarities in structure, anatomical location, and function (Huising and Flik, 2005).

Thus CRH reflects many aspects of evolutionary principles. It displays remarkable conservation of structure, related to the selective constraints due to its functional signaling modalities. At the same time it has given rise to substantial variation in producing descendant molecules that diverged through gene duplication hundreds of millions of years ago.

Over this vast time period, the CRH hormone family appears to have continued to add new functions to its signaling pathways. In mammals, the CRH hormone family signaling pathways are widespread, with virtually all tissues that have been examined (e.g., skin, heart, stomach, and intestine) being shown to express some member of the CRH hormone family. Primates appear to have evolved a novel function for CRH signaling, synthesizing CRH in the placenta, which acts on the fetal adrenal to regulate and maintain pregnancy and fetal development (Power and

Schulkin, 2006). This is another hallmark of evolution: the co-opting of existing structure, be it anatomical or molecular, to perform diverse functions.

Hormones and Endocrine Glands

The body contains many different organs, each with specialized functions. The actions of these organs are coordinated and regulated; the living body is more than the sum of its parts. The central nervous system is the prime regulator and coordinator of the body's organs. It sends and receives information through the nervous system and also through chemical messengers (information molecules) including but not limited to steroid and peptide hormones. The endocrine gland system synthesizes and excretes these hormones in response to internal and external stimuli. Some glands

TABLE 7.1 *Some endocrine systems in which information molecules are secreted*

Endocrine gland	Major hormones secreted
Anterior pituitary	Prolactin, adrenocorticotropic hormone (ACTH), luteinizing hormone (LH), thyroid-stimulating hormone (TSH), follicle-stimulating hormone (FSH)
Neurointermediate lobe/ posterior pituitary	Oxytocin, arginine vasopressin (AVP), endorphins, enkephalins
Pineal	Melatonin
Thyroid gland	Thyroxin, calcitonin
Parathyroid gland	Parathyroid hormone (PTH)
Heart	Atrial natriuretic factor
Adrenal cortex	Glucocorticoids, mineralocorticoids, androgens
Adrenal medulla	Epinephrine, norepinephrine, dopamine
Kidney	Renin, 1,25 dihydroxy vitamin D
Pancreas	Insulin, glucagon, pancreatic polypeptide, amylin, enterostatin
Stomach and intestines	Ghrelin, leptin, corticotropin-releasing hormone (CRH), urocortin, cholecystokinin (CCK), gastrin-releasing peptide, peptide YY, bombesin, somatostatin, obestatin
Liver	Insulin-like growth factor, angiotensinogen, 25-hydroxy vitamin D
Gonads: ovary	Estrogens, progesterone
Gonads: testis	Testosterone
Macrophage, lymphocytes	Cytokines
Skin	CRH, vitamin D
Adipose tissue	Leptin, adiponectin, androgens, glucocorticoids, cytokines

TABLE 7.2   *Some important peptides and neuropeptides*

	Synthesized in brain	Peripheral organ(s)
β-endorphins	Yes	
Dynorphin	Yes	
Enkephalin	Yes	
Somatostatin	Yes	
Corticotropin-releasing hormone	Yes	Colon; skin
Urocortin	Yes	Stomach; heart
Atrial natriuretic factor	Yes	Heart
Bombesin	Yes	
Glucagon	?	Pancreas
Vasoactive intestinal polypeptide	?	
Vasotocin	Yes	
Substance P	Yes	
Neuropeptide Y	Yes	Adipose tissue
Neurotensin	Yes	
Galanin	?	
Calcitonin	Yes	Thyroid gland
Cholecystokinin	Yes (?)	Intestine
Oxytocin	Yes	Mammary gland
Prolactin	Yes	
Vasopressin	Yes	
Angiotensin	Yes	Kidney
Interleukins	Yes	Adipose tissue
Thyrotropin-releasing hormone	Yes	
Gonadotropin-releasing hormone (GRH)	Yes	
Luteinizing-hormone-releasing hormone (LHRH)	Yes	
Neurotropin	Yes	
Calretinin	Yes	
Leptin	No	Adipose tissue; stomach
Ghrelin	Yes (?)	Stomach
Insulin	No	Pancreas

appear to perform primarily or even totally an endocrine function (e.g., adrenal glands, parathyroid gland). However, all organs have regions of endocrine or exocrine cells that synthesize and secrete hormones (Table 7.1). The notion of an endocrine gland has become greatly expanded (see chapter 9 for a more detailed discussion).

Steroid hormones cross the blood-brain barrier, whereas peptide hormones typically do not. Some peptides enter the brain via transport mechanisms that move them across the blood-brain barrier (e.g., leptin, insulin). Other peptides appear to be restricted to acting on areas of the brain outside of the blood-brain barrier (the circumventricular organs). Many peptide hormones are produced in both the periphery and also in the central nervous system. In the periphery they are peptides; in the central nervous system they are called neuropeptides (Table 7.2). In some of these instances there are interactions between a neuropeptide and peptide and steroid hormones produced in the periphery. The peripheral peptide affects the secretion of the steroid, which crosses the blood-brain barrier and affects the synthesis and secretion of the neuropeptide.

In some cases the neuropeptide and the peptide perform complementary functions, aligning the central response with the peripheral response to a challenge. Behavior and physiology are coordinated. For example, in response to sodium loss or deprivation, peripheral endocrine responses that serve to conserve water and sodium are paralleled by central responses to motivate salt-seeking and ingesting behavior. There are separate renin-angiotensin systems in periphery and brain. The peripheral renin-angiotensin system regulates sodium conservation and redistribution by the body; the central system generates salt-seeking and ingesting behavior (Schulkin, 1991).

The Digestive Endocrine System

In chapter 2 we discuss the human alimentary tract in gross terms. In this chapter we discuss the human digestive system as a functioning, coordinated set of diverse organs. This is a very good analogy to the whole body: diverse sets of organ systems acting in a coordinated fashion to achieve viability of the organism. A major player in the regulation and coordination of those diverse organ systems is the brain.

The digestive system is divided into different regions that serve different functions. Within each region enteroendocrine cells synthesize and secrete peptides that act locally, systemically, and centrally to influence digestive processes, metabolism, and feeding behavior. Accessory organs such as the liver, gall bladder, and pancreas also secrete peptides in response to food ingestion. These secretions affect and are affected by gut secretions either directly or indirectly, often by central nervous system mechanisms.

Certainly a prime function of the digestive tract is to receive, digest, absorb, and eventually eliminate ingested material. However, it is also an active player in metabolism and immune responses as well. To eat food is to deliberately introduce foreign substances into the body; certainly necessary for viability of the organism, but also a challenge to homeostasis and the immune system. The paradox of eating is that eating is both necessary for homeostasis and a threat to homeostasis at the same time (see chapter 10). The gastrointestinal tract is a key barrier protecting the internal milieu. Viability often requires challenging homeostasis. The extent of the perturbation must remain within tolerable limits, however. One function of the gut-brain axis is to regulate the flow of nutrients into and out of circulation.

When food enters the different compartments of the digestive tract a postprandial phase of peptide secretion begins. Many enteroendocrine cells have properties similar to those of taste-receptor cells of the tongue. For example, human duodenal L cells and mouse intestinal L cells express both the sweet taste receptor T1R3 and the taste G protein gustducin (Jang et al., 2007; Margolskee et al., 2007). These gut taste sensors allow the enteroendocrine cells to regulate their peptide secretions in response to the nutrient properties of the food (Cummings and Overduin, 2007). These secretions regulate gut motility, gastric acid and digestive enzyme secretion, and pancreatic secretions, and stimulate the vagus nerve; some end up in circulation where they can act on other organ systems including brain. For example, both sugar and artificial sweeteners increase intestinal expression of the sodium-dependent glucose transporter isoform 1 (SGLT1), which enhances glucose absorption (Margolskee et al., 2007). Glucagon-like peptide-1 (GLP-1) secretion, which affects appetite, insulin secretion, and gut motility, is regulated through these taste transduction elements (Jang et al., 2007). The effects of these peptides increase the efficiency of digestion and absorption of nutrients but also initiate physiological cascades that will end feeding. In other words, many of these peptides act to reduce food intake.

Gut-Brain Peptides

The brain and the gut are linked through the nervous system and through a multitude of gut-brain peptides (Table 7.3). Many of these peptides have

TABLE 7.3   *Gut-brain peptides involved in regulating food intake*

Peptide	Site of synthesis	Sites of action relevant to feeding			Function(s)
		Hyp	Hindbrain	Vagus	
Ghrelin	Stomach	X	X	X	Stimulates feeding
Leptin	Stomach	X	X	X	Inhibits feeding
GRP	Stomach		X	X	
NMB	Stomach		X	X	
CCK	Small intestine	X	X	X	Terminates feeding
APO AIV	Small intestine	X		X	Is secreted in response to fat absorption
GLP1	Small intestine; colon	X	X	X	Delays gastric emptying
Oxyntomodulin	Small intestine; colon	X			
PYY	Small intestine; colon	X		X	Delays gastric emptying
Amylin	Pancreas	X	X		Inhibits gastric emptying, gastric acid, and glucagon secretion
Enterostatin	Pancreas			X	Is secreted in response to fat ingestion
Glucagon	Pancreas				
Insulin	Pancreas	X			
PP	Pancreas		X	X	

been shown to be produced in both gut and brain; some appear to be produced only in the gut, but are transported to brain areas where they bind to receptors. Some of these peptides act primarily on vagal afferents.

These peptides have diverse functions and effects on different tissues. They influence gut motility, other gut secretions, peripheral metabolism, and central signaling that results in changes in feeding behavior. They are ancient information molecules that have evolved diverse functions in physiology, metabolism, and behavior. For example, the gut-peptide ghrelin is produced in the stomach and binds to the growth hormone secretogue receptor and stimulates release of growth hormone (Kajima et al., 2001). Ghrelin also stimulates feeding, the only gut peptide so far known to do so, at least partly by up-regulating neuropeptide Y (NPY) and agouti-related protein (AgRP) in the arcuate nucleus (Kamegai et al., 2001).

Gut-brain peptides are not just involved in feeding behavior. Some of them apparently have functions and effects on reproduction as well (Gosman et al., 2006). For example, elevated ghrelin inhibits luteinizing hormone and stimulates prolactin secretion (Arvat et al., 2001). Peptide YY (PYY) and NPY are members of the pancreatic polypeptide-fold family (see below). Excessive NPY secretion is associated with hypogonadism in rodent models; PYY inhibits NPY secretion and thus might influence reproductive potential (Gosman et al., 2006). Upon reflection it is not surprising that feeding and reproduction should be linked.

Urocortin and CRH are gut-brain peptides. Urocortin is expressed in the stomach along with the CRH type 2 receptor; urocortins and CCK act synergistically to delay gastric emptying and reduce food intake in mice resistant to dietary-induced obesity (Gourcerol et al., 2007). CRH, urocortin, and both CRH type 1 and type 2 receptors are expressed in the colon (Tache and Perdue, 2004). Exogenous urocortin and CRH delay gastric emptying but enhance colonic motility (Martinez et al., 2002). The CRH signaling system in the brain is associated with fear and distress. The effects of urocortin and CRH on gut function are consistent with fear or distress delaying gastric emptying and stimulating colonic emptying. Thus distressful stimuli would result in digesta being excluded from the intestinal tract where it would stimulate blood flow and energy expenditure necessary for digestion, allowing blood flow and energy expenditure to be prioritized to brain and muscle tissue to respond to the perceived threat (Power and Schulkin, 2006). Fear really does empty the bowels, for an adaptive purpose.

Pancreatic Polypeptide-fold Family

The pancreatic polypeptide-fold family is an excellent example of related information molecules that have evolved multiple and diverse functions in different tissues. The family consists of the paralogous peptides neuropeptide Y (NPY), peptide YY (PYY), and pancreatic polypeptide (PP). PYY has two active forms, PYY_{1-36} and an active cleaved product, PPY_{3-36}.

There are five receptors for these ligands (Y1R, Y2R, Y4R, Y5R. and y6R) that couple to inhibitory G proteins (Cummings and Overduin, 2007). An NPY-preferring receptor (Y3R) has been suggested to exist based on pharmacological studies, however it has not been cloned. Evidence suggests that the pharmacological binding that was attributed to Y3R may be better explained by tissue-specific effects on the other receptors (reviewed in Berglund et al., 2003). Receptor y6 is typically given a lower case designation as it appears to be nonfunctional in humans and pigs, is absent in the rat, and has very different pharmacological properties in mouse and rabbit, two species in which it appears to be functional (Wraith et al., 2000; Berglund et al., 2003). The ligands have different affinities for the different receptors. These ligands are primarily synthesized and secreted by different organs: NPY by brain, PYY by intestine, and PP by pancreas.

NPY is a potent orexigenic (increasing or stimulating appetite) molecule. NPY mRNA is up-regulated in the arcuate nucleus in response to food deprivation (Brady et al., 1990), and intracerebroventricular (icv) injection of NPY stimulates food intake, often with a preference for carbohydrate-rich foods, in a number of animals including fish, reptiles, birds, and mammals (reviewed in Berglund et al., 2003). Recent research has shown that NPY has direct peripheral effects as well as central effects.

In animal models PYY and PP have both orexigenic and anorectic (diminishing appetite) effects depending on their site of action, probably due to the differing actions of the different receptors. For example, both PYY and PP reduce food intake if administered peripherally; PP and PYY$_{3-36}$ reduce food intake if administered to the arcuate nucleus, probably by activating Y4R and Y2R, respectively (Cummings and Overduin, 2007). However, diffuse icv injections of PYY$_{3-36}$ or PP increase food intake, probably via the YR5 receptor in the hypothalamus (Batterham et al., 2002; Cummings and Overuin, 2007). Ligand and receptor systems are indeed complex and flexible.

The pancreatic polypeptide-fold family likely derives from a single ancestral ligand-receptor pair that underwent gene duplication events. The evidence supports the hypothesis that NPY and PYY were the result of the first gene duplication, and that PP is the result of a duplication of the PYY gene. Fish lack PP, but have NPY and PYY orthologs, implying that the PYY-PP split occurred after the tetrapods had diverged from fish (Berglund et al., 2003). Interestingly, however, fish express a third ligand (PY) that appears to be the result of a gene duplication of PYY in fish (Cerda-Reverter et al., 1998). This peptide is closer to fish PYY than fish

NPY, but is equally distant from mammalian NPY and PYY. Thus it does not appear to be an ortholog of PP. Instead, two independent PYY gene duplication events seem to have occurred after the tetrapod-fish split, one in each lineage (Berglund et al., 2003).

The genes for receptors Y1, Y2, and Y5 are located in a cluster on human chromosome 4 (location 4q31). Based on an examination of the human, mouse, and pig receptors, the first duplication was for Y1 and Y2; the other receptors (Y4, Y5, and y6) appear to be duplications of the Y1 receptor, with Y4 and y6 becoming translocated to human chromosomes 10 and 5, respectively (Wraith et al., 2000). Interestingly, there is substantial conservation in receptors Y1, Y2, and Y5 among human, mouse, and pig, with significantly more variability in receptors Y4 and y6 among these species.

The Leptin Story

Our final example of a gut-brain peptide is leptin. Although leptin is usually associated with adipose tissue, leptin is also a gut peptide. Leptin is secreted by the stomach but not by the rest of the GI tract so far as we know (Bado et al., 1998). However, the intestines do express the receptor Ob-Rb (Cammisotto et al., 2005), and leptin protein has been detected in the intestinal tract (Cammisotto et al., 2006). On the other hand, recombinant leptin placed into acidic and peptidase conditions such as are found in the stomach is rapidly degraded. Leptin secreted by the cells of the stomach is bound to the soluble short-form receptor (Ob-Re), which appears to protect leptin from degradation by gastric acid and peptidases (Cammisotto et al., 2006). Leptin released by the stomach may act through Ob-Rb receptors in the intestine to regulate feeding (Peters et al., 2005) and also nutrient absorption (reviewed in Picó et al., 2003).

Leptin has been identified in all mammals studied to date, including the Australian carnivorous marsupial the dunnart (Doyon et al., 2001). If the dunnart is used as the outgroup, a molecular phylogeny of eutherian mammals based on the amino acid sequence of leptin molecules fits with well-established phylogenies based on morphological, fossil, and DNA evidence. Primates group together; carnivores group together; rodents form another unit; and sheep, cattle, and pigs form a group with the Beluga whale, in accordance with the evidence that whales are related to the artiodactyls (Gingerich et al., 2001). The rate of amino acid substitutions does not differ among the mammalian lineages characterized so far.

What are leptin's functions? Well, that depends on the tissue, the state of the organism, especially the expression of other information molecules such as insulin, glucocorticoids, and CRH, and also on the age of the organism. Leptin has many functions, which change with time, tissue, and circumstances.

Fat mass has been linked to appetite and food intake for over 50 years (e.g., Kennedy, 1953). The peptide leptin is one the main mediators of that connection. Leptin is primarily synthesized and released by adipose tissue and circulates in the bloodstream in direct proportion to adipose tissue (Havel et al., 1996).

Even before the technique of producing animal knockout models was invented, careful selective breeding had produced strains of rodent models with abnormal expression of certain genes. An obese mouse model (the Ob strain) developed over 50 years ago provided strong evidence that there were one or more unidentified humoral factors important to the regulation of food intake, and the lack of these factors was associated with the propensity to become obese (Hervey, 1959). The peptide leptin was predicted many years before it was isolated and characterized based on experiments on two different kinds of obese mutant mouse models. The ob/ob and db/db mouse models both became obese. Parabiosis experiments, in which the circulatory systems of two mice are connected, provided evidence that ob/ob mice lacked a circulating factor (Hervey, 1959), and db/db mice lacked the receptor (Coleman, 1973). Briefly, if a wild type mouse was partnered with an ob/ob mouse, the ob/ob mouse reduced food intake and lost weight. If ob/ob and db/db mice were partnered the ob/ob mouse lost weight but the db/db mouse was unaffected. Thus Coleman (1973) concluded that ob/ob mice lack a circulating factor that inhibits food intake; db/db mice lack a functional receptor.

In the early 1990s the missing humoral factor responsible for this particular model of obesity was identified as a 16-kilo Dalton protein hormone secreted by adipose tissue (Zhang et al., 1994) and was named leptin after the Greek *leptos*, meaning thin. The receptor was cloned shortly thereafter and shown to be present in the ventral and arcuate nuclei of the hypothalamus (Tartaglia et al., 1995; Mercer et al., 1996; Fei et al., 1997).

Leptin is implicated in body weight regulation through control of appetite and possibly of energy metabolism. It acts through a specific leptin receptor (with at least five isoforms) located on target tissue cells. Leptin receptor B (Ob-Rb), the longest form of the receptor, is able to activate all signaling pathways so far known (Cammisotto et al., 2005). Ob-Rb is highly expressed in the hypothalamus, where it appears to play an important role in the regulation of food intake, both through effects on appetite and on the hedonic perception of food (Isganaitis and Lustig, 2005). There is a short, soluble form of the receptor, a cleavage product of the longer forms, that appears to act as a binding or carrier protein or both (Ahima and Osei, 2004).

Mouse models produced with defective leptin receptors also displayed obesity (e.g., Cohen et al., 2001). Restoring leptin to mice that were leptin deficient produced substantial weight loss but had no effect in mice with defective leptin receptors (Halaas et al., 1995). Leptin crosses the blood-brain barrier via a saturable active transport system (Banks et al., 2000). Central infusions of leptin reduce food intake (reviewed in Schwartz et al., 2000). There are multiple proposed mechanisms for this anorexigenic effect, including that leptin is thought to influence the hedonic perception of food. For example, Leptin has been suggested to function as part of a "lipostat," a system that attempts to maintain total adiposity within a stable range, at least partly by regulating appetite in relation to fat stores. This engendered great excitement, especially in the popular press. The "silver bullet" for weight loss had been found.

Biology, however, is rarely so simple. Giving pharmacological doses of leptin to obese animals that are not leptin deficient does not reliably lead to reduced appetite and weight loss. One consideration raised by the evolutionary perspective is that obesity in our or other animals' evolutionary past was probably rare. Abnormally high leptin levels due to excess adipose tissue are unlikely to have occurred very often. The more likely scenario is leptin levels were often low due to low fat stores caused by food shortage. Leptin may serve as an indicator of peripheral fat stores to the central nervous system, but its evolved, adaptive function more likely is as an indicator of low energy stores cuing either increased foraging effort or strategies to conserve energy, rather than as a signal to reduce food intake due to obesity.

Circulating leptin also is regulated; it does not simply circulate in proportion to body fat. Leptin levels drop following fasting and return to normal following refeeding (Mizuno et al., 1996; Mars et al., 2005). These changes are far out of proportion to any change in adipose tissue. There is also a circadian rhythm to circulating leptin, with the highest levels generally being during the habitual sleep time for humans and rodents (Licinio et al., 1998).

Certain physiological states are associated with leptin resistance, that is, hyperleptinemia without subsequent reduction in appetite or food intake. Pregnancy and obesity are instructive examples, as the resistance to the anorexigenic effects from these two states appears to arise from different mechanisms. Leptin is produced by the placenta, and the excess leptin in maternal circulation during pregnancy likely is of placental origin. The soluble, short form of the leptin receptor is also up-regulated during pregnancy. Thus, although maternal circulating leptin is increased, much of the leptin is likely bound to the soluble, short-form receptor and thus inactivated (Henson and Castracane, 2006). The increased circulating leptin during pregnancy does not result in a decrease in appetite; pregnancy is a state of hyperphagia not hypophagia.

Obesity is also considered a condition of leptin resistance. The large amounts of adipose tissue result in high circulating leptin, but this hyperleptinemia does not appear to reduce appetite. At some level of circulating leptin, the ability of leptin to reduce appetite is compromised. One potential mechanism involves the active transport of leptin across the blood-brain barrier (Banks et al., 2000). At some peripheral serum leptin concentration, this system becomes saturated, and increases in circulating leptin concentration no longer result in increases of leptin transport into the central nervous system. Increased peripheral leptin will not change central leptin signaling. This is further evidence that the central and behavioral functions of leptin may be more related to low levels rather than the high levels seen in obesity.

Leptin is not just about food intake and adipose tissue homeostasis, however. Leptin levels may also act as a signal of minimal stores necessary for reproduction. Indeed, we know that leptin has significant effects on reproduction. The leptin-deficient obese mice were also infertile, both males and females. Restoring leptin reversed the infertility (Chehab et al., 1996). The reproductive functions of leptin include an association with the onset of puberty, and a role in fertility for males and females, ovarian folliculogenesis, and implantation of the fertilized ovum. Giving leptin to mice resulted in their attaining sexual maturity at a significantly earlier age (Chehab et al., 1997). Circulating leptin has a transitory increase at the onset of puberty in boys (Mantzoros et al., 1997). Leptin is expressed by the placenta, the umbilical cord, and other fetal membranes (Ashworth et al., 2000). Leptin receptors are widespread in fetal tissues, and leptin may play a role in fetal development (Henson and Castracane, 2006). Spermatozoa secrete leptin (Aquila et al., 2005). Leptin appears to have many functions beyond any potential "lipostatic" function.

The Curious Case of Chicken Leptin

Leptin is an ancient molecule, perhaps as ancient as CRH. Leptin orthologs have been found in birds (Taouis et al., 2001; Kochan et al., 2006), fish (Johnson et al., 2000; Huising et al., 2006), lizards (Spanovich et al., 2006), and amphibians (Boswell et al., 2006; Crespi and Denver, 2006). The leptin molecule and receptor has been characterized in poultry, both in chickens and turkey (Taouis et al., 2001). Chicken and turkey leptin show substantial homology with each other and with mammalian leptins; they are greater than 80% identical to mammalian leptins so far characterized. There was some controversy regarding the chicken leptin sequence initially (Taouis et al., 2001) due to the fact that it showed about 95% identity with mouse and rat leptin! The chicken leptin sequence has been verified by independent researchers, along with the puzzlingly high degree of identity with rodent leptin. The high degree of identity between the leptin molecules from these two quite distant lineages appears to be a case of parallel or convergent evolution (Doyon et al., 2001).

As in mammals, leptin in chickens has a strong relationship with fat. However, in chickens the primary site of leptin synthesis is the liver (Taouis et al.,1998; Ashwell et al., 1999), although adipose tissue does also synthesize and release leptin (Taouis et al., 2001). Liver is very lipogenic in chickens, and fat plays an increased role in avian energy metabolism. The liver is also the primary site of leptin synthesis for carp (Huising et al., 2006), indicating that perhaps hepatic expression of leptin is the ancestral condition that has been lost in mammals.

In other ways leptin functions in chickens as it does in mammals.

Circulating leptin increases after feeding and decreases with fasting; providing exogenous leptin decreases food intake (Denbow et al., 2000). Young hens given leptin attained sexual maturity earlier (Paczoska-Eliasiewicz et al., 2006). Leptin has potent effects on the ovaries in chickens (Paczoska-Eliasiewicz et al., 2003; Cassey et al., 2004) and is involved in the nutritional regulation of reproduction (Cassey et al., 2004). Thus the links between fat, leptin, and female reproduction are likely ancient in origin.

Trophic Functions of Leptin

Leptin has different functions at different times in the life of an organism. Leptin signaling in fetal tissue and in neonates has very different effects from leptin signaling in adults. Leptin is a developmental hormone for young animals. For example, leptin receptor (Ob-Rb) is expressed in the gastrointestinal tract, from the esophagus to the colon, in human fetuses from about 8 weeks (Aparacio et al., 2005). Leptin protein was detected as well, although leptin mRNA was not detected until the 11th week. Amniotic fluid contains significant concentrations of leptin, probably of placental origin. Fetal swallowing of amniotic fluid begins at about this time, and the early presence of leptin before leptin mRNA was detected may have reflected leptin from swallowed amniotic fluid (Aparacio et al., 2005). These data suggest that leptin, acting through its receptor, has a role in growth and maturation of the GI tract.

Breast milk contains leptin (Casabiell et al., 1997; Housenechte et al., 1997; Smith-Kirwin et al., 1998), and leptin receptors are present in the GI tract after birth (Barrenetxe et al., 2002). Indeed, leptin receptors are present in the adult GI tract (Cammisotto et al., 2005). The maturation of the GI tract is not complete until at least 1 year of age, and leptin may play an important role from fetal through adult. However, leptin-deficient mice do not display any developmental problems with their GI tracts (Cammisotto et al., 2005). Thus either there are multiple, compensatory mechanisms in the developmental process of the GI tract, or perhaps leptin's role is permissive but not necessary.

Leptin displays trophic properties in other species. A leptin ortholog has been characterized in the South African clawed frog, *Xenopus laevis*

(Crespi and Denver, 2006). The function of leptin appeared to change over metamorphosis. Leptin functions to regulate appetite in postmetamorphic frogs and even in late metamorphosis, but it has no effect on appetite for tadpoles in early metamorphosis. However, it does speed the metamorphic development, causing the hind limbs to grow more quickly and hastening the development of the toes (Crespi and Denver, 2006).

Leptin also appears to have trophic actions on the neonatal rodent brain. Interestingly, leptin has no affect on appetite in neonatal rats or mice. Leptin-deficient mice do not differ from wild type mice in size or fatness for the first two weeks of life, after which they rapidly diverge in fatness due to an exaggerated appetite (Bouret and Simmerly, 2004). Circulating leptin increases dramatically in the first and second weeks of life in rodents. This time period corresponds to a key brain developmental period; during this time the arcuate nucleus makes connections with other hypothalamic nuclei. These connections appear to be stimulated by leptin. Leptin-deficient mice have poorly developed connections between the arcuate and other hypothalamic nuclei (Bouret and Simmerly, 2004).

Thus leptin function not only differs by tissue, but also by age. Leptin participates in key developmental periods of an animal's life; interestingly, many of the tissues leptin acts on during development are also ones it acts on in the adult. The GI tract, the hypothalamic nuclei, and the gonads are all targets of leptin signaling in development and afterward. The nature of leptin's effects differ, however, with leptin stimulating and regulating development in early life.

Silver Bullet or Lead Buckshot?

When leptin was first characterized, and shown to be the circulating factor that, when missing, led to obesity in rodents, there was great enthusiasm and hope that a "silver bullet" in the battle to control appetite had been found. There was skepticism and wariness as well, since other molecules had been touted in the past without fulfilling expectations (e.g., CCK). The characterization of leptin has indeed greatly increased our understanding of the regulation of appetite and the pathophysiology of obesity. However, very few obese people have known defects in the leptin signaling system. Most obese people have high circulating leptin and do not have receptor defects. Although obese humans are often characterized as exhibiting leptin "resistance," mechanisms for this hypothesized resistance that leads to overeating have not been demonstrated. The biology of leptin supports the hypothesis that evolutionarily low leptin levels were far more common than high circulating leptin. The leptin signaling system appears much better adapted to reacting to circumstances of low leptin (e.g., low amounts of adipose tissue or periods of low or no food intake) rather than circumstances of excess. For example, the fact that the transport system that carries leptin from blood across the blood-brain barrier becomes saturated at circulating levels found in obesity suggests that regulation of feeding behavior by leptin is compromised at high circulating levels.

The adaptive functions of leptin may be more heavily weighted toward responding to food adversity rather than food plenty. Low circulating leptin levels may motivate animals to increase feeding behaviors; low circulating leptin may make food more salient in the environment. Animals are always faced with a number of behavioral choices. They can't do everything, therefore some behaviors have priority. Circulating leptin may be one means of giving feeding behaviors relative priority over other behaviors. If leptin concentration is low then feeding becomes relatively more important; if leptin concentration is high then other behaviors may be perceived to have relatively higher priority. Of course in the modern environment the tradeoffs are much reduced; food is ubiquitous and the costs of engaging in feeding behavior (in terms of having to forgo other behaviors) are low. We can often eat while we do other things (e.g., drive, read, or even write books).

Summary

Physiology and metabolism are complex, multidimensional phenomena. Coordination among organs of the body is necessary; that is a fundamental concept. Cross talk among organs is accomplished by the nervous system and signaling molecules. The gut-brain connections are important for understanding feeding behavior.

Just as species have evolved, so have molecules and physiological pathways. A finite set of signaling molecules functions to coordinate and regulate metabolism. The incredible diversity of function that is created by these information molecules is due in part to these potent molecules having diverse functions in diverse tissues, and tissue- and context-specific regulation and regulatory function. Many of the original molecules have "split" via gene duplication to form families of signaling molecules and pathways. Information molecules have diverse functions in diverse tissues, and their effects depend on complex interactions with other signaling molecules.

CHAPTER EIGHT

Appetite and Satiety

....

We eat food; we need nutrients. When we eat, how much, how often, what kind of food depends on many things. But regardless of when, how much, and so on, one aspect of our eating is to satisfy our nutrient requirements.

An organism has specific nutrient requirements, and evolution has produced adaptations to enhance intake of some, but not all, of those nutrients when they are scarce. The most general, hunger, motivates an animal to eat food; thirst motivates an animal to drink water. A specific appetite for salt (sodium) has been well documented in many species (Richter, 1936; Denton, 1982; Schulkin, 1991; Fitzsimons, 1998). Briefly, sodium deficiency (or significant water loss) results in the stimulation of the renin-angiotensin system in the kidney, which acts peripherally to conserve water and sodium, and to induce release of the steroid aldosterone from the adrenal gland. Aldosterone crosses the blood-brain barrier and induces central angiotensin, which, through various brain circuits, results in motivated behaviors to acquire water and salt. The effects include changing the hedonic perception of salt, so foods and solutions that taste salty become preferred (see Rozin and Schulkin, 1990; Schulkin, 1991; Fitzsimons, 1998 for reviews).

Sodium appetite is a nice example of a number of important concepts in feeding biology. It is an example of communication and coordination between peripheral organs (in this case the kidney) and the brain; physiology and behavior are in synch; need is matched to motivation and behavior. It is an example of the same peptide (angiotensin) having complementary functions in periphery (regulating physiology to conserve sodium) and brain (motivating behavior to find and ingest sodium). The number of information molecules is large, but it is a finite number and could even be considered small compared with the complexity of living organisms. These information molecules have varied functions in diverse tissues under diverse conditions, and it is that diversity of function that enables the complexity if life.

The idea of specific appetites for nutrients has intellectual appeal, but, except for sodium, has been difficult to demonstrate (Richter, 1957; Rozin, 1976). For instance, calcium deficiency can result in an appetite for calcium, but it also induces a sodium appetite that in many ways appears more robust than the preference for calcium (Schulkin, 2001). In another example, although rats fed a thiamine-deficient diet readily chose a thiamine-replete alternative diet when offered it, the evidence suggests that rats have no inherent ability to detect thiamine. Rather, the behavior of the thiamine-deficient rats is best explained as a combination of a learned aversion to the deficient diet and a learned association with recovery from malaise for the new diet (Rozin and Schulkin, 1990). Rats that have been made thiamine deficient can learn to avoid the taste of the thiamine-deficient food and prefer the taste of the thiamine-replete food, even though the tastes are arbitrary. If the flavorings are switched then the rats will, for a time, prefer the thiamine-deficient food and avoid the thiamine-replete food (Rozin, 1976), although eventually they will once again learn to avoid the flavor of the thiamine-deficient diet.

Animals have many strategies to guide their food-choice behavior, with the strategies differing among species and among nutrients (Rozin, 1976). Time scale is an important consideration. The murkiness of the calcium story may be due in part to the fact that calcium stores (bone) are large compared with need. The consequences of short-term dietary calcium deficiency are much less serious than an equal time period of sodium or water insufficiency. The acute response to calcium deficiency primarily involves conservation and mobilization of stores; only after the deficiency becomes chronic does a behavioral component appear (Schulkin, 2001).

Hunger in human beings, as a response to an energy deficit, probably falls in between these two time scales: not as immediate as a response to sodium depletion or serious water loss, but more rapid than in the case of calcium deficiency. We are relatively large animals, so we can store a significant amount of our energy requirement. However, the manner in which the energy deficit is incurred will impact the physiological and behavioral response as well. The hunger caused by a short- to medium-term fast, such as overnight sleep, may differ from the hunger caused by a sustained period of high exertion, and both will differ from a prolonged period of food deprivation.

There are different time scales over which to consider the regulation of food intake. The shortest is the regulation at the start and end of a meal. There are physical, physiological, and hormonal signals that are involved in this process, all generally involving the intestinal tract, the vagus nerve, and brain.

Food intake is regulated. It is rare that we eat as much as we are physically capable of ingesting. There are multiple paradigms within which to approach the investigation of regulation of food intake, for the simple reason that there are multiple motivations for feeding. Nutrition is indeed a fundamental aspect of food intake, but it is not the sole driving force. There are hedonic aspects to feeding, with some similarities to addictions. Also, animals that are subjected to unpleasant, "stressful" circumstances experience changes in food intake and preference patterns. Many eating disorders in humans are associated with an adverse psychosocial environment (Dallman et al., 2003).

Satiation, Satiety, and Appetite

The regulation of feeding rests on the opposing (or perhaps complementary) concepts of appetite and satiety. These regulate the duration of feeding, the frequency of feeding, the amount of food consumed, and the type(s) of food consumed. The gut and brain coordinate and interact via these processes. The interaction between gut and brain begins before food is ingested. The sight, smell or even just anticipation of food triggers physiological cascades that prepare the gut to receive and digest food and other organs to metabolize the absorbed nutrients. (These cephalic-phase responses are examined in detail in chapter 9.)

Satiation and satiety are two related but distinct concepts. Satiation regulates short-term feeding; when an animal becomes satiated it stops eating. Satiety regulates feeding frequency, or the time between feeding bouts. So satiation refers to the processes that terminate a meal (defined in this context as a feeding bout), and satiety refers to the processes that regulate the number of meals and the time between meals (Cummings and Overduin, 2007).

Cholecystokinin (CCK) is the first intestinal satiation peptide that was described (Gibbs et al., 1973). It is produced in the small intestine and the brain. There are two known CCK receptors: The type 1 receptor (CCK1R) is expressed predominantly in the gut and the type 2 receptor (CCK2R) is expressed predominantly in the brain (Moran and Kinzig, 2004; Cummings and Overduin, 2007). Peripheral injection of CCK reduces meal size in a dose-dependent manner (e.g., Gibbs et al., 1976). The anorectic effects of endogenous CCK are potent and rapid; they are short-lived, however, lasting less than 30 minutes postinjection (reviewed in Gibbs et al., 1993). For example, an intravenous infusion of CCK given to rhesus macaques led to a significant reduction in food intake (Gibbs and Smith, 1977; Figure 8.1a). However, after 15 minutes the rate of food intake was not different; thus at the end of 3 hours after the monkeys received CCK they had eaten less, but the total decrement in food intake was within the first 15 minutes (Figure 8.1b).

Thus CCK acts to produce satiation, but it is not a satiety signal. In other words, it terminates feeding but has little effect on the frequency of meals or on total daily food intake. Indeed, chronic administration of CCK results in rats eating many small meals per day, with total food intake unchanged from controls (e.g., West et al., 1984; reviewed in Moran and Kinzig, 2004). In rats without expression of CCK1R, meals were of greater duration and consisted of greater amounts ingested, however, the number of meals per day decreased. The net effect in this experiment was an overall greater daily food intake and eventual obesity (Moran and Kinzig, 2004). Thus a disruption of CCK signaling can lead to obesity, although an increase in CCK signaling does not reliably lead to weight loss. This kind of asymmetry may be very common.

There are many gut-brain peptides that have satiation actions. Some are relatively specific to metabolic substrate. For example, the glycoprotein apolipoprotein A-IV (APO AIV) is secreted from the intestine in response to fat absorption. It is also synthesized in the hypothalamic arcuate nucleus. Exogenous administration of APO AIV decreases food intake (Tso and Liu, 2004). The anorectic gut peptide PYY is secreted propor-

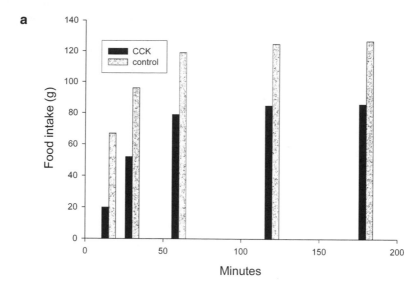

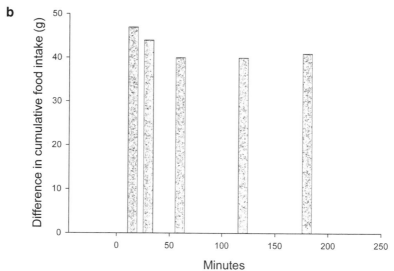

FIGURE 8.1. (a) Rhesus macaques ate less food when they were given an intravenous infusion of CCK. (b) The entire difference in food intake was accounted for by the first 15 minutes of feeding. Data from Gibbs and Smith, 1977.

tionally to caloric load, though the potency of the response is greatest for lipid, then carbohydrate, and finally protein (Degen et al., 2005; Cummings and Overduin, 2007).

Currently we know of only one gut-brain peptide that has orexigenic (appetite inducing) action: ghrelin. Ghrelin was initially isolated from rodent gastric epithelium (rats, Kojima et al., 1999; mice, Tomasetto et al., 2000). Ghrelin was shown to induce increased adiposity in rodents both by increasing food intake and by affecting fat metabolism (Tschop et al., 2000). The effects on food intake are mediated by activation of NPY neurons in various brain areas, including the arcuate nucleus (reviewed in Gil-Campos et al., 2006).

Ghrelin is synthesized and secreted by gastric mucosal cells in the stomach and proximal small intestine. Ghrelin increases gut motility, decreases insulin secretion, and powerfully increases food intake in many species (reviewed in Gil-Campos et al., 2006). Ghrelin secretion is enhanced by fasting and is suppressed by feeding (Wren et al., 2001a, b; Inui et al., 2004). Circulating levels of ghrelin surge prior to meals. Indeed, the timing of ghrelin increases can be entrained by consistent meal times (e.g., Cummings et al., 2001). Thus the preprandial secretion of ghrelin appears to act as a cephalic-phase or anticipatory response to stimulate feeding behavior (see chapter 10 for a further discussion).

It is uncertain whether ghrelin is produced in the brain. Levels of cytoimmunological staining of rodent brains for ghrelin have been barely above baseline. However, by comparing staining of brains of wild type mice with ghrelin knockout mice, a low ghrelin signature was detected (Sun et al., 2003). So either ghrelin is produced in low levels in the brain, or there is some transport of ghrelin across the blood-brain barrier. Intriguingly, the ghrelin knockout mice did not differ from wild type in feeding behavior, or indeed in any measured parameter (Sun et al., 2003). The ghrelin-null mice responded to exogenous ghrelin in the same way as the wild type animals, so ghrelin functional signaling was intact in the knockout mice, just presumably absent due to the deletion of the gene for the ligand. There are multiple redundant circuits that regulate feeding behavior; ghrelin appears not to be necessary for normal food intake and growth, despite its potent effects on feeding behavior. There have been, to our knowledge, no studies that have tested whether the ability to antici-

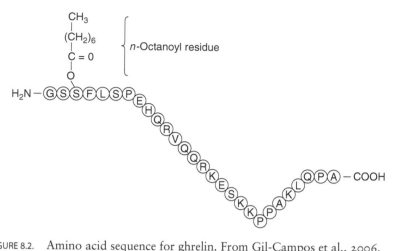

FIGURE 8.2. Amino acid sequence for ghrelin. From Gil-Campos et al., 2006.

pate meals due to entrained circadian rhythms or to conditioning is compromised in ghrelin null mice.

Ghrelin circulates in an acylated and nonacylated form; in the acylated form a medium-chain fatty acid, usually n-octanoic acid, is attached to serine at position 3 (Figure 8.2). The acylated form is the active form for the central actions on appetite. The nonacylated molecule appears to have peripheral actions including cardiovascular effects, though the mechanisms are uncertain (reviewed in Gil-Campos et al., 2006).

Signals That Regulate Appetite

The importance of food intake to survival suggests that it is quite reasonable that there would be multiple signals regulating appetite and motivation to feed, and that the signals probably act on different time scales. Body size could be an important determinant of the relative importance between the signals associated with short-term versus long-term energy status. Small animals simply can't afford to go without feeding for any length of time, because they cannot store a very high proportion of their need in either glycogen or fat. Larger animals have much longer starvation times and therefore have more options to deal with short-term needs. One might then infer that larger animals would have evolved more mechanisms for dealing with energy needs and energy regulation that use long-term signals, that is, relevant to fat and adipose tissue.

Rats are small-bodied animals. There is good evidence that feeding behavior after fasting is associated with liver energy status in rats (Ji and Friedman 1999). Rats were fasted for 24 hours and then offered food ad lib at the beginning of the daylight period, a period during which they normally eat very little. Fasted rats ate dramatically more than the control rats during the light period. Glycogen contents of the liver of fasted rats was significantly decreased relative to that of fed rats but it rapidly returned to normal and indeed overshot the values from fed animals upon refeeding (Ji and Friedman 1999). However, liver ATP content remained below normal for a longer period of time. Food intake behavior roughly paralleled the changes in ATP and glycogen content of the liver, with food intake being very high while liver glycogen and ATP were lower than basal levels, and then declining to match the rate of food intake in control animals after liver energy status returned to normal (Ji and Friedman 1999).

Leptin has potent effects on feeding behavior. Leptin levels typically reflect total adipose tissue, though not in all circumstances. For example, in very lean populations of people, circulating leptin concentrations are not associated with adiposity (Bribiescas, 2005; Kuzawa et al., 2007). Levels of leptin in wild baboon populations tend to be low as well, and to be more related to age than to body weight (Banks et al., 2001). Of course these wild populations are leaner than captive baboons. At low levels of circulating leptin (reflecting low amounts of adipose tissue) leptin concentration appears to be reflective of other processes. One possibility is that circulating leptin concentration reflects fatty acid flux through adipose tissue in addition to total adipose tissue. In fasted animals the concentration of circulating leptin decreases to a much greater extent than would be predicted from any change in adipose tissue (Weigle et al., 1997; Chan et al., 2003; MacLean et al., 2006). Similarly, after a meal leptin is elevated beyond what could be expected from any change in adipose.

At its core, eating is a motivated behavior (Richter, 1953; Stellar, 1954; Berridge, 2004). A simple but important concept is that animals cannot perform an unlimited number of behaviors at any given time; they have to "choose." If they choose to eat then they are by default choosing not to do other things. There will always be competing motivations that must be resolved. One adaptive consequence of hunger and satiety is that these perceptions will influence whether eating food will be preferred relative to other behavioral options. The motivation to eat will be relative to other motivations and sensory inputs; both hunger and satiety can be overridden.

The central nervous system is the organ that coordinates and prioritizes behavior; it will be involved in regulating feeding behavior for all species, from flies to humans. In the case of flies the signaling system is very simple (Dethier, 1976). When a fly lands on food, it begins to feed; there is an excitatory reflex that responds to food detection with ingestive behavior. An inhibitory signal shuts off feeding when the fly's stomach becomes full. If the inhibitory reflex is disrupted by cutting sensory nerves from the gut then the fly will continue to eat until it bursts its stomach and dies (Dethier, 1976). This is a simple and elegant system that works well for fly feeding behavior; mammals, and all other vertebrates, have more complex and varied neural circuits that influence feeding. Hunger and satiety are not the same in mammals and flies (Berridge, 2004).

Hunger and satiety are complex perceptions that involve the integration of stimuli from many neural sites as well as peripheral organs (Figure 8.3). The peripheral tissues of primary importance are the digestive tract (from the mouth/tongue to the stomach to the small and large intestines), the pancreas, the liver, muscle, and adipose tissue. In the brain all levels of neural organization are involved, from brain stem to cortex. Certainly there are some key areas that research has shown to be critically involved in feeding behavior (e.g., parabrachial nucleus in brain stem; arcuate nucleus in the hypothalamus; prefrontal cortex), but neural systems that underlie feeding behavior are distributed throughout the brain. Even the relatively straightforward behavior of induced salt appetite is encoded by neural fiber systems that are widely distributed in brain stem and the lim-

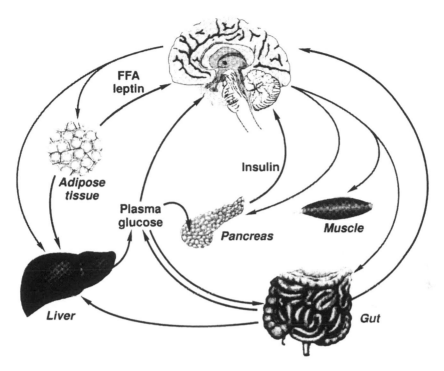

FIGURE 8.3. Physiology and behavior act in a coordinated, regulated manner. Brain and peripheral organs all communicate with each other via ancient, conserved information molecules. *Note:* FFA = free fatty acids.

bic areas of the forebrain (Fitzsimons, 1998). Feeding behavior in mammals is centrally regulated by complex, integrated, distributed networks that include all neural levels of organization.

Meal-related information is transmitted to the brain via several modalities. Areas of the brain within the blood-brain barrier receive inputs from peripheral organs via at least four mechanisms: (1) direct nerve connections, (2) steroid hormones that can pass the blood-brain barrier, (3) peptide hormones and other molecules that cannot passively cross the blood-brain barrier but are actively transported into the brain via saturable and often regulated molecular transport systems, and (4) the circumventricular organs. The circumventricular organs are outside the blood-brain barrier, and thus can respond to inputs from peripheral organs via peptide secretions as well as nerve impulses. In addition they can respond to circulating metabolic products. In essence they can "taste" the extracellular fluid. They are thus well situated to regulate behavior and physiology in response to peripheral inputs and to inform the other brain regions.

The weight of evidence supports the hypothesis that caudal brain stem networks can regulate and support at least some semblance of normal ingestive behavior without forebrain input (Grill and Norgren, 1978; Grill and Kaplan, 2002). Decerebrate rats display appropriate oral/facial responses to oral infusions of various tastants. For example, Grill and Kaplan (2002) showed that decerebrate rats display positive hedonic responses to glucose (sweet) and aversive responses to quinine (bitter). Meal size, as measured by the volume of a sucrose solution ingested, also is similar between normal and decerebrate rats. Injections of CCK reduce meal size in a similar fashion in normal and decerebrate rats (Grill and Smith, 1988). However, decerebrate rats do not display completely normal ingestive responses (Grill and Kaplan, 2002). They do not regulate their intake relative to need. They cannot form learned taste aversions (Grill and Norgren, 1978).The regulation of feeding behavior in decerebrate rats is, not surprisingly, indeed even comfortingly, compromised. Normal feeding behavior takes a complete brain.

The nucleus of the solitary tract in the brain stem integrates afferent signals from the tongue and the gastrointestinal tract (Norgren, 1995; Travers et al., 1987). The information then travels anteriorly through the brain stem to the hypothalamus among other forebrain areas. The arcuate nucleus and the paraventricular nucleus are important areas in the regulation of food intake (Woods et al., 1998; Bouret and Simerly, 2004). Two neuronal types in the arcuate nucleus have critical and opposite roles in the regulation of feeding behavior; NPY and proopiomelanocortin (POMC) neurons are involved in stimulating or inhibiting feeding, respectively. NPY neurons express NPY and agouti-related protein (AgRP), both potent orexigenic peptides. Central infusion of NPY or AgRP increased food intake and led to obesity in rats (Sahu, 2004). NPY/AgRP neurons express ghrelin receptors, and one mechanism for ghrelin's orexigenic effects is via stimulating NPY/AgRP neurons (Zigman and Elmquist, 2003). Leptin appears to act through the NPY and POMC neurons, inhibiting NPY neurons and stimulating POMC neurons, thus inhibiting food intake. Interestingly, insulin also acts on NPY neurons (inhibits) and POMC neurons (stimulates) to inhibit food intake. Leptin and insulin activate some of the same intracellular pathways within the same neuronal populations, though not identically in all cases (Berthoud and Morrison, 2008).

In contrast to leptin's effects on adult mice, exogenous leptin given to neonatal mice does not result in decreased food intake, at least for the first few weeks postnatal (Mistry et al., 1999; Bouret and Simerly, 2004). This is consistent with the fact that leptin-null mice do not differ in size or fatness for the first few weeks after birth; only then do they begin to diverge from the wild type pattern (Bouret and Simerly, 2004). Leptin receptors are expressed in the hypothalamus of neonatal rats; peripheral administration of leptin does modify NPY and POMC mRNA expression in the arcuate nucleus of neonatal rats (Proulx et al., 2002). So leptin signaling is apparently intact in neonatal mice and rats. However, arcuate nucleus projections to other hypothalamic nuclei are immature at birth and begin to develop after the first week postnatal (Bouret et al., 2004). There is a surge in leptin secretion at this same time in mice and in rats (Ahima et al., 2000; Bouret and Simerly, 2004). In adult leptin-deficient mice, exogenous lepton resulted in rapid rewiring of the arcuate nucleus (Pinto et al., 2004). Leptin appears to have a critical role in the development of the appetite control circuits in the hypothalamus (Bouret and Simerly, 2004).

As important as hypothalamic circuits and the arcuate nucleus may be to the regulation of feeding behavior, other brain areas are important as well. For example, leptin receptors are found throughout the brain; even the brain stem is a target for leptin signaling (Hosoi et al., 2002). Taste circuits run through the brain stem to the solitary nucleus and then to the parabrachial nucleus. From the parabrachial nucleus there are two circuits. One runs to the amygdala and the bed nucleus of the stria terminalis; the other runs to the gustatory cortex through the gustatory thalamus (Norgren, 1995). Feeding is a complex behavior. It requires interaction and coordination between peripheral organs and multiple brain regions (see Figure 8.3). Neural circuits important to feeding behavior are distributed throughout the brain.

Cortex certainly is involved in feeding behavior. We decide when and what to eat, and we can decide not to eat. The hypothalamic feeding circuits do not control our feeding behavior; they do provide input into our motivation (Berridge, 2004). Obese women were shown to have significantly less activation of the left dorsolateral prefrontal cortex in response to eating a meal than did lean women or women who had formerly been obese (Le et al. 2007). One role of prefrontal cortex is to inhibit behavior, especially behavior that is no longer appropriate. Prefrontal cortex is thought to be important in decision-making (Heekeren et al., 2004, 2006) and may be important for the regulation of feeding by inhibiting orexigenic brain regions (Gautier et al., 2001). In a study of men and women fed a liquid meal, prefrontal cortex and hypothalamic activation was associated with serum increases of the meal-terminating gut peptide glucagon-like peptide 1 (Pannacciulli et al., 2007).

Thus caudal brain stem, hypothalamic brain regions, and cortex, among others, have been shown to be involved in regulation of feeding. Feeding behavior is coded at all levels and is generally consistent with a hierarchical view of behavioral organization. The British neurologist John Hughlings Jackson (1835–1911) concluded from his clinical observations of patients with brain lesions that most psychological functions involved brain hierarchies; importantly, function is represented at all levels of neural organization. For example, the ability to smile requires a motor pattern encoded in the brain stem; damage to that area of brain stem would cause facial muscle paralysis and a smile could not be formed. However, the desire to smile or the knowledge that a smile was appropriate would still be intact. On the other hand, specific damage to motor cortex in one of Hughlings Jackson's patients resulted in one side of the patient's face being unable to smile voluntarily; however, when told a joke the patient would smile on both sides of his face. The lower forebrain areas that coded the emotion that made the joke funny still worked, as did the brain stem regions that coded the muscle movements for a smile. What was lost was the voluntary control, coded by the motor cortex. Thus function is coded at all levels, but there is a general hierarchy in which forebrain directs brain stem function.

This hierarchy is not absolute; neocortex does not rule all other brain areas in all cases. Strong emotional reactions can overrule voluntary regulation (Berridge, 2004). Indeed, each level of brain hierarchy is semiautonomous (Gallistel, 1980). Many actions and behaviors can take place with minimal cortical involvement. For example, cortex is certainly important in deciding when to eat, but once food is placed in the mouth the brain stem doesn't need cortical direction to chew and swallow. And brain stem sends signals up to forebrain that informs and affects neural function. For example, the taste or texture of the food might cause the forebrain to direct the brain stem to stop chewing, definitely not swallow, and instead spit that food out.

Hierarchy within the forebrain is even more problematic (Berridge, 2004). Forebrain neural circuits are rarely linear top-down architecture; they are truly circuits, with complex connections and feedback loops. The various areas of the brain interact to create behavior. In addition, information molecules (steroid and peptide hormones) are released from peripheral organs and can contribute to forebrain signaling directly, with or without being involved in brain stem signaling. The simple model of bottom-up signaling and top-down control does not describe feeding biology.

Neural circuits that regulate feeding behavior are distributed throughout the brain. Inputs and outputs travel from brain stem to the limbic system to cortex. These circuits are ancient and appear to be operative in most if not all mammals; much of our knowledge of them has come from tracing the neural connections involved in feeding in the laboratory rat. The insights gained from investigating rat feeding neurobiology have generally held true when extended to other mammals including ourselves.

The most important difference between humans and other mammals arguably is the large increase in cortex in humans. Of course we have other unique features; all extant animals have their specific adaptations. We also share a great deal of our biology with other mammals, especially other primates. If we wish to investigate what aspects of our feeding behavior differ from other mammals, an appropriate place to start is with the increased corticalization of function in humans.

Cortex is traditionally thought to serve an important inhibitory function on behavior. The limbic system is thought to be important in emotion and motivation, the brain stem in involuntary behavior and in delivering inputs up to forebrain and outputs from forebrain to the periphery. Of course the brain works together; the division given above is a gross oversimplification.

Consider our feeding behavior, past and present. In the past, external factors often limited food availability. The question of whether food is available is usually not an issue these days; stores and restaurants are often open 24 hours a day, 7 days a week. Food can be stored easily and accessibly. There are still constraints on our feeding behavior, but they are more likely to be internally driven. We decide when we eat. Being hungry is a powerful motivation, but we can delay eating, consciously choosing when to eat. There are also social aspects to eating. These can be as important as the nutritional aspects of eating, especially in the timing of eating. Even if a person is hungry she may refrain from eating because she is at work or engaged in another activity that she doesn't want to interrupt. And a person who is not hungry may eat something because of the social setting, a lunch meeting with a business associate or a friend, for example. People really do decide when to eat.

Metabolic Models

Peripheral physiology both affects and is affected by feeding behavior; the gut, liver, adipose tissue, and other endocrine organs in the body send and receive signals that assist in the regulation of food intake. Metabolic signals and processes may be as important as are endocrine signals.

The word *metabolism* is used in different contexts. The most appropriate dictionary definition in the context of this book is "the chemical changes in living cells by which energy is provided for vital processes and activities and new material is assimilated." In this definition multiple concepts are involved that are relevant to different organs over varying time scales. Providing energy, vital processes, and activities and assimilating new materials all reflect different metabolic pathways, organ systems, and physiological conditions. Metabolism is regulated by a number of signals. Some are generated by metabolism itself; metabolic signals that are by-products or end products of metabolism. Others will be generated by exogenous signals from the environment, transduced through the brain.

Energy metabolism is the prime focus of this book. Energy metabolism certainly doesn't encompass all the metabolic aspects of feeding biology and susceptibility to obesity, but in the end, obesity does represent excess stored energy, and thus energy metabolism systems will be involved. Energy comes from metabolic fuel, the main components being carbohydrates, lipids or fats, and proteins.

Metabolic flexibility, the ability to switch between metabolic fuels (i.e., up- or down-regulate carbohydrate, fat, or protein oxidation) varies among individuals and groups. This variation complicates the concept of energy balance and possible lipostatic mechanisms to regulate appetite. It raises questions such as: If the energy from ingested food is stored rather than metabolized even though there is a metabolic demand for that energy, does that result in "excess" food intake? In the following section we explore the idea that food and energy intake as measured by a researcher may not reflect what the body senses.

Metabolism and Obesity

In Sprague-Dawley rats, there is variation in the susceptibility to gaining weight on a high-fat diet. Several investigators have looked at different aspects of metabolism to try to characterize this variation in order to be able to predict which rats will be susceptible and which resistant. Ji and Friedman (2003) showed that rats that were less able to up-regulate fat oxidation were more prone to gaining weight on a high-fat diet (Figure 8.4a and b).

This is evidence for the metabolic substrate theory of regulation of food intake and subsequent weight gain (Friedman and Stricker, 1976). This model predicts that when fed a high-fat diet, rats that could rapidly and efficiently up-regulate their fat oxidation would be able to provide the body organs, especially the liver, with a sufficient rate of energy (e.g., molecules of glucose, glycogen, or ATP). Those rats that were poor at fat oxidation would deliver a lower rate of energy; because they could not up-regulate fat oxidation sufficiently to cope with the influx of fatty acids they would end up storing more of the fat directly in their adipose tissue. The metabolic fuel theory of food regulation predicts that sensors or signals, likely from the liver, would thus indicate to the brain that the available energy for metabolism was insufficient. In other words, there is a deficit in oxidizable fuel due to the poor ability to oxidize fat. The behavioral response would be to cue further food intake. Thus the energy that was stored in adipose tissue instead of being oxidized to support metabolism is effectively replaced by further food intake instead of mobilized from adipose. The total energy intake increases, with a resulting increase in stored energy in fat.

This model makes several important points to be considered when

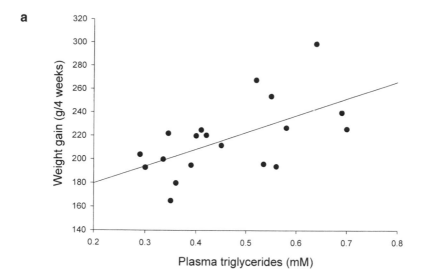

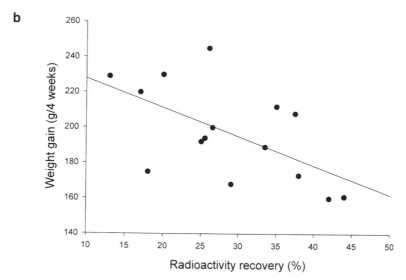

FIGURE 8.4. (a) Rats were fasted for 18 hours; fasting triglycerides were positively correlated with the amount of weight gain over 4 weeks when the rats were fed a high-fat diet. (b) Recovery of radioactive CO_2 from rats fed labeled palmitic acid was negatively associated with weight gain over 4 weeks on a high-fat diet. Thus, rats who were less able to up-regulate fat oxidation gained more weight. Data from Ji and Friedman, 2003.

investigating causes of weight gain and obesity. Although the body can undoubtedly tell whether there is a net increase or decrease in available energy and the relative proportion of energy coming from different metabolic fuels (fat, carbohydrate, and protein), it can't tell from which fuel any particular ATP molecule was derived, or whether that ATP molecule was created from ingested food (exogenous source) or from mobilized stores (endogenous source). Physiology may be able to detect that a source of ATP is desired, but it can't necessarily dictate whether that source of oxidizable substrate will come from stored energy as opposed to new food intake. In fact an evolutionary argument can be made that in the presence of available food, ingestion might generally be favored over mobilizing energy stores. Energy stored on the body can and very likely will be used at a later time, but in the short term it may become effectively invisible to aspects of the energy metabolism and appetite regulating systems.

Other investigators looked at the same phenomenon of obesity-prone and obesity-resistant rats but focused on circulating leptin, the hormone synthesized by adipose tissue. They showed that the rats that were prone to gaining weight on a high-fat diet showed an increased leptin response to feeding (Leibowitz et al., 2006). This finding would seem to be counterintuitive. Since giving exogenous leptin usually decreases appetite, why would an exaggerated leptin response to feeding result in greater food intake? And the obesity-prone rats did indeed have higher food intake.

The answer may lie in how leptin secretion from adipose tissue is regulated. Although leptin is often considered a reliable predictor of total adipose tissue, that result reflects basal or steady-state circulating leptin concentrations. Circulating leptin concentrations are certainly generally associated with adiposity, however leptin concentration also varies over the day, due both to a circadian rhythm and to eating patterns. In general, leptin concentrations are highest late at night and lowest early in the morning just after awakening (Licinio et al., 1998). The association with adiposity can be further affected by how long it has been since the animal ate. For example, after acute fasting, circulating leptin levels decrease dramatically, far in excess of actual changes in total adipose tissue. Thus although leptin in the steady state condition appears to be highly correlated with total adipose tissue, under a catabolic condition such as fasting, the drop in leptin secretion is far greater than would be expected from changes in amount of fat. A hypothesis to explain these various findings is that leptin synthesis in and secretion from adipose tissue is at least partly regulated by whether fat is being stored or mobilized. A net mobilization of fat results in decreased leptin secretion; a net deposit of fat results in increased leptin secretion.

If we now look at the characteristics of obesity-prone versus obesity-resistant Sprague-Dawley rats from the various studies we can see that the data are consistent. One inference from combining the findings that obesity-prone rats are both poor at fat oxidation and have an exaggerated leptin response to eating a high-fat diet is that being poor at fat oxidation results in a larger proportion of ingested fat being stored in adipose tissue; this in turn results in a greater leptin secretion from adipose tissue. Thus obesity-prone Sprague-Dawley rats appear to be poor at increasing their oxidation of fat, which means that when fed a high-fat diet they have an exaggerated flow of lipid into adipose tissue, which triggers an exaggerated leptin response.

This suggests that leptin is not the key metabolic signal regulating appetite for this particular mechanism for developing obesity, though it is possible that rats with an exaggerated leptin response to eating may also have an exaggerated decline in circulating leptin in response to fasting. This is intuitively attractive. A poor ability to oxidize fat would result in high fasting triglyceride levels (which are seen in the obesity-prone rats) due to a lower rate of oxidation of the lipid released from adipose tissue during fasting. This would also mean a lower rate of ATP production, unless carbohydrate or protein metabolism is up-regulated to compensate. In either case it makes sense that signals that would favor further release of lipid from adipose tissue would be enhanced, leading to a greater decline in leptin secretion, and, when access to food is finally obtained, a greater appetite. However, these data also suggest that there are other metabolic signals, possibly linked more directly with ATP production and concentration in organs such as the liver and muscle, that are driving the increased appetite that leads these rats to overeat and gain weight.

Metabolic flexibility has become an important consideration in studying vulnerability to obesity in people. The evidence supports the idea that there is considerable variation in the ability of people to shift between fat and glucose oxidation. Men and women differ in the ability to up-regulate fat oxidation (see chapter 12). There also appears to be significant variation among racial and ethnic groups (see chapter 13).

Summary

Food intake is certainly regulated; animals and people rarely eat as much as they are physically capable of doing. Animals make decisions about what, when, and how much to eat.

Feeding is a complex behavior. It requires interaction and coordination among peripheral organs and multiple brain regions. Neural circuits important to feeding behavior are distributed throughout the brain. Inputs arrive at the brain via the nervous system through the brain stem, but also via steroid and peptide hormones; these information molecules can reach the brain by crossing the blood-brain barrier, via the circumventricular organs of the brain, which are outside the blood-brain barrier, or by traveling up the vagus nerve.

There is a general hierarchy of brain organization; function is coded at all levels of brain organization, but cortex can "rule" the limbic system and brain stem. This hierarchy is not complete; the brain works as a whole and it is impossible to say that any particular area controls another in a simple top-down model. However, the expanded cortex of humans has increased the behavioral flexibility of our feeding biology. We can inhibit the appetite and satiety signals coming from periphery and other brain areas and thereby refrain from eating when hungry and eat when we aren't hungry.

Finally, metabolic signals may play as important a role in feeding behavior as do endocrine signals. At the basic biological level, food intake is about acquiring metabolic fuel that is oxidized to provide metabolic energy to perform the functions of life. Variation in the ability to up-regulate fat oxidation has been shown to be a risk factor for dietary-induced obesity in rats; there is documented variability among people in the ability to switch between glucose and fat oxidation. Metabolic flexibility may be a key component in the vulnerability to obesity on a Western diet.

CHAPTER NINE

Getting Ready to Eat

E ating requires a complex set of coordinated physical, physiological, and behavioral actions. Of course when we eat we don't really
have to think about it. We can do it easily and naturally. But when we
consider it as scientists, we can see that eating is a very complex process.
It is also episodic. We don't eat all the time. We don't even think about
eating all the time.

However, if you hadn't eaten for a while and you smelled an enticing
aroma wafting from the kitchen, your body would react. Your physiology
would begin to gear itself up to receive food. Your mind would turn to
eating. And perhaps you would get out of your chair and go into the
kitchen to find out when dinner was ready. Already your body would have
changed its metabolic state. It would be ready for dinner, regardless of
what the cook says. This anticipatory response to the smell, sight, and
taste of food was first discovered by Pavlov in the late 19th century.

Pavlov Revisited

In 1904 Ivan Petrovich Pavlov received the Nobel Prize in Physiology for
his work on the digestive system. A key facet of his work was the demonstration that the gut and the central nervous system (CNS) work together
in the process of digestion. Digestive secretions, saliva, gastric acid, and
so forth could be stimulated by the brain. Thus even before the peptide
revolution and the identification of the many gut-brain peptides, the concept of a link between central and peripheral feeding physiology was
proposed.

The concept of *factory* is important in understanding Pavlov's approach toward and contribution to science: many investigators sharing resources and technologies in the pursuit of knowledge about a common theme, performing related but not identical investigations. Pavlov established a factory for the study of physiology (Figure 9.1). He also conceived of the digestive system as a complex chemical factory, with many parts working in concert to provide the nutrients necessary for life (Todes, 2002). The orderliness of his mind was reflected in both his methods of investigation and his conception of physiology. (For more details on Pavlov's life and accomplishments see *Pavlov's Physiology Factory* by Daniel P. Todes [2002]).

At the time of Pavlov's early investigations there was controversy over whether the nervous system contributed to digestive secretions. The consensus among physiologists at that time was that the central nervous system was not involved (Todes, 2002). Pavlov disagreed. In a series of experiments at his factory, each building upon and expanding the results of the ones before, Pavlov was able to demonstrate the connection between

FIGURE 9.1.   The vivarium at Pavlov's laboratory in Saint Petersburg, Russia.

the brain and the gut by investigating the phenomenon now known as the cephalic-phase response in digestive physiology (Smith, 1995).

In this chapter we examine the role of cephalic-phase responses in the regulation of food intake. We examine cephalic-phase responses from an adaptive perspective. We focus on the functions of various cephalic-phase responses, their adaptive value, and possible selective pressures that have influenced their evolution.

Cephalic-Phase Responses

We eat food; we need nutrients. Nutrients are essential, but many are also toxic in high concentrations, and food contains many things, good and bad, besides nutrients. The function of our digestive physiology is to render food into nutrients safely and efficiently.

The term *cephalic-phase response* refers to anticipatory physiological regulation related to food and feeding. It refers to digestive and metabolic responses to food cues generated by the central nervous system that act to prepare the organism to ingest, digest, absorb, and metabolize food (Pavlov, 1902; Powley, 1977; Smith, 1995). These anticipatory physiological responses increase the efficiency with which an organism turns food into nutrients. One result is an increase in the amount of food that can be processed at a given time—an advantage in our evolutionary history, but possibly one source of our species' susceptibility to obesity in the modern milieu. For certain foods (e.g., those high in simple sugars) maybe we are too efficient. Recent evidence suggests that physiological responses that serve to end feeding also can have a cephalic phase. Thus, by the first bite of food, or even before, physiological processes have been set in motion that will influence the duration of the meal and the amount of food eaten.

The phrase originally used by Pavlov was *psychic secretions* (see Powley, 1977; Todes, 2002), to emphasize the role of the "psyche" in digestive secretions. The change in terminology to *cephalic-phase responses* distanced the concept from any association with the mystical notions associated with the word *psychic*, and reflected the fact that some cephalic-phase responses are not secretions (e.g., gastric motility, thermogenesis).

Pavlov initially studied pancreatic and gastric secretions. His first success was demonstrating that the vagus nerve was linked with pancreatic secretions (Smith, 1995; Todes, 2002). The demonstration of a cephalic-phase gastric secretion response took longer, but his team of investigators were able to use the surgically treated dog models Pavlov developed to show that gastric acid secretion begins within minutes of tasting, or even seeing, food, as long as the vagus nerve is intact. Severing the vagus nerve abolishes the cephalic response. A connection with the brain is required for these peripheral physiological responses.

For many people the name Pavlov is associated with salivating dogs. For Pavlov, salivary secretions were digestive secretions, serving the same intrinsic function as gastric and intestinal secretions—enabling the animal to utilize food for bodily needs. He demonstrated in dogs that salivation varied with the food ingested. For example ingesting dry food stimulated greater salivation than ingesting wet food (Todes, 2002). He also, famously, demonstrated that salivation can occur in anticipation of feeding, and that dogs can be trained to associate many external signals with the imminent appearance of food (Pavlov, 1902). And yes, the sound of a bell was one of the stimuli he used.

Pavlov termed these responses to arbitrary stimuli *conditional responses*. The secretory response was in anticipation of feeding. Thus digestive secretions are anticipatory as well as reactive. The gut and the brain are intimately linked, working together to acquire and utilize the nutrients necessary for life.

The general concept of cephalic-phase responses has changed little since the original demonstrations. It was revitalized by Powley (1977), with special emphasis on the cephalic-phase insulin response, an important metabolic response to feeding. In functional terms, cephalic-phase responses are anticipatory changes in physiology and metabolism that serve to prepare the digestive tract to digest food and absorb nutrients, and to prepare other organ systems (e.g., liver, adipose tissue) to metabolize and store the absorbed nutrients. What has changed since Pavlov's day is that the list of secretions and other responses that appear to have a cephalic phase has continued to expand (Table 9.1). Cephalic-phase responses can be physical (e.g., gut motility), secretory (e.g., digestive enzymes, peptide hormones), or metabolic (e.g., thermogenesis). They can affect digestion, metabolism, and behavior. Recent evidence suggests that cephalic-phase responses not only prepare an animal to digest, absorb,

TABLE 9.1 *Some known cephalic-phase responses*

Cephalic-phase response	Organ(s)	Function(s)
Salivation	Mouth	Lubricates food; begins digestion of starch; dissolves food particles (essential for taste)
Gastric acid secretion	Stomach	Hydrolysis of food
Gastrin	Stomach	Stimulates gastric acid secretion
Lipase	Stomach; pancreas	Fat digestion
Gastric emptying	Stomach	Regulates food passage
Intestinal motility	Intestine	Regulates food passage
Bicarbonate	Intestine	Neutralizes stomach acid
Cholecystokinin (CCK)	Small intestine	Terminates feeding
Insulin	Pancreas	Regulates blood glucose
Pancreatic polypeptide	Pancreas	Regulates pancreatic and gastrointestinal secretions
Digestive enzymes	Pancreas	Aids in digestion of protein, carbohydrates, and fat
Bile	Gall bladder	Fat emulsification
Leptin	Adipose tissue; stomach	Reduces appetite
Ghrelin	Stomach	Stimulates appetite; stimulates GH secretion, fat absorption
Thermogenesis	Many	Represents the increase in energy metabolism due to digestive and physiological processes related to eating

and metabolize food, but they may also play a role in appetite and satiety via stimulating endocrine secretions. Thus cephalic-phase responses may affect the beginning and end of a meal.

Importance of Anticipatory Responses in Regulatory Physiology

Animals anticipate. Behavior, physiology, and metabolism are not merely reactive. The senses convey information about the external environment to the central nervous system. The central nervous system interprets this information within the constraints of experience (knowledge); intrinsic, evolved tendencies (phylogeny); and current conditions (social setting, nutritional status, and so forth). Is there a threat? Is there an opportunity? The central nervous system sends messages to the appropriate peripheral organs to begin the physiological cascades that prepare the organism to respond to the anticipated challenge. The animal changes its state in advance of the potential need.

These anticipatory physiological changes can be in response to circumstances or can reflect internal, clocklike rhythms. For example, there are circadian rhythms in the secretion of many hormones (e.g., cortisol, leptin, ghrelin), which allow an animal to be in the most appropriate physiological state at different time points. Moore-Ede (1986) suggested the term *predictive homeostasis* for these anticipatory changes in physiology. Anticipatory responses are associated with the central coordination of physiology and the interplay of physiology and behavior. Some authors (e.g., Schulkin, 2003; Sterling, 2004) have emphasized how the central coordination of physiology, anticipatory physiological responses, and the interplay of physiology and behavior appear to be given short shrift by the classic homeostatic paradigm of physiological regulation. The concept of allostasis, defined as the "process by which an organism achieves internal viability through bodily changes of state," has been proposed as an alternative (Schulkin, 2003, p. 21). Many of the examples of allostatic regulation provided by Schulkin (2003) involve the concept that the hormones that regulate peripheral physiology in response to a challenge are also involved in changing central motive states of the brain and thus inducing behaviors that aid the animal to meet the challenge (see also Epstein, 1982; Herbert, 1993; Smith, 2000). The regulation of food intake is a paradigmatic example of the linkage of peripheral physiology with central motive states and the concept that behavior and physiology act together to preserve viability.

The concept of allostasis and especially allostatic load (McEwen, 2000; Schulkin, 2003) is applicable to the health consequences of obesity. Obesity is excess adipose tissue. Adipose tissue is very metabolically and endocrinologically active. Many of the health consequences of obesity may result from the overexpression of normal physiology.

Importance of Anticipatory Responses in Feeding Biology

We are a species that feeds in meals. We have discrete times when we eat, separated by substantial time periods when we do not eat. In consequence, food enters our bodies in pulses. This is true of most, but not all, animals. For example, ruminants, such as cows, attempt to maintain at least a minimal rumen fill at all times, so that their digestive tract is never empty and is always transferring nutrients to the bloodstream. Most nonhuman primates feed more by grazing than in what could be called meals.

In the wild, nonhuman primates may refrain from feeding because of external or internal constraints. For example, golden lion tamarins, small New World monkeys, feed extensively from the time they awake. However, when they reach a fruit tree that has abundant fruits they will often cease to eat after a short period of feeding and will engage in social behavior. After approximately 20 to 30 minutes, social behavior stops and the animals begin feeding again. The resumption of feeding is preceded by a rain of seeds being eliminated from the monkeys' digestive tracts (MLP, personal observation). In other words, under conditions of high abundance of a fruit the monkeys' fruit intake can exceed their digestive capacity. They are physically forced to stop eating!

Humans rarely experience complete digestive bulk fill these days. It is certainly possible that in our distant past there may have been times when gorging was an adaptive strategy, but humans usually don't refrain from eating because their stomachs are full. We deliberately refrain from eating until certain, usually socially determined, times.

This pattern of feeding in meals has probably been present for a substantial part of our evolutionary history. It reflects more than a simple response to food intake exceeding the rate at which the digestive system can process food. Meals reflect our adaptation of cooperative food gathering and sharing. Meals have social significance as well as nutritional value. Meals may have been a significant behavioral adaptation in our evolution.

Because we are meal-feeders, our internal milieu is not strictly constant. The state of the digestive system and numerous other organ systems (e.g., liver, kidney, adipose tissue) constantly change to accommodate conditions of nutrient excess (feeding) followed by potential nutrient deficits (between meals). Nutrients continually flow into and out of storage pools. Meals result in perturbations of the internal milieu that must be accommodated (Woods, 1991; see also chapter 10). The nutrients that enter the blood in pulses from the digestive tract during feeding must be metabolized or transported to and sequestered in the appropriate storage pools.

Sometime later, when the gastrointestinal tract is largely empty, nutrients reenter the blood from nutrient stores. The secretory responses that regulate absorption, storage, and mobilization of nutrients constantly change over the day.

These changes take time. Cephalic-phase responses, anticipatory physiological responses to cues that feeding is imminent, allow organisms to get "ahead of the curve." They improve the efficiency with which animals digest food, and absorb, metabolize, and store the liberated nutrients. They also prime the organism to meet the resulting homeostatic challenges presented by the influx of nutrients, such as changes in blood pH and electrolytes.

Cephalic-phase responses will occasionally result in digestive and metabolic secretions that are not followed by feeding. Sometimes prey escapes, or other factors (predators, conspecifics) interfere with feeding. However, the advantages of anticipating feeding appear to have outweighed any costs of wasted digestive secretions.

Evidence for Cephalic-Phase Responses

There is a considerable literature on cephalic-phase responses, going back to Pavlov (Todes, 2002). Cephalic-phase responses have been demonstrated in a wide range of mammals, including humans, nonhuman primates, dogs, cats, sheep, rabbits, and rats (Powley, 1977). Some of these cephalic-phase responses are general, and some appear to be specific to the nutritional properties of the tastant, that is, responses to sweet substances differ from those to bitter substances or to high-fat substances. It has long been established that sensory contact with food stimulates cephalic-phase digestive responses that result in increased secretion of saliva (Pavlov, 1902), gastric acid (Pavlov, 1902; Farrell, 1928), and pancreatic secretions, including enzymes, proteins, and bicarbonate (Pavlov, 1902; Preshaw et al., 1966). Even the sight of food within sealed plastic containers stimulated gastric secretions in humans (Feldman and Richardson, 1986). Adding smell and taste to the sensory experience increased the response (Feldman and Richardson, 1986).

In humans, dogs, and rats it has been repeatedly shown that the palatability of the offered food is positively correlated with the extent and

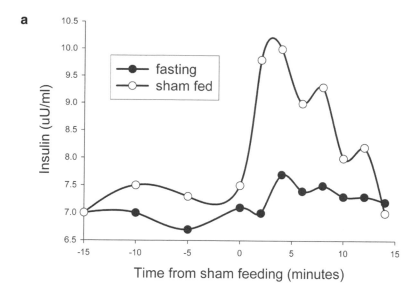

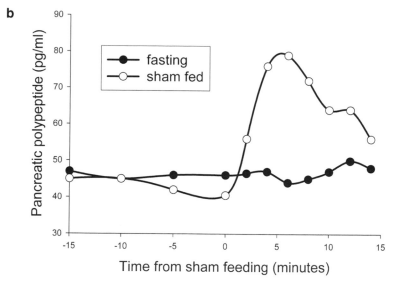

FIGURE 9.2. Human response to sham feeding: (a) cephalic-phase insulin and (b) pancreatic polypeptide. Data from Teff, 2000.

magnitude of the cephalic salivary and gastric secretions (reviewed in Powley, 1977). Thus appetite, or the psychological state of wanting food, directly affects the physiological processes of digestion and metabolism (Pavlov, 1902; Powley, 1977).

Cephalic-phase responses can be demonstrated using the technique of sham feeding. In humans, this consists of masticating the test diet or tastant, but not swallowing it. In animal models, the use of fistulas at different parts of the GI tract allows an animal to masticate and swallow, but keeps the food from the GI tract below the fistula. Thus food sensory cues can be restricted to sections of the alimentary canal. Pavlov (1902) used esophageal fistulas, restricting sensory input to the mouth and tongue. He showed that the gastric secretions stimulated by sham feeding exceeded those stimulated by the sight of food. Many experimenters have preferred gastric fistulas, which mean that potential effects of the exposure of food to the stomach must be considered. In both cases metabolic changes due to absorbed nutrients are minimal if not nonexistent.

Sham feeding stimulates a number of changes in the gastrointestinal tract, in the bloodstream, and in behavior. For example, sham feeding increases gastric acid secretion in dogs (e.g., Pavlov, 1902), rats (e.g., Martinez et al., 2002), and humans (e.g., Goldschmidt et al., 1990). Sham feeding induces the secretion of peptides from the pancreas, resulting in an anticipatory rise in blood insulin and pancreatic polypeptide concentration (Teff, 2000; Figure 9.2a and b).

Pavlov (1902) demonstrated that food placed directly into the stomach of dogs was poorly digested; however, if sham feeding preceded the intragastric intubation, then digestion was enhanced. Several clinicians in the 1800s and early 1900s independently discovered that patients with fistulas that were by necessity fed by gastric intubation fared much better if they were allowed to chew and taste food prior to the actual meal being delivered to their gastrointestinal tract. Appetite was much better, and patients were better able to maintain body weight. One patient insisted on swallowing the masticated food, even though it was shortly regurgitated from his esophageal pouch (reviewed in Powley, 1977). These observations reinforce the importance of the brain in coordinating digestive responses.

Another method to demonstrate cephalic-phase versus reactive responses is by showing that the physiological changes occur before post-

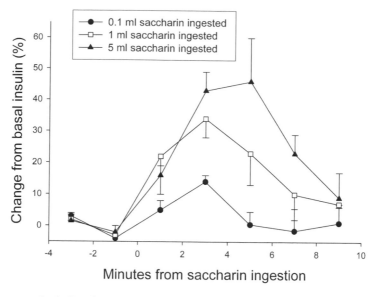

FIGURE 9.3. Cephalic-phase insulin response to the ingestion of different volumes of 0.15% sodium-saccharin solution in rats. Plasma glucose levels did not change. Data from Powley and Berthoud, 1985.

ingestional effects. For example, an initial pulse of insulin secretion in response to food ingestion occurs within 10 minutes (peak value at 4 minutes postingestion) in normal-weight men, and this is before any change in blood glucose concentration due to absorption of nutrients could occur (Teff et al., 1991). The cephalic-phase insulin response can also be triggered by nonnutritive sweet-tasting substances, such as saccharin (Powley and Berthoud, 1985).

Not all experiments have been able to demonstrate a cephalic-phase insulin response. For example, humans differ from rats in that merely tasting a sweet substance does not appear to be sufficient to generate the cephalic-phase insulin response in humans. Oral infusion of a glucose solution stimulates a cephalic-phase insulin response in rats (e.g., Berthoud et al., 1980). Rats that ingested saccharine solutions displayed a reliable and dose-dependent cephalic-phase insulin response (Powley and Berthoud, 1985; Figure 9.3); ingesting sweetened solutions or sucking on sweetened tablets do not reliably initiate insulin secretion in human subjects (Bruce et al., 1987; Abdallah et al., 1997). A study in which human subjects tasted but did not swallow sweet-tasting liquids found no cephalic-phase insulin response or effects on blood glucose (Teff et al., 1995). However, in the same study sham feeding on apple pie produced a reliable insulin response.

Most studies using the sham feeding paradigm with human subjects have reliably found a cephalic-phase insulin response. The complexity of the food stimulus appears to affect the cephalic responses in humans; the more modalities involved, the greater the response (Feldman and Richardson, 1986). Perhaps the stimulation of taste alone, without the associated motor and other secretory involvement present in eating (e.g., chewing, salivation), is not sufficient to cue the full cephalic metabolic responses in humans. A sweet taste alone may not ensure the expectation that food is to be ingested, and it is that expectation that underlies the functional aspect of the cephalic insulin response.

The Role of Taste

The mouth is the "clearinghouse of the organism" (Pavlov quoted in Smith, 1995). It is the proximal end of the alimentary tract and the first stage of digestion. Food is masticated and mixed with saliva, which begins the digestive process. Food is also tasted.

Taste plays a variety of roles in appetite, food intake, and digestion (Norgren, 1995). Food has hedonic qualities, and these influence food choice, both in type and in amount. But taste also has a role in anticipatory physiological regulation. Taste is a direct cue to the digestive tract that feeding has commenced and that food to be digested is entering the system. Nutrients from the digested food will then begin to flow into the bloodstream. Nutrients come in pulses (meals) for humans, not in more or less continuous flows as for ruminants. We must switch between mobilizing energy stores (acquiring the food; energy expenditure) to energy distribution to stores (digestion, absorption, and storage). A number of physiological events must occur for food to be digested and for the nutrients to be absorbed and then deposited in the appropriate storage organs or depots. Salivary secretion, containing amylase for initial starch breakdown, is increased; acid secretion in the stomach is increased; proteolytic enzymes are secreted into the small intestine; and bile for fat digestion is released by the pancreas, among other changes that prepare the digestive tract to digest food. Insulin is released into the bloodstream even before blood glucose levels rise. These cephalic-phase responses (Powley, 1977) anticipate the incoming nutrient load and prepare the body to absorb and assimilate the pulse of nutrients. The physiology of digestion, absorption, and assimilation of nutrients is anticipatory as well as reactive, and taste plays a key role in this anticipatory regulation.

There are a number of so-called basic tastes. The exact number is in some dispute and can vary depending on definitional decisions. For example, astringency is considered a taste by some. Humans are capable of discriminating at least five basic tastes: sweet, sour, salty, bitter, and umami (also called savory). Evidence suggests that there might be a fat taste as well. Taste has two main functions: to promote or inhibit ingestion, and to prepare the body to utilize or metabolically respond to ingested materials. Cephalic-phase responses generally prepare an animal to digest, absorb, and then store nutrients that enter the body through feeding. However, cephalic-phase responses can also serve to inhibit feeding and prepare an animal to deal with toxic or tainted food. For example, a bitter-tasting substance can decrease gastric motility (Wicks et al., 2005). Some tastes and smells are nauseating, intrinsically or from learned associations. The reactions to these stimuli precede any actual physiological or metabolic effects of their ingestion.

Sweet taste generally stimulates ingestion in rats and humans and stimulates cephalic-phase digestive and metabolic responses. This is true of nutritive and nonnutritive sweet-tasting substances (e.g., saccharin; see Figure 9.3). In nature, a sweet taste is associated with a high concentration of simple sugars. For omnivorous or frugivorous species, animals that will naturally feed on ripe fruits, nectars, and other plant parts that contain significant concentrations of simple sugars, the ability to detect sweet taste and having a preference for sweet-tasting foods makes intrinsic sense. Some animals do not appear to be able to detect sweet tastes, however. Strict carnivores such as felines appear to have lost the ability to detect sweet (Li et al., 2006).

Sour taste is associated with acid. It can generate aversive responses; it is not known if any cephalic-phase responses are generated. Umami is associated with fermented foods that contain high concentrations of free glutamates. It is also associated with high-protein foods and the ability to detect amino acids. The food additive monosodium glutamate (MSG) produces the umami taste by stimulating oral glutamate receptors. Taste sensitivity for MSG is associated with an increased preference for high-protein foods in both men and women (Luscombe-Marsh et al., 2007). Cephalic-phase pancreatic secretions (Ohara et al., 1988), including a cephalic-phase insulin response (Niijima et al., 1990), are produced by oral infusion of MSG solutions. Insulin is important for the metabolism of amino acids.

Salty taste is associated with sodium. Sodium deficiency can generate a robust appetite for salty-tasting substances (Richter, 1936; Denton, 1982; Schulkin, 1991; Fitzsimons, 1998). There is a large literature on salt appetite. Oral infusion of a sodium chloride solution in conscious beagles produced a significantly lower pancreatic response than did either sucrose or MSG (Ohara et al., 1988).

Bitter taste is generally a feeding deterrent. Many toxic substances found in plants (e.g., alkaloids) taste bitter. There is significant genetic variation in humans in the ability to detect certain "bitter" substances. Bitter taste can slow gastric motility (Wicks et al., 2005).

Is There a Fat Taste?

Humans generally like fat in their food. Fat in foods is likely detected through multiple mechanisms (Mattes, 2005). Texture is thought to provide cues to fat content in food. This is referred to as "mouth feel" in the food industry. However, texture alone does not appear to explain the ability of humans to perceive fat content (Mattes, 2005). Rats appear to be able to detect fatty acids by olfaction, but the evidence for this ability in humans is mixed at best (Mattes, 2005). There is growing evidence that certain chemical receptors in the mouth can detect fatty acids. One such suggested receptor is CD36 (also known as fatty acid translocase, or FAT), a transmembrane protein that binds lipids, including long-chain fatty acids, and is capable of transporting them across plasma membranes. CD36 has many functions in cellular lipid transfer. It is also expressed in rat and mouse lingual gustatory papillae (taste buds), with about 16% of taste bud cells expressing CD36 (Laugerette et al., 2005).

Recent investigations in mice have indicated that CD36 may serve a fat taste transduction mechanism. Mice were shown to prefer solutions and diets with linoleic acid, but CD36-null mice that do not express CD36 were not able to make this distinction (Laugerette et al., 2005). The CD36-null mice were not different from wild type mice in their preference for sucrose solutions and avoidance of quinine solutions, indicating that CD36 may be involved in fat-specific responses (Laugerette et al., 2005).

Whether the cues are textural, olfactory, or a fatty taste, oral fat exposure initiates a series of cephalic-phase responses. Gastric lipase is secreted; GI transit is modulated; endocrine and exocrine secretions from the pancreas are stimulated; and stored lipid from enterocytes is mobilized, among other responses (reviewed in Mattes, 2005). However, a cephalic insulin response generally is not stimulated by oral exposure to fat. Thus the cephalic responses appear to be to fat, and not to feeding or food in general. The responses do not have to be reflected in conscious judgments about the food being eaten. In one study (Crystal and Teff, 2006), although young women were unable to distinguish consciously between a high-fat and low-fat cake, after sham feeding on the cakes, the stimulated secretion of pancreatic polypeptide was significantly greater for the high-fat cake.

Certain fats appear to initiate a cephalic-phase response in humans. Men and women who sham fed on high-fat foods containing either olive oil or linoleic acid had increased circulating triglycerides and nonesterified fatty acids. They also reported greater feelings of satiety (Smeets and Westerterp-Plantenga, 2006). There is a storage theory of fat absorption that hypothesizes that a significant amount of fat from a meal remains in the gut lumen or enterocyte and enters circulation only after a subsequent meal (Jackson et al., 2002; Mattes, 2002). The evidence supports a taste component to the release of fat into circulation from this storage compartment (Mattes, 2002; Tittelbach and Mattes, 2001).

Evidence for Central Nervous System Contribution

Responses to food cues can be both intrinsic and learned (Booth, 1972; Rozin, 1976). Conditioned taste preferences and conditioned taste aversions provide strong proof that responses to food cues can be learned and modified. For example, most animals readily learn to avoid food sources that render them ill, a special visceral learning linked to food ingestion

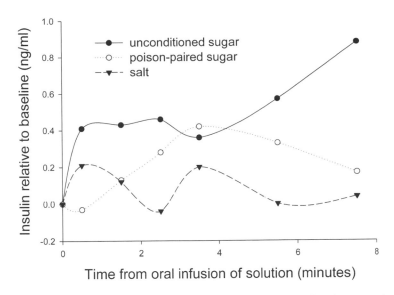

FIGURE 9.4.   The cephalic-phase insulin response to an orally infused sugar solution was significantly attenuated in rats that had been conditioned by pairing the sugar solution with a poison that caused gastrointestinal upset. There was no consistent insulin response to a sodium chloride solution. Data from Berridge et al., 1981.

(Garcia et al., 1974; Rozin, 1976). Bait shyness linked to poison is well known and an important adaptation (Richter, 1953).

Rats can be conditioned to react aversively to sweet solutions by rendering them ill after ingestion. They will decrease their intake of the sweet solutions when they are exposed to them again; they also emit species specific oral/facial rejection responses as opposed to the normal positive oral/facial responses associated with ingestion (Berridge et al., 1981). Importantly, the cephalic-mediated insulin response is now decreased to orally infused sweet solution (Berridge et al., 1981; Figure 9.4).

Cephalic-phase responses can also be stimulated by operant conditioning; that is, animals can learn to associate arbitrary sensory stimuli with the availability of food and then will react as if the food itself has been perceived. Animals have been conditioned to time of day (Woods et al., 1970; Dallman et al., 1993), sounds, tastes in water, and visual cues (e.g., Pavlov, 1902; Woods et al., 1977), among others. For example, insulin secretion in anticipation of feeding can be associated with environmental stimuli such as time of day, sounds, visual cues, or tastes (e.g., Woods et al., 1977). Rats fed at a certain time every day begin to secrete insulin time-locked to a circadian clock (Woods et al., 1970; Dallman et al., 1993).

The vagus nerve is thought to be a main pathway for cephalic-phase responses. The vagus nerve innervates the gastrointestinal tract from the esophagus to the colon. Vagal afferents below the diaphragm integrate meal-related gastrointestinal signaling (Schwartz and Moran, 1996). Truncal vagotomy largely eliminates gastric acid secretions and pancreatic enzyme and bicarbonate secretions induced by tasting food, including the insulin cephalic-phase response (Powley, 2000). Blocking cholinergic inputs also blocks cephalic-phase responses. For example, infusion of atropine eliminates the increase in gastric acid secretion due to sham feeding in humans (Katschinski et al., 1992).

Bypassing the mouth and placing food directly into the stomach also eliminates most cephalic-phase responses (reviewed in Powley, 1977). In rodent models in which the pancreatic β cells have been destroyed and new β cells transplanted such that they lack innervation by the vagus nerve, the reactive insulin secretion to increases in blood glucose is still intact, but the cephalic-phase insulin secretion is absent (e.g., Berthoud et al., 1980).

The gut and the brain work in synchrony to prepare the organism to digest and metabolize food in anticipation of its being ingested. Signals from the CNS produce peripheral responses that feedback to the CNS via humoral signals that cross the blood-brain barrier and via vagal nerve stimulation. These responses are integrated within the brain to regulate feeding and food intake.

Many cephalic-phase responses are intrinsic. They appear to require only brain stem function and are not dependent on forebrain structures. Decerebrate rats obviously are incapable of forming learned associations; however, they have been shown to produce competent cephalic-phase insulin responses to glucose infused into the oral cavity (e.g., Flynn et al., 1986; Figure 9.5). Decerebrate rats display appropriate oral/facial responses to oral infusions of various tastants. For example, Grill and Kaplan (2002) showed that decerebrate rats display positive hedonic responses to glucose (sweet) and aversive responses to quinine (bitter).

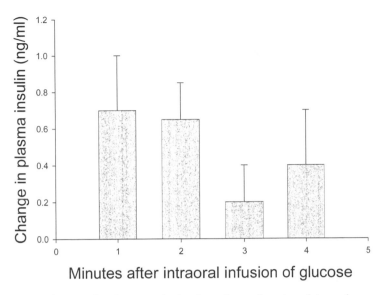

FIGURE 9.5. Infusion of a glucose solution into the oral cavity of decerebrate rats stimulated an immediate (cephalic-phase response) increase in insulin secretion. Data from Flynn et al., 1986.

Both hypothalamic brain regions and caudal brain stem have been shown to be involved in regulation of feeding. The brain stem has been proposed to respond to what Smith (2000) has termed direct feeding controls. These are signals generated by the mouth, tongue, stomach, and intestines that reflect inputs such as taste, gastric distension, and responses to the properties of the food within the alimentary tract. In this model the forebrain responds to what are termed indirect feeding controls, which include blood-borne metabolic signals of deprivation or excess. Of course the forebrain is also involved in learned food associations.

Thus the central control axis for regulation of feeding behavior consists of ventral forebrain networks for neuroendocrine signals and caudal brain stem networks for behavioral organization and reflex control (Grill and Kaplan, 2002). Both can regulate cephalic-phase responses. The neural networks that underlie cephalic-phase responses and the regulation of ingestive behavior have been reviewed in Zafra et al. (2006).

Cephalic-Phase Insulin Response

Glucose is a prime example of a nutrient whose blood concentration is actively regulated by both reactive and anticipatory physiology. If blood glucose falls below a critical concentration it can lead to rapid brain damage and then death. However, high concentrations of blood glucose, which are potentially toxic, are associated with macular degeneration, brain cell death, and higher mortality after stroke (Williams et al., 2002; Gentile et al., 2006). A number of mechanisms have evolved to resist changes in blood glucose concentration and keep it within safe levels. Glucose is constantly being shuttled between different pools, or perhaps more precisely, the energy contained in glucose is shuttled between these pools.

Insulin is the primary peptide regulating glucose metabolism. Insulin increases glucose storage (in the form of glycogen) in liver and muscle, decreases lipolysis and gluconeogenesis, and increases fatty acid synthesis by adipose tissue (Porte et al., 2005). The net result is to lower blood glucose concentration by increasing the conversion of glucose to other energy storage molecules (glycogen and fat) and decreasing the production of glucose from the liver.

In humans and rats there is a robust cephalic-phase insulin response (reviewed in Powley, 1977; Powley and Berthoud, 1985; Teff, 2000). In response to masticating and tasting food, the pancreas rapidly begins to secrete insulin. This initial pulse of insulin secretion is followed by a larger, more sustained insulin secretion in response to the absorption of digested nutrients (Teff, 2000). The cephalic-phase insulin response thus anticipates and mimics, at an attenuated level, the postabsorptive insulin response to changes in blood glucose concentration (Teff, 2000).

Although the magnitude of the cephalic-phase insulin response is lower than the postprandial response, it appears to have significant physiological effects (Ahren and Holst, 2001). Preventing the cephalic-phase response, for example through infusion of trimethaphan, which inhibits neurotransmission across autonomic ganglia, results in both significantly higher peak blood glucose concentration and impaired reduction of glucose within the first hour postprandial (Ahrens and Holst, 2001). Thus the absence of a cephalic-phase insulin response compromises glucose control and can even lead to hyperinsulinemia (Berthoud et al., 1980). Administration of insulin immediately prior to or at the beginning of a meal, that is, during the preabsorptive period, improves glucose control in obese humans (Teff and Townsend, 1999) and type 2 diabetics (Bruttomesso et al., 1999).

Obese individuals often have a blunted or even absent cephalic-phase insulin response (e.g., Teff & Townsend, 1999). It is not clear whether a blunted cephalic-phase insulin response contributes to obesity, obesity contributes to a loss of the cephalic-phase response, or both.

Summary

Appetite, food preferences, and the regulation of food intake are key aspects of energy balance and weight homeostasis. There is an ever-growing list of peptides, receptors, and other gene products associated with appetite and the regulation of food intake, and investigating their interrelated roles is an important process for understanding the proximate mechanisms by which people decide when and what to eat, and when to stop eating.

Anticipatory, feed-forward systems are vital to regulatory physiology. Physiology is not merely reactive. Anticipatory physiological regulation is an adaptive strategy that enables animals to respond faster to physiologic and metabolic challenges. The cephalic-phase responses are anticipatory responses that prepare animals to digest, absorb, and metabolize nutrients. They enable the sensory aspects of the food to interact with the metabolic state of the animal to influence feeding behavior, including the size of the meal. Cephalic-phase responses increase digestive efficiency and aid in controlling the resulting elevation of metabolic fuels in the blood due to feeding (Woods, 2002). Thus digestion, metabolism, and appetite are regulated in a coordinated fashion.

Cephalic-phase responses are a fundamental concept in regulatory physiology, a paradigmatic example of anticipatory physiological responses. Cephalic-phase responses need to be integrated across the diverse regulatory information molecules that are being discovered. They need to be viewed in an adaptive, evolutionary perspective. They represent, in our opinion, the results of feed-forward evolutionary pressures, an "arms race," if you will, between the imperatives of increasing the rate of nutrient acquisition and of defending the internal milieu.

CHAPTER TEN

The Paradox of Feeding

The acquisition of food is a necessity for animals. Strong selective pressures have acted to produce the anatomy, physiology, and behavior that serve to enhance an animal's ability to ingest, digest, absorb, and ultimately metabolize the nutrients necessary for survival and reproduction. But animals do not eat constantly, even if food is always present. What are the selective pressures and adaptive functions that have shaped the physiological responses that stop an animal from feeding? In other words, what are the costs of feeding, and under what circumstances should animals refrain from feeding even when food is available?

In much of the literature on appetite and regulation of food intake an implicit or even explicit assumption is that food intake is regulated to maintain energy balance, which will result in weight/adipose tissue homeostasis. And indeed there is a substantial body of empirical data indicating that animals, including humans, will maintain a relatively constant body weight under ad lib food conditions (Woods et al., 1998; Havel, 2001). At the same time, the epidemic of human obesity and the constant problem of overweight and obese captive animals in zoos and laboratory colonies also indicate that weight homeostasis is often violated, and thus people and animals are often in chronic positive energy balance.

From an evolutionary perspective, the important questions are to what extent and in what ways is energy balance a target of selection? After all, as valuable and insightful as the concepts of energy and energy balance are, they remain human constructs. Animals cannot measure their energy intake or expenditure directly; insofar as they behave as if they can it is due to mechanisms that measure and monitor proxy measures of energy balance (e.g., circulating levels of leptin and insulin that are directly correlated with the amount of adipose tissue). In addition, for many species, seasonal periods of hyperphagia, hypophagia, or both are routine. These species, such as hibernating bears and migrating birds, spend a significant part of their adult lives in sustained positive or negative energy balance. Although humans do not hibernate or seasonally migrate (for the most part), it still behooves scientists concerned with understanding the regulation of food intake in humans to consider carefully the extent and manner in which the maintenance of energy balance has been an important, adaptive aspect of our evolution. The case for adaptations to avoid sustained negative energy balance is reasonably self-evident; the arguments for avoiding moderate positive energy balance are less clear.

Time scale is an important aspect of understanding the function of the regulation of food intake. Much of the literature frames the regulation of food intake in terms of energy balance. However, the appropriate time scale of energy balance for humans is days, not meals. People do not generally start and stop eating a meal due to being in negative or positive energy balance. There are short-term satiety signals that regulate meal size and frequency. There are also social, cultural, and psychological factors that influence how much and how often a person eats (Rozin, 2005). For example, portion size of meals in restaurants differs dramatically between France and the United States; portion sizes are larger in the United Sates. People in France, on average, spend more time eating a meal than do people in the United States, yet they consume fewer calories per meal (Rozin, 2005).

Thus there are at least two aspects to the regulation of food intake with different time scales: total food intake over a day or set of days (probably strongly influenced by energy balance), and the type and amount of food eaten during a meal, which are influenced by physiological and psychological processes in addition to energy balance or weight/adipose tissue homeostasis. Whether a person is in positive or negative energy balance undoubtedly affects meal choices (e.g., duration, type of food, and so forth), but many other aspects of physiology and psychology are also operant. Indeed, there are anticipatory physiological responses to feeding, the cephalic-phase responses, that set in motion endocrine cascades that directly and indirectly regulate meal size and duration. These cephalic-phase responses can be conditioned, so experience, learning, and hence social and cultural factors, can and do play a role in their expression.

Cephalic-phase responses serve postabsorptive metabolism and physiology as well as digestion (see chapter 9). They prepare the organism to assimilate the ingested nutrients. This is a key adaptation, because although feeding is necessary for survival, it also presents a significant challenge to homeostasis, in what has been termed "the paradox of eating" (Woods, 1991).

The homeostatic paradigm has guided thought and research on physiology for over a hundred years. From the work of Claude Bernard (1865) to Walter Cannon (1932) to modern physiology, the concept that stability of the internal milieu is required for health and survival has been a central tenet. Woods (1991) eloquently presented the fundamental paradox of feeding from this physiological perspective. Organisms must consume food in order to survive, but the act of consumption brings exogenous substances into the body and presents a challenge to the stability of the internal milieu. Nutrients are required, but many nutrients are also toxic; they have maximal as well as minimal levels for concentration in the blood. The influx of nutrients from feeding can have negative as well as positive effects and requires metabolic adaptations to return the intercellular fluid to its homeostatic set points.

Although both of us have questioned the primacy of the homeostatic paradigm (Schulkin 2003; Power, 2004), neither of us denies the importance of homeostasis in understanding functional and adaptive aspects of physiology. We have merely argued that homeostasis does not represent all of physiological regulation and that often animals are required to abandon homeostasis to serve the goal of being a viable organism (defined in the evolutionary sense as capable of passing on its genetics to future generations). Viability, not stability, is the parameter of evolutionary importance (see Power, 2004). In the case of the challenges presented by digesting, absorbing, and metabolizing nutrients, however, the homeostatic paradigm provides insight into the selective pressures driving the evolution of complementary digestive and metabolic adaptations. In particular it highlights the inherent contradictions between adaptive changes that enhance the efficiency of digestion and absorption of nutrients with the necessity of maintaining the bloodstream within critical parameters for many of the absorbed nutrients.

The process of feeding can be divided into multiple phases. For our purposes we have proposed three phases: (1) foraging, (2) ingestion and digestion, and (3) absorption and metabolism. In the first phase the food is completely external to the animal. This is the foraging phase, where the animal is searching for, identifying, and finally acquiring the food. This phase consists of appetitive behaviors (Craig, 1918). Although the food is completely external to the animal, sensory cues about the food (visual, olfactory, and so forth) and learned associations already begin to stimulate salivary, gastric, and other secretions that prepare the animal for the next phase: ingestion and digestion.

In phase two, ingestion and digestion, the food is consumed and becomes internal to the animal, yet it is still apart from the internal milieu. The alimentary tract still presents a barrier. This barrier is vital, as the flow of ingested material into the bloodstream must be regulated, and some materials in the food are better not absorbed. For one thing, although we eat food, what we actually need are nutrients. After food is eaten it must be digested, the material broken down into its constituents, and the nutrients absorbed and then transported to storage depots for later use. The efficiency with which an organism can utilize food depends on the coordination of these processes. Adaptive changes in one process provide both opportunities and selective pressures for changes in others.

Humans are omnivores. We eat many types of foods, with quite varied nutrient composition. Our food can vary substantially in its proportions of fat, carbohydrate, and protein, among other constituents, and can present quite different digestive and absorptive challenges. Different foods require different digestive and absorptive responses. Animals, or at least omnivorous ones (e.g., humans and rats), appear to have developed mechanisms to detect these constituents in ingested food, and thus anticipate the required digestive and other exocrine and endocrine secretions. Our digestive tract is not continuously in the same state. Digestive secretions are not constant but rather change in accord with feeding, with the characteristics of the ingested food, and even with the anticipation of feeding. Cephalic-phase responses account for a significant portion of the secretions induced by eating (Katschinski, 2000).

These anticipatory digestive responses to food cues increase the rapidity and efficiency with which food is transformed into nutrients and thus can be absorbed through the intestinal walls. This increased efficiency in absorption presents both advantages and challenges to the organism. The main advantage is obvious: The digestive tract can process a greater amount of food per unit time, and thus the rate of nutrient transfer from the environment to the animal is greater. This allows, among other possible effects, greater total food intake, shorter latency time between meals, shorter total feeding/foraging time to meet requirements, an increased ability to meet requirements on foods of lower quality or that are scarce in the environment, or some combination of these effects. However, the more quickly and efficiently an organism can digest and absorb ingested food, the greater the potential disruption to the homeostatic conditions of the internal milieu. This requires correspondingly rapid and efficient metabolic responses in order to accommodate the pulse of nutrients entering the bloodstream, keep the concentrations within tolerable limits, and eventually return the intercellular fluid to the "normal" range. This is the third phase of feeding: regulation of the absorption and metabolism of digested nutrients. Anticipatory metabolic responses to sensory contact with food or food cues that aid this phase of feeding have been demonstrated, most notably the cephalic-phase insulin response (Powley and Berthoud, 1985).

Digestive and metabolic cephalic-phase responses likely would have evolved in concert. The advantages of rapid and efficient digestion and assimilation of nutrients are tempered by the subsequent greater perturbation of homeostasis due to the rapid influx of these efficiently acquired nutrients. Anticipatory, cephalic-phase metabolic responses serve to ameliorate the challenge to homeostasis and thus allow increases in digestive and absorptive efficiency. The net effect of the coordination of these processes establishes many of the constraints on feeding, such as maximal meal size and frequency, types of foods that can be eaten, the efficiency with which nutrients can be incorporated into the body, and so forth.

There are trade-offs and interrelated constraints between these phases of feeding. A foraging strategy that provides more food than the digestive system can process may not be adaptive (unless food caching is possible). A rapid and efficient digestive strategy that routinely delivers pulses of nutrients above the rates that metabolic adaptations can process may present greater selective disadvantages than advantages.

Consider the example of a snake and a small mustelid mammal (e.g., a weasel) that both feed on small mammals (e.g., mice). The weasel will, of necessity, catch many more mice per day than will the snake. The weasel will also digest ingested mice more rapidly. However, the snake will incorporate mouse nutrients into its body more efficiently than will the weasel. The snake will put on more mass per gram of eaten mouse than will the weasel.

There would be no advantage to the snake to have the weasel's foraging and digestive strategies. Its metabolism doesn't require that rate of nutrient input and likely could not maintain homeostasis if faced with such a high rate of nutrient input. The weasel, of course, would die if it had to maintain its metabolic rate using the snake's foraging and digestive strategies. The relatively slow feeding strategy of the snake is efficient at turning food into growth. The relatively fast feeding strategy of the weasel includes using a much higher proportion of ingested food for metabolism. These two strategies represent very different, but equally successful, solutions to the ultimate problem: surviving and reproducing.

The fact that mammalian physiology requires that a high proportion of ingested food fuel metabolism generates interesting ideas regarding cephalic-phase responses and the concept of satiety. Cephalic phase responses that increase the rate of nutrient absorption may be required by the generally high mammalian metabolic rate but may also contribute to it. There might be a necessary "inefficiency" in the mammalian feeding strategy in which some proportion of ingested nutrients is "burned off" in defense of homeostasis. Thermogenesis, the increase in metabolism and body temperature induced by feeding, may in part be an adaptation to defend homeostasis by metabolically removing glucose and fatty acids from circulation. Thermogenesis appears to have a cephalic phase (Diamond and LeBlanc, 1988; Soucy and LeBlanc, 1999). Short-term satiety signals likely serve the same ultimate purpose: to aid the animal to ameliorate the perturbation to the internal milieu due to the flow of nutrients from feeding by limiting meal size.

To what extent are satiety signals reflective of longer-term energy balance as opposed to shorter-term considerations of defending homeostasis? It is interesting that ingesting fructose produces a much less robust insulin response than does an isocaloric dose of glucose (Teff et al., 2004). Of course the absorption of fructose into cells is not insulin dependent, relying on GLUT5 as opposed to GLUT4 transporters. However, if insulin plays a significant role in satiety, then foods high in fructose, such as foods sweetened with high-fructose corn syrup, potentially will have a low satiety-to-calories response. Such foods are believed to play a significant role in weight gain among some people (Isganaitis and Lustig, 2005). For example, women who self-reported being restrained eaters (i.e., consciously refraining from eating certain foods for health and weight reasons) reported higher hunger scores on the day they were given a high-fructose beverage for breakfast compared with days when they were give an isocaloric high-glucose beverage. The day following the high-fructose beverage condition, when offered food ad lib, they consumed more fat than they had consumed the day after the high-glucose condition. Unrestrained eaters showed no differences in either hunger scores or fat intake the following day (Teff et al., 2004), so this response to fructose varies among people.

A Role for Cephalic-Phase Responses in Appetite?

Cephalic-phase responses have been suggested to play a role in appetite and satiety (Powley, 1977; Woods, 1991). Palatable foods generally result in more robust cephalic-phase responses than do less preferred foods. Preventing cephalic-phase responses results in animals and humans eating smaller meals (reviewed in Woods, 1991). This is an example of the short-term effects of cephalic-phase responses and the role of defense of homeostasis in appetite and satiety. Cephalic-phase responses would appear to allow larger meals, presumably due to their ability both to stimulate digestive processes and to address the challenge to homeostasis of the subsequent absorbed nutrients.

It has also been suggested that cephalic-phase responses are linked to motivation to feed and thus may actually play a more direct role in determining meal size and total daily food intake beyond the permissive one of ameliorating negative consequences of feeding. For example, the cephalic-phase insulin response biases metabolism toward directing metabolic fuels into storage, preparing the organism for ingested nutrients. The fuel oxidation theory of appetite (Friedman & Stricker, 1976) predicts that one result is that fuel available for oxidation by the liver is reduced, and this aids in development of an appetite. Saccharin ingestion has been shown to increase subsequent food ingestion in rats, a response abolished by hepatic vagotomy (Tordoff & Friedman, 1989; Figure 10.1).

So far only one gut peptide has been identified as having orexigenic activity. Ghrelin is secreted into the bloodstream by the stomach and in-

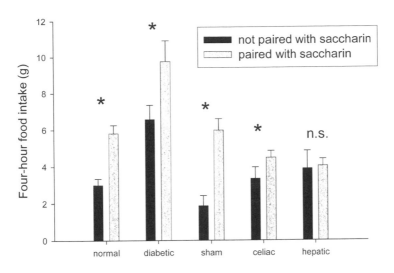

FIGURE 10.1. Rats given a saccharin solution before being offered food ate more food except for those that had a hepatic vagotomy. Data from Tordoff and Friedman, 1989. Key: normal = unmanipulated rats; diabetic = rats made diabetic using streptozotocin; sham = sham-operated rats; celiac = rats with celiac vagotomy; hepatic = rats with hepatic vagotomy; * = p < .05.

testines. Exogenous ghrelin rapidly increases food intake in rats (Wren et al., 2001a) and humans (Wren et al., 2001b). Indeed, ghrelin is as potent at stimulating feeding as neuropeptide Y. In rats, starvation increases plasma ghrelin concentration and refeeding rapidly decreases circulating ghrelin (Ariyasu et al., 2001). Ghrelin has been suggested to act to initiate feeding (Cummings et al., 2001).

In human volunteers provided meals on a fixed schedule, plasma ghrelin concentration displayed a consistent pattern of being low immediately after a meal, slowly increasing, and then having a rapid increase in concentration immediately prior to the next meal (Cummings et al., 2001; Figure 10.2). The pattern of ghrelin concentration was roughly opposite to the pattern of insulin concentration, but in phase with the circadian cycle of leptin concentration, though much more variable. Both ghrelin and leptin were lowest immediately after breakfast. Leptin concentration gradually rose throughout the day, with small declines immediately after each meal, and reached a zenith roughly during the middle of the sleep period. Ghrelin followed the same pattern, except for the pronounced

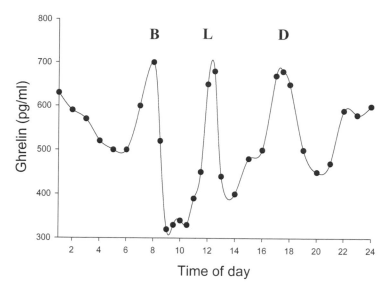

FIGURE 10.2.   Mean plasma ghrelin in 10 human subjects (9 women, 1 man) over a 24-hour period consuming breakfast (B), lunch (L), and dinner (D) at set times (8:00, 12:00, and 5:30). Subjects were aware of when meals were to be provided. Data from Cummings et al., 2001.

surges immediately prior to a meal, followed by equally dramatic declines immediately after the meal (Cummings et al., 2001; Figure 10.2). Because the meals were provided at known, fixed times, this evidence supports the hypothesis that the surge in ghrelin secretion prior to meals is a cephalic-phase response that serves to initiate feeding, prepare the person to digest and metabolize food, or both.

   In a study of real and sham feeding in human volunteers, serum ghrelin concentration increased prior to the meal in identical fashion. After both real and sham feeding, serum ghrelin concentration initially declined; the decline continued for the real feeding condition, but serum ghrelin concentration began to rise 60 minutes after sham feeding (Arosio et al., 2004). Thus ghrelin secretion appears to have a cephalic phase for both the anticipatory rise before a meal and the decline with the initiation of feeding.

   Further evidence for a cephalic-phase component to ghrelin secretion patterns comes from a study involving rats fed either ad lib or at a fixed time. Freely feeding rats had a peak of ghrelin secretion just before the dark phase of their light cycle; meal-trained rats had a significantly higher peak of ghrelin secretion at the start of the 4-hour period of the light phase during which they had become accustomed to having access to food (Drazen et al., 2006).

These results are consistent with the pattern of plasma ghrelin in 6 humans (3 men and 3 women) accustomed to eating three meals per day who fasted over a 33-hour period. Plasma ghrelin increased on both mornings (around 8:00), the middle of the day (12:00 to 13:00) and in the early evening (17:00 to 19:00). These increases were followed by a spontaneous decline in plasma ghrelin despite the fact that the subjects were fasting (Natalucci et al., 2005; Figure 10.3).

These results in rats and humans suggest that appetite can be trained to a circadian clock. It suggests that hunger can occur in anticipation of food being available, but if the food is not delivered, then the system re-

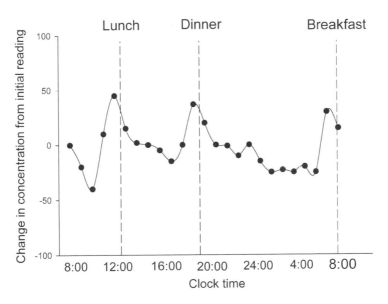

FIGURE 10.3. Subjects (3 men, 3 women) were accustomed to eating breakfast, lunch, and dinner at approximately 8:00, 12:30, and 18:30; they then fasted for approximately 32 hours, starting at midnight. Blood was collected via catheter every 20 minutes starting at 8:00 for 25 hours when they were fed breakfast. The pattern of circulating ghrelin concentration showed a consistent pattern of rising before the expected meal time and declining afterward, even when no food was ingested. Data from Natalucci et al., 2005.

verts to a "nonfeeding" state. Perhaps readers are familiar with the phenomenon of being hungry but not able to eat; after several hours the feelings of hunger recede. Of course the scent and sight of food can bring them back intensified, but otherwise, a person will no longer be "hungry" despite having not fed. Again, this is consistent with energy balance not being the only determinant of hunger and food intake. We can be trained to be hungry at certain times.

A Role for Cephalic-Phase Responses in Satiety?

Leptin has been suggested to regulate food intake. Leptin is secreted by adipose tissue; plasma leptin concentration is highly correlated with fat mass. Leptin acts on neurons in the hypothalamus and arcuate nucleus. It appears to act in opposition to neuropeptide Y, and in concert with insulin and corticotropin-releasing hormone (CRH) to reduce food intake. Leptin and insulin are thought to influence the hedonic perception of food, both centrally (reviewed in Isganaitis and Lustig, 2005) and via taste receptors. Leptin modulates sweet taste sensation through actions on sweet taste receptor cells via the leptin receptor. Increased leptin decreases firing of sweet taste cells in mice (Kawai et al., 2000).

Leptin also is synthesized and secreted by the gastric mucosa, and appears to be secreted during meals (Bado et al., 1998). Vagal stimulation results in gastric mucosal leptin secretion but no increase in plasma leptin concentration (Sobhani et al., 2002), suggesting that gastric leptin is secreted during the cephalic phase of gastric secretions and acts in a paracrine fashion. Leptin receptor mRNA is present in vagal afferent neurons that innervate the stomach, suggesting that leptin may have direct stimulatory effects on vagal afferents (Peters et al., 2005). Infusions of leptin into the celiac artery, but not the jugular vein, significantly reduced intake of a sucrose solution by rats, and this result was abolished by vagotomy (Peters et al., 2005).

Some of the leptin secreted by the gastric mucosa appears to survive the gastric acid and travels intact to the intestine where it is thought to perform multiple functions regulating the absorption of lipid, carbohydrate, and protein (reviewed in Picó et al., 2003). The functional form of the leptin receptor (Ob-Rb) is expressed in human jejunum and ileum

(Morton et al., 1998; Barrenetxe et al., 2002). Leptin has been shown to inhibit D-galactose absorption (Lostao et al., 1998) and increase the intestinal absorption of small peptides (Morton et al., 1998). Leptin also stimulates CCK (cholecystokinin) secretion (Guilmeau et al., 2003). Leptin and CCK appear to form a positive feedback loop. Plasma CCK stimulates gastric leptin secretion (Bado et al., 1998), and duodenal infusion of leptin in rats increased plasma CCK comparable to the effects of feeding (Guilmeau et al., 2003). Leptin and CCK appear to act synergistically to activate vagal afferent neurons (Peters et al., 2004), and leptin and CCK infused via a celiac catheter acted synergistically to decrease intake of a sucrose solution (Peters et al., 2005).

CCK has direct effects on meal size. Giving exogenous CCK to rats resulted in meals that were shorter on average than the meal duration of control rats. However, the treated rats ate a greater number of meals per day, and total daily food intake did not differ between treated and control rats (West et al., 1984). In rats without expression of CCK receptor type A, meals were of greater duration and consisted of greater amounts ingested. The number of meals per day decreased, however, the net effect was an overall greater daily food intake, and eventual obesity (Moran and Kinzig, 2004).

Cholecystokinin is another good example of an information molecule with multiple functions. It is a neurotransmitter and a gut peptide. Although there is a single CCK gene, CCK has multiple molecular forms derived from post-translational or extracellular processing. There are two receptors that bind the different molecular forms with different affinities. In addition to its role in satiety, CCK in the gut regulates intestinal motility, gastric emptying, pancreatic enzyme secretion, and release of bile from the gall bladder. Central CCK has been associated with anxiety, sexual behavior, learning and memory (Moran and Kinzig, 2004).

Thus, in addition to its role in long-term energy balance, leptin has been suggested to play a role in short-term satiety signals, either directly via vagal afferents or indirectly through stimulation of CCK secretion. Indeed, leptin and CCK appear to act synergistically to reduce both long- and short-term food intake (Matson and Ritter, 1999; Barrachina et al., 1997).

Diverse Functions of Information Molecules

Leptin provides an example of a key concept we believe needs emphasizing in regulatory physiology and that we have emphasized throughout this book. More and more, physiologically important peptides, steroids, and other information molecules are being shown to have multiple functions in diverse tissues. Their actions and regulation can be tissue and context specific. Leptin, for example, is secreted by placenta and appears to have important functions in fetal development (Bajari et al., 2004; Henson and Castracane, 2006). Leptin also acts on the gonads, and acts to regulate sexual maturity and fertility, especially in females (Bajari et al., 2004). In many ways leptin serves as much as a reproductive hormone as it does a hormone of energy balance.

Ghrelin also appears to be an ancient regulatory molecule. Its structure is highly conserved among mammals, and it has been detected in chickens, fish, and bullfrogs (Tritos and Kokkotou, 2006). Ghrelin is a potent secretogue of growth hormone from the pituitary through binding to the GHS (growth hormone secretogue) receptor (Takaya et al., 2000). Circulating ghrelin exists in both an acylated and nonacylated form. The nonacylated form does not activate the GHS receptor, but does appear to have effects on glucose homeostasis, lipolysis, adipogenesis, cell apoptosis, and cardiovascular function, suggesting that an additional, as yet undetected, receptor might exist (Tritos and Kokkotou, 2006). Ghrelin is produced by post-translational cleavage of a prepropeptide of 117 residues; an alternative cleavage of this prepropeptide produces obestatin (Zhang et al., 2005). Intriguingly, obestatin appears to suppress food intake (Zhang et al., 2005), although this affect was not replicated by other investigators (Gourcerol et al., 2007). Thus the preproghrelin gene appears to produce at least two distinct peptide hormones that may have opposing actions. These facts highlight the importance of post-translational mechanisms in understanding gene effects.

Evolution is limited in that it can only work on existing variation. There are many examples of ancient regulatory molecules, like leptin and ghrelin, which, over time, became adapted and co-opted to serve multiple, diverse functions. These regulatory molecules serve as "information molecules," transmitting information among organ systems and coordinating the responses of peripheral organs and the central nervous system to external and internal challenges to an organism's viability. An evolutionary perspective predicts that these molecules will have multiple and diverse functions and that their regulation, function, and mode of action will vary between different taxa, and between different tissues within taxa.

Appetite, Satiety, and Energy Balance

The combination of meal size and frequency has a strong influence on total daily energy intake, but it is not clear that regulation of energy intake is the function of all appetite and satiety signals. Motivation to begin feeding and to stop feeding does not always track energy or other nutrient requirements exactly, especially in the short term. Different foods have different physiological and metabolic consequences. The motivation to eat, or to stop eating, certain food types may have more to do with these metabolic consequences and their effects than with energy balance per se.

For example, Dallman and colleagues have proposed that eating high-fat, high-sugar foods (referred to as "comfort foods") results in physiological and metabolic responses (e.g., increases in insulin and glucocorticoids) that stimulate pleasure-associated areas of the nucleus accumbens (Dallman et al., 2005). The motivation to eat comfort foods has been documented in humans and rats subjected to unpleasant, challenging, or stressful circumstances. This behavior is proposed to represent, in part, a form of self-medication, with the ingestion of food serving as more of a medicinal or therapeutic function than a nutritional one; eating comfort foods is one way to ameliorate hormonal signals that increase arousal. This effect has been shown to be stronger in human females compared with males (Zellner et al., 2006). People who increase food consumption due to stress report both that the foods they overeat are the foods that they normally avoid for reasons of health or weight loss, and that they eat these foods to feel better (Zellner et al., 2006).

Stressful and challenging circumstances will increase CRH in many brain areas. Adrenalectomy will also change the basal levels of CRH, increasing CRH in the paraventricular nucleus of the hypothalamus and decreasing CRH in the central nucleus of the amygdala (e.g., Swanson and Simmons, 1989). Interestingly, sucrose ingestion will normalize CRH in these brain regions in adrenalectomized rats (Dallman et al., 2003; Figure 10.4). In another study (Peciña et al., 2006) rats were trained to associate a sound with the availability of sucrose pellets if they pressed a bar; injection of CRH into the nucleus accumbens increased cue-triggered bar pressing for the sucrose pellets (Figure 10.5). This can be interpreted as CRH increasing the salience of external cues and the motivation of the rats to respond. It can also be interpreted as the rats reacting to the increased neural CRH with a higher motivation for sucrose because of its "comforting" attributes; that is, the rats were self-medicating to alleviate the arousal due to the artificially up-regulated central CRH.

In addition, the same peripheral endocrine profiles can result in different behaviors among individuals. For example, a breakfast accompanied by a high-fructose beverage resulted in a lower increase in blood glucose and insulin than the same breakfast accompanied by an isocaloric high-glucose beverage (Teff et al., 2004). The subsequent increase in circulating leptin and the decrease in circulating ghrelin were also smaller following the high-fructose beverage. All these endocrine patterns are consistent with fructose producing a lower satiety than glucose. Despite the fact that the endocrine responses were not different between restrained

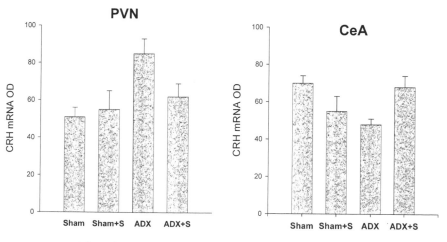

FIGURE 10.4. Adrenalectomized rats had higher basal CRH in the paraventricular nucleus (PVN) of the hypothalamus and lower basal CRH in the central nucleus of the amygdala (CeA). Sucrose ingestion normalized CRH levels. Data from Dallman et al., 2005.

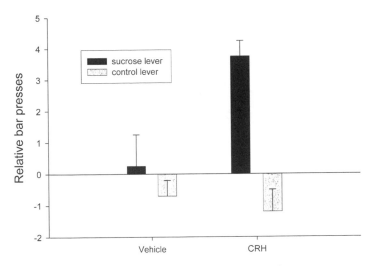

FIGURE 10.5. Rats injected with CRH in the nucleus accumbens were more responsive to an auditory cue signaling that sucrose pellets were available by bar pressing; they increased their pressing of the sucrose bar and decreased pressing of the control bar. The relative bar presses are the number of presses within 2.5 minutes of the auditory cue minus the number of presses in the 2.5 minutes prior to the cue. Data from Peciña et al., 2006.

and unrestrained eaters, only in the restrained eaters were hunger scores higher the day of the high-fructose beverage and fat intake higher the day after. This is consistent with there being multiple signaling mechanisms that influence food intake.

Energy balance and body weight are indeed regulated. The regulation appears to be asymmetric, however. Loss of body weight (sustained negative energy balance) is defended against more tightly than gain of body weight (sustained positive energy balance). This makes intrinsic evolutionary sense. For much of our evolutionary history, food intake was controlled by external rather than internal factors. Food supply could vary dramatically, and in times of plenty the original food storage method was body fat. Thus our energy balance regulatory system appears to have evolved a bias in favor of moderate positive balance and the storage of excess energy in the form of body fat (Schwartz et al., 2003).

This asymmetry can be seen in the neural endocrinological basis of appetite and feeding, as well. There appear to be redundant mechanisms to stimulate food intake. Both neuropeptide Y (NPY) and agouti-related peptide (AgRP) have been shown to have strong orexigenic effects. However, mice that are deficient in one or both of these neuropeptides are not anorexic and respond to starvation with hyperphagia just as do controls (reviewed in Schwartz et al., 2003). In contrast, leptin deficiency reliably leads to hyperphagia and obesity. Mice that were both leptin and NPY deficient were less hyperphagic and less obese than leptin-deficient mice, but still hyperphagic and obese compared with normal mice (Erickson et al., 1996).

Summary

Physiological and metabolic systems serve the survival and reproductive capabilities of the organism (fitness). In real-world situations animals are constantly balancing competing imperatives. It can be argued that the resolution and prioritization of those competing imperatives is a principal function of the central nervous system. Another principal function of the CNS is the coordination of responses to and in anticipation of challenges. The central nervous system is intimately involved in physiological regulation.

A principal function of satiety is meal termination; however, meal termination is only indirectly related to energy balance. Cephalic-phase responses serve to increase meal size/duration (efficiency of digestive and metabolic responses) and thus increase food intake per meal. They also begin endocrine cascades to terminate a meal. There are multiple reasons, in addition to energy balance and adipose tissue homeostasis, to terminate a meal. Defense of homeostasis (sensu Woods, 1991) is an important consideration; elevated levels of blood glucose, amino acids, and insulin contribute to a loss of appetite.

Another simple, and perhaps little-considered, factor is that animals have many necessary functions to perform to be viable. Strong incentives and redundant neural circuits reinforce feeding behavior. There have to be equally strong mechanisms to stop feeding in the presence of available food or animals would be constantly feeding, regardless of other imperatives. There are many constraints on animals, but time is a universal one. Animals, to be viable, must apportion their time among the various activities necessary for survival and reproduction.

To what extent is the adaptive value of terminating feeding related to conserving time for other activities? And to what extent does the current human obesity epidemic relate to the fact that the amount of time needed in the past to acquire sufficient calories far exceeds the amount of time today needed to acquire a gross surfeit of calories? The evolution to regulate meal size and the evolution to regulate energy balance have independent aspects. Therefore the regulation of food intake on the time scale of the meal can be decoupled from the regulation of energy balance, and this has significant implications for understanding the pathogenesis of obesity.

CHAPTER ELEVEN

The Biology of Fat

···

Fat is an essential part of our bodies. Fats, or lipids, perform many functions: nutritional, hormonal, even structural. For example, myelin, which sheaths the axons and increases the speed with which nerve impulses travel, is 80% lipid. Certain fatty acids are essential for proper brain development. Indeed, brains are high-fat organs. This makes brains energetically expensive to build and maintain. It also makes them a high-quality food. Many predators will first eat high-energy organs such as the liver and brain from their prey. In times of plentiful prey these may be the only things eaten by the primary predator, with the rest of the carcass left for scavengers.

Lipids perform many other vital functions in living tissue. Cell membranes are composed of phospholipids, glycolipids, and steroids. And of course the cholesterol-based steroid hormones such as estrogens, testosterone, and glucocorticoids perform vital functions necessary for life. Fat is essential and adaptive. Without fat in our diet and on our bodies we would die.

The above examples reaffirm that not all fat is bad; indeed fat is necessary and has multiple and diverse functions within the body. In this book we are mostly concerned with the nutritional, metabolic, and hormonal aspects of fat. The main depots of fat on the body are in adipose tissue, and it is adipose tissue that is relevant to the subject of obesity. Adipose tissue performs the nutritional and metabolic functions of fat in the body and is involved in many of the hormonal functions.

Obesity isn't excess weight; it is excess adipose tissue. To understand the health effects of obesity we need to explore the adaptive functions of adipose tissue. Fat is a good thing, in our diets and on our bodies, in moderation. Too much of this "good thing" becomes maladaptive. Normal physiology pushed beyond adaptive function becomes pathology, a concept referred to as allostatic overload. In this chapter we investigate the functions of adipose tissue and some of the reasons why too much adipose tissue can be harmful.

Adipose Tissue

Adipose tissue is loose connective tissue that contains large numbers of cells called adipocytes. Adipocytes contain a single internal fat droplet surrounded by a thin rim of cytoplasm. Adipose tissue can be found under the skin (subcutaneous fat), around the visceral organs (visceral fat), and in muscle (intramuscular fat). Adipose tissue is thought to perform a variety of adaptive functions, such as being a source of stored energy, insulation to reduce heat loss, padding to protect the internal organs. Most fat on the adult human body is contained in white adipose tissue. Newborns also have brown adipose tissue. Brown fat, which contains a higher number of mitochondria, is capable of releasing substantial amounts of heat energy (nonshivering thermogenesis) via uncoupling protein reactions that are important in thermal regulation (Nichols and Rial, 1999; Nichols, 2001). In general brown fat is minimal in adult humans (Haney et al., 2002; Cohade et al., 2003), but it remains important throughout life in many other, mainly small, mammals. Brown fat is especially active in cold-adapted animals, being capable of generating heat at up to 60 times the rate of an equal mass of liver tissue (Nichols and Rial, 1999). White adipose tissue is not as metabolically active, though it is not merely a passive store of energy. It has endocrine, immune, and metabolic function. Adipose tissue has important functions in energy metabolism.

White adipose tissue can make up a substantial portion of body weight; in healthy women 20% or more of their weight is white adipose tissue. Adipose tissue serves as a source of fatty acids stored as triglycerides. When energy intake exceeds energy expenditure there will be a net deposit of lipid into adipocytes; when energy intake is insufficient, fatty acids will be released from adipocytes to provide the metabolic fuel for life.

Fat contains more than twice the metabolizable energy per gram of carbohydrate or protein. This is why it is such an efficient means of storing energy. The average-weight human stores enough energy in fat to satisfy minimal energy requirements for more than a month.

The ability to store substantial amounts of energy in adipose tissue has many adaptive advantages. It buffers the organism from the effects of unpredictable and variable food supply. It allows available excess energy to be consumed and then used at a later time. It allows an animal to go longer between feeding. The ability to store fat increases the behavioral flexibility and the potential feeding strategies an animal can employ. Our ancestors' ability to increase their adipose stores probably served multiple adaptive functions.

Adipose tissue is not just about storing energy, however. Although fat has many important functions in the body and is the most efficient energy storage medium, it is also toxic (Schrauwen and Hesselink, 2004; Slawik and Vidal-Puig, 2006). Lipid droplets that accumulate in cells and organs cause pathology (e.g., fatty liver). To prevent the adverse effects of lipotoxicity, fatty acids must be either oxidized or sequestered. Fat oxidation releases a significant amount of energy; fat is certainly an important energy source for living organisms. However, fat can be, and these days for humans frequently is, ingested in larger quantities than can be oxidized. There are limits to metabolism and to the rate that animals can process energy. Fat is toxic to cells in high levels. If fat cannot be oxidized it must be safely stored. Thus another basic function of adipose tissue is to sequester excess fat to prevent lipotoxicity (Slawik and Vidal-Puig, 2006).

Adipocytes are cells specially adapted for fat storage; fat is preferentially stored in adipose tissue, and thus less fat accumulates in muscle, liver, heart, and so forth, where it would cause morbidity. So adipocytes store energy and prevent lipotoxicity. Adipocytes store fat for the positive benefit associated with its energy value; storing energy that can be used far away in space and time from the act of ingestion. Adipocytes also store fat to avoid the negative metabolic consequences of excess cellular lipid.

Based on these functions, adipose tissue is acting as a storehouse to protect other tissue and to allow energy ingested at one time to be used at a later time. These are very important adaptive functions. But adipose tissue is not just a passive organ. It actively regulates metabolism through multiple pathways. Indeed, adipose tissue contains more than just adipocytes. A large number of nonfat cells are also found in adipose tissue, including fibroblasts, mast cells, macrophages, and leukocytes (Fain, 2006). Both adipocytes and these nonfat cells produce, regulate, and secrete active peptides and steroids (Kershaw and Flier, 2004), as well as immune-function molecules (Fain, 2006). The biology of fat has now been shown to be much more complex and integrative than our previous conception. Adipose tissue is an active component of regulatory physiology.

It is the metabolically active functions of adipose tissue that are related to the development of much of the pathology associated with obesity. The conception of adipose tissue as an endocrine and immune-function organ gives insight into why excess adipose tissue can have important effects on physiology and metabolism. Just consider for a moment the expectations you would have if an animal's liver or adrenal gland doubled in size. That would have metabolic and health consequences. Obesity is an increase in adipose tissue well beyond the functional range that was typically experienced during our past. The secretions from adipose will likely be out of balance with the other organ systems. In some aspects the metabolic consequences of obesity are analogous to the endocrine dysfunction any hyperplasia of an endocrine organ is likely to cause.

The Endocrine System

Originally, the endocrine system was considered to consist of eight endocrine glands (adrenals, gonads, pancreas, parathyroid, pineal, pituitary, thymus, and thyroid), organs whose primary purpose appeared to be the synthesis of endocrine hormones that were secreted into the blood. These hormones travel in the bloodstream to reach their target end-organ systems where they bind to receptors and initiate their particular signaling cascades. The concept of an endocrine organ underwent significant changes as we learned more about physiology, metabolism, and the organs of the body. The endocrine glands were not unique, except perhaps in that endocrine signaling was their primary function. Organs with other primary functions also have endocrine functions. For example, in 1983 Forssmann and colleagues published a paper titled "The Right Auricle of the Heart Is an Endocrine Organ." Since then the list of tissues that synthesize and secrete hormones has expanded to include virtually every tissue in the body. The old concept of an endocrine organ has been forced to adapt to the biological reality that many tissues can and do act in an endocrine fashion; that is, they produce substances that act on other end-organ systems to regulate physiology. The digestive system discussed in chapter 7 is an excellent example; heart, skin, and adipose tissue also have been shown to have endocrine functions.

The original concept of endocrine hormones was "action at a distance." In other words, a hormone secreted into circulation traveled to the appropriate end-organ where it had its effect. Another conceptual change was the realization that many of these hormones act locally as well as at a distance. Nowadays, secreted hormones are said to have endocrine, paracrine, or autocrine hormone function depending on whether they act on other tissue (endocrine), locally on nearby cells and tissue (paracrine), or on the very cells doing the secreting (autocrine). The concept of an endocrine organ and endocrine function has become more complex as our understanding of physiology and metabolism has become more complete.

Adipose Tissue and Endocrine Function

Adipose tissue is far more metabolically active than was once believed (Kershaw and Flier, 2004). The original notion of adipose tissue as a relatively metabolically inert store of energy has been replaced by a concept of adipose tissue as a metabolically active player in many physiological and endocrine processes. Adipose tissue is both a means to store energy and an endocrine organ important in whole-body physiology. Adipose tissue stores, synthesizes, and secretes hormones, as well as enzymes that regulate hormones. These molecules affect both local and systemic concentrations of peptides and steroid hormones. Adipose tissue also expresses many receptors for peptide and steroid hormones. The hormones secreted by adipose tissue can act locally and on other end-organ systems in autocrine, paracrine, or endocrine fashion.

Adipose tissue functions as an endocrine gland in three different ways. It stores and releases preformed steroid hormones. Many of these hormones also are metabolically converted from precursors in adipose tissue, or the active hormones are converted to inactive metabolites. Adipose tissue expresses many enzymes involved in steroid hormone metabolism

TABLE 11.1 *Bioactive molecules produced by adipose tissue*

Hormone	Function(s)	Changes in obesity
Leptin	Effects on food intake, onset of puberty, bone development, immune function	Circulating leptin increased
Tumor necrosis factor α (TNF-α)	Represses genes involved in uptake and storage of nonesterified fatty acids (NEFAs) and glucose	Adipose tissue expression of TNF-α increased
Adiponectin	Enhances insulin action	Circulating adiponectin lowered
Interleukin 6 (IL-6)	Involved in regulation of insulin signaling; central effects on energy metabolism	Circulating IL-6 is increased; expression of IL-6 greater in visceral fat
NPY	Angiogenesis in adipose tissue; regulates leptin secretion	Uncertain; but increased NPY and Y2 receptor function associated with increased visceral fat
Resistin	Effects on insulin action; linked with insulin resistance	Serum resistin elevated in rodent obesity models
Aromatase	Converts androgens to estrogens	No change, but increased fat mass results in greater total conversion
17β-hydroxysteroid hydrogenase	Converts estrone to estradiol and androstenedione to testosterone	No change, but increased fat mass results in greater total conversion
3α-hydroxysteroid hydrogenase	Inactivates dihydrotestosterone	
5α-reductase	Inactivates cortisol	
11β-hydroxysteroid dehydrogenase type 1	Converts cortisone to cortisol	Increased activity in adipose tissue

(Table 11.1). For example, estrone is converted to estradiol in adipose tissue. Indeed, most if not all circulating estradiol in postmenopausal women comes from their adipose tissue (Kershaw and Flier, 2004). Adipose tissue expresses 11β-hydroxysteroid dehydrogenase type 1 (11β-HSD1), which converts cortisone to cortisol, and 5α-reductase enzymes, which convert cortisol to 5α-tetrahydrocortisol (5α-THF). Thus adipose tissue regulates the local concentrations of glucocorticoids and contributes to the metabolic clearance of glucocorticoids (Andrew et al 1998).

Finally, adipose tissue produces and secretes a large number of bioactive peptides and cytokines. These peptides are referred to as adipokines to emphasize the role of adipose tissue in their synthesis and secretion. The list continues to grow as our understanding increases. Below we discuss the function and regulation of some of the better understood information molecules produced by adipose tissue.

A Vitamin That Is Actually a Steroid Hormone

Vitamin D is an interesting example of how the definitions of an endocrine organ and an endocrine hormone have to be flexible. First of all, despite its name, vitamin D is not a nutrient for most animals; rather, vitamin D_3 (the animal form; vitamin D_2 is found in plants) is a steroid hormone produced in the skin via exposure to ultraviolet B (UVB) radiation. The exceptions to this rule include obligate carnivores such as cats (Morris, 1999) and polar bears (Kenny et al., 1999), which appear to have lost this photosynthetic capability; for them vitamin D is truly a vitamin. However, primates, including humans, have a robust vitamin D_3 photosynthetic system in their skin. It has been estimated that several days per week of ten to fifteen minutes per day full sunlight exposure to the hands and face is sufficient to provide adequate vitamin D for people, at least during the summer months (Holick, 2004). In winter the amount of UVB radiation that penetrates the atmosphere due to absorption by ozone and other molecules can be very low at latitudes far from the equator. For example, almost no UVB radiation penetrates the atmosphere at the latitude of Boston, Massachusetts, between the end of October and the end of February (Holick, 1994). Vitamin D production is seasonally variable above 30 degrees latitude, with a very short season indeed in the Arctic Circle. Of course the ozone hole of the Southern Hemisphere has undoubtedly increased the vitamin D production season far south of the equator.

Vitamin D itself, in either the plant or animal form, has low biological activity. To become the active hormone it must go through two hydroxylation steps. The first occurs in the liver, where it is hydroxylated to 25-hydroxy vitamin D (25-OH-D). This hydroxylation step is unregulated. Circulating 25-OH-D is the best measure of vitamin D status; it is the storage form of the vitamin but has limited biological activity itself. The most biologically active metabolite of vitamin D is 1,25-dihydroxy vitamin D (1,25-OH$_2$-D). This metabolite is produced primarily in the kidney via a tightly regulated hydroxylation of 25-OH-D. The primary nutritional functions of 1,25-OH$_2$-D are to regulate the circulating levels of ionized calcium and phosphorus (DeLuca, 1988) by acting on the intestine (regulates calcium and phosphate absorption), the kidney (regulates calcium and phosphate excretion), and bone (affects bone remodeling and thus whether there is a net flow of calcium and phosphate into or out of bone).

Thus the endocrine hormone 1,25-OH$_2$-D is the product of metabolism from three different organs: skin, liver, and kidney. The 1,25-OH$_2$-D produced in the kidney acts on distant tissues and on the kidney not only to regulate calcium excretion but also to down-regulate production of the 1α hydroxlase enzyme that converts 25-OH-D to 1,25-OH$_2$-D. Thus it acts in a paracrine or autocrine fashion as well. Recently, evidence suggests that 1,25-OH$_2$-D is also produced in skin, from a hydroxylation of circulating 25-OH-D. Thus vitamin D leaves skin, enters circulation, is hydroxylated in the liver to 25-OH-D, and then some small fraction of the 25-OH-D diffuses back into skin to be further hydroxylated to the active form. In skin 1,25-OH$_2$-D acts in a paracrine fashion, an autocrine fashion, or both to regulate skin cell differentiation and maturation (Holick, 2004).

Vitamin D and Adipose Tissue

Adipose tissue acts as a storage depot for vitamin D metabolites and other fat-soluble molecules. This is one reason why vitamin D deficiency takes a long time to develop, and why humans far from the equator can remain vitamin D sufficient through the winter even with low dietary vitamin D. High sun exposure during the summer months would result in several months' supply of vitamin D and its metabolites stored in body fat. This mechanism is potentially important for infant health as well. Infants are born with several months' supply of vitamin D transferred in utero and stored in adipose tissue. Even though breast milk is deficient in vitamin D (Hillman, 1990) a solely breast-fed infant has enough stored vitamin D to avoid frank deficiency for many months. And of course if the infant is exposed to intense enough sunlight (or another source of UVB radiation) endogenous photosynthetic production will be sufficient. Over the course of our evolution, vitamin D was mostly acquired through sunlight, which explains why humans did not evolve physiological mechanisms to concentrate vitamin D metabolites in milk. Infants usually had plenty of sun exposure. Thus in our past, vitamin D deficiency and its associated diseases (e.g., rickets) would have been largely absent. Vitamin D deficiency is a disease of modern humanity.

Obesity is excess adipose tissue. Obesity is also associated with lower circulating concentrations of 25-OH-D and increased risk of vitamin D deficiency (Wortsman et al., 2000; Arunagh et al., 2003; Hyppönen and Power, 2006). Diet and sun exposure do not explain this association. It appears that a large mass of adipose tissue leads to excess vitamin D (and possibly other fat-soluble vitamins) becoming sequestered in adipose tissue. It appears that the net flux of vitamin D in obese people tends to be into adipose tissue, and thus circulating levels are decreased.

Thus obesity is associated with vitamin D deficiency. Is there any evidence that vitamin D has any affect on the vulnerability to obesity? Low calcium intake is associated with weight gain and higher fat mass in both people and animal models (Heaney et al., 2002; Zemel, 2002). Vitamin D metabolism is partly regulated by calcium intake. A number of studies suggest that vitamin D status and obesity are related. Lipid metabolism in adipose tissue is regulated by $1,25\text{-OH}_2\text{-D}$ (Sun and Zemel, 2004); an up-regulation of $1,25\text{-OH}_2\text{-D}$, due, for example, to low calcium intake appears to increase fat deposition into adipose tissue. Increased calcium intake decreases circulating $1,25\text{-OH}_2\text{-D}$ and reduces fat deposition by inhibiting lipogenesis (Sun and Zemel, 2004; Zemel, 2004).

Steroid Hormones and Fat

Steroids and fat have long been associated. For one thing, steroids are created from cholesterols, which are fat. Thus they are fat soluble and can diffuse into adipose tissue from circulation. Steroids are found in adipose tissue, including steroid hormones, for example, estrogens, progesterone, testosterone, and glucocorticoids. These steroid hormones are stored in adipose tissue and then released when triggered by other physiological cascades. Adipose tissue is both a sink and a source of steroids, retaining or releasing them in response to physiological cues. But adipose tissue is not just a passive storehouse of steroids. It is metabolically active. We now know that steroids are both produced and inactivated in adipose tissue through the actions of enzymes synthesized by adipocytes. Adipose tissue is an important regulator of steroid hormones.

Excess adipose tissue can result in dysregulation of steroid hormone metabolism. There are several well-documented examples of obesity being associated with either higher or lower circulating levels of steroids. Vitamin D is one such example.

Glucocorticoids are also affected by the amount of adipose tissue. Both the local concentration and systemic levels of cortisol are affected by conversions between cortisol and corticosterone and by metabolism of glucocorticoids by enzymes expressed in adipose tissue (see Table 11.1). Obesity is associated with both increased adrenal glucocorticoid production and higher glucocorticoid metabolic clearance, which appears to result in normal plasma concentrations. In obese individuals, 11β-HSD1 activity is reduced in liver and the inactivation of cortisol by 5α-reductase is enhanced (Andrew et al., 1998; Stewart et al., 1999). But 11β-HSD1 activity is enhanced in adipose tissue (Rask et al., 2001).

The sex hormones can also be affected by adipose tissue. Aromatase enzymes in adipose tissue convert androgens to estrogens. Obese men, especially those with central obesity, are hyperestrogenic and hypoandrogenic. The increased adipose tissue mass of obese men contributes directly to their increased circulating estrogens and decreased testosterone (reviewed in Hammoud et al., 2006).

In an ironic parallelism, obese women have an increased risk of hyperandrogenism. The most common hyperandrogenic disorder found in women is polycystic ovary syndrome (PCOS), which is strongly associated with obesity. However, obesity is generally associated with sex hormone imbalance in women (Pasquali et al., 2003). Hyperinsulemia may lead directly to androgen excess in women through insulin's effects on circulating sex hormone–binding globulin (SHBG), which affects metabolic clearance and conversion of steroids, and by stimulating ovarian androgen secretion (Pasquali et al., 2006).

Leptin

Adipose tissue produces many peptides as well. The list of known adipokines (peptides produced by adipose tissue) is long and continues to grow. For example, adipose tissue produces a number of interleukins, including interleukin 1β (IL-1β), IL-6, IL-8, IL-10, IL-18, and interleukin-1 receptor antagonist (IL-1Ra) (Juge-Aubrey et al., 2003; Fain, 2006). In the next sections we briefly discuss four adipokines: leptin, tumor necrosis factor α (TNFα), adiponectin, and neuropeptide Y (NPY). We start with leptin.

Leptin was not the first adipokine to be discovered, but it is in many ways the canonical adipokine. It is secreted primarily by adipocytes. Leptin is a 167-amino-acid peptide with structural homology to cytokines (Kershaw and Flier, 2004). In general, leptin secretion is directly proportional to adipose tissue mass. However, leptin synthesis and secretion are regulated by a number of factors. Insulin, glucocorticoids, and estrogens, among others, increase leptin secretion; androgens, free fatty acids, and growth hormone, among others, decrease leptin secretion (Kershaw and Flier, 2004). Leptin secretion also varies by the type and location of the fat; subcutaneous fat generally has higher leptin secretion than does visceral fat (Fain et al., 2004; Kershaw and Flier, 2004).

Leptin has diverse functions in the body, but a major function would appear to be as a signal of adiposity, or the level of energy stores on the body. Centrally, leptin, insulin, NPY, cortisol, and corticotropin-releasing hormone (CRH) all interact, sometimes in opposition and sometimes in concert, to regulate appetite and food intake. These information molecules serve to coordinate peripheral physiology with central motive states, linking energy metabolism, perception of energy requirements (e.g., temperature, social factors, reproductive state, illness, circadian rhythms), and perception of energy stores with motivated behaviors that regulate energy intake and expenditure. These molecules affect the hedonic perception of food, the willingness to feed, the motivation to search or work for food, and, ultimately, the amount of food ingested both on a daily basis and over longer time scales.

Leptin has other important endocrine effects, including regulating tissue development and reproductive function (see chapter 7). It is also important in immune function. Leptin performs diverse functions in diverse tissues in addition to its role in adipose tissue regulation.

Leptin and Pregnancy

Because leptin is strongly associated with a measure of maternal nutritional status (fat mass), it is a plausible candidate for being an important metabolic signal for fertility and the maintenance and duration of pregnancy. Low leptin levels are associated with pregnancy loss in humans (Laird et al., 2001). Leptin levels may be abnormally high in pregnancies complicated by conditions such as diabetes mellitus and preeclampsia (Hendler et al., 2005; Ategbo et al., 2006). Although the evidence does not indicate that leptin is a primary signal for either puberty or pregnancy, the evidence does imply that it may function as one among many metabolic signals that maternal condition is satisfactory for reproduction.

Leptin is produced by the placenta. Placental weight is correlated with placental leptin mRNA (Jakimiuk et al., 2003). Cord serum leptin was correlated with placental leptin mRNA, with maternal serum leptin, and with fetal mass (Jakimiuk et al., 2003). In humans, maternal serum leptin concentration is highest at midgestation, after which it declines. Pregnancy is considered to be a state of hyperleptinaemia with leptin resistance; that is, high maternal leptin does not decrease food intake. Maternal circulating leptin levels drop precipitously at parturition, presumably due to the loss of placental leptin. Serum leptin concentrations still are correlated with maternal fat mass during pregnancy. Figure 11.1 shows regression equations for fasting serum leptin against fat mass during pregnancy and six months postpartum (from Butte et al., 1997). The lines are parallel, implying a consistent effect of fat mass on serum leptin, but the values during pregnancy are shifted upward, suggesting a consistent placental component.

Leptin is associated with insulin, insulin-like growth factor, and growth hormone, but appears to be an independent predictor of fetal size in humans. Large-for-gestational-age fetuses have higher than normal leptin, small-for-gestational-age fetuses have lower leptin. In twin pregnancies, the larger twin has higher circulating leptin (Sooranna et al., 2001). In humans, cord blood leptin is associated with both length and head cir-

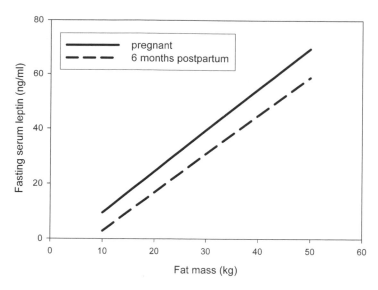

FIGURE 11.1. Maternal serum leptin concentrations in relation to maternal fat mass during pregnancy and six months postpartum. Equations for regression lines from Butte et al., 2001.

cumference of neonates. Evidence supports the hypothesis that most fetal leptin is of placental origin, though some is produced by fetal adipose tissue. Interestingly, maternal circulating leptin levels are higher in mothers of female compared to mothers of male infants (Al Atawi et al., 2005).

Leptin receptors are found in placenta, and the placenta secretes the short, soluble form sOB-R. This may partly explain why the increased maternal serum concentration of leptin during pregnancy does not result in a decrease in appetite, as much of the placental leptin is bound. Human data are lacking, but in rodents, leptin receptors are found in many if not most fetal tissues in addition to adipocytes (e.g., hair follicles, cartilage, bone, lung, pancreatic islets cells, kidney, testes, and so forth). Leptin is suspected of having endocrine, autocrine, and paracrine effects in placental and fetal tissues. It is hypothesized that leptin has important functions in regulating fetal growth and development (Henson and Castracane, 2006). Leptin may be a signal/marker of growth and development. Leptin receptors are found in fetal lung tissue in the baboon, and may play a role in lung maturation (Henson et al., 2004). Leptin receptors are also found in the intestinal tract (Lostao et al., 1998; Barrenexte et al., 2002).

Tumor Necrosis Factor α

Tumor necrosis factor α (TNFα) is also a cytokine. Like leptin, expression of TNFα is greater from subcutaneous fat relative to visceral fat (Fain et al., 2004; Kershaw and Flier, 2004). TNFα acts both locally (autocrine and paracrine) and systemically (endocrine). In adipose tissue it regulates the expression of other adipokines such as adiponectin and interleukin-6. It down-regulates the expression of genes necessary for uptake and storage of nonesterified fatty acids (NEFA) and glucose, and for adipogenesis.

TNFα also acts on the liver to regulate genes important to metabolic pathways, decreasing glucose uptake and metabolism and fatty acid oxidation and increasing cholesterol and fatty acid synthesis. It activates serine kinases that increase serine phosphorylation of insulin receptor substrates, increasing their rate of degradation. This results in a reduction of insulin signaling (Kershaw and Flier, 2004).

Adiponectin

Adiponectin is the most abundant of the circulating adipokines, with concentrations approximately 1,000 times higher than leptin. Adiponectin differs from most other factors secreted by adipose tissue in that its circulating levels are decreased in obesity (Arita et al., 1999). Low adiponectin is associated with insulin resistance, and high circulating levels appear to be protective against type 2 diabetes (Lihn et al., 2005; Trujillo and Scherer, 2005). A main target of adiponectin bioactivity is the liver, and adiponectin enhances hepatic insulin sensitivity (Trujillo and Scherer, 2005).

Adiponectin forms complexes; these high-molecular-weight molecules are the more active forms (Lihn et al., 2005; Trujillo and Scherer, 2005). There are important post-translational events that affect adiponectin signaling; hydroxylation and glycosylation of the adiponectin molecule are required for the high molecular weight complexes to form. In a sense, molecular metabolism must occur. Thus adiponectin produced by bacterial phages in which the post-translational changes do not occur is not as effective as adiponectin produced by mammalian cells (Trujillo and Scherer, 2005). This illustrates an important complexity about biological systems; although they are inherently genetic at their base (DNA to RNA that is then transcribed to a peptide) post-translational events can be extremely important.

Circulating adiponectin concentration is higher in females compared to males; however, paradoxically, adiponectin appears to be suppressed by estrogens. Maternal circulating adiponectin declines over pregnancy (Catalano et al., 2006; O'Sullivan et al., 2006), possibly due to the estrogenic environment of pregnancy. Maternal circulating adiponectin concentration is even lower during lactation, possibly due to the suppressive effect of prolactin. Interestingly, although circulating concentration of adiponectin was lower in male compared to female mice, mRNA synthesis did not differ between males and females (Combs et al., 2003). This indicates that post-transcriptional mechanisms may be responsible for the sex difference in circulating adiponectin protein.

Adiponectin appears to be an important adipokine that should be carefully studied. It has significant effects on metabolism, including energy metabolism. The sex differences in adiponectin metabolism may be very important for understanding sex differences in disease risk due to obesity.

NPY

Neuropeptide Y (NPY) is an important orexigenic molecule that acts centrally to regulate appetite. Like most, if perhaps not all neuropeptides, NPY has actions in the periphery as well; it is found in the pancreas and also adipose tissue. NPY is expressed and secreted by adipose tissue (Kos et al., 2007), as well as the type 2 receptor Y2 (Kuo et al., 2007).

Mice exposed to cold or aggression had increased expression of NPY and Y2 in abdominal fat (Kuo et al., 2007); NPY and Y2 activation stimulated angiogenesis (Lee et al., 2003) and led to an increase in visceral fat in mice exposed to the challenges of cold or conspecific aggression when the mice were fed a high-fat high-sugar diet (Kuo et al., 2007; Figure 11.2). Environmental and social stressors, such as cold and conspecific aggression, have been linked to a preference for so-called comfort foods; that is, foods high in energy from fat and simple sugars (Dallman et al., 2003). It appears that NPY may be involved in both the motivation to eat such foods and in the increase in visceral fat from the combination of a

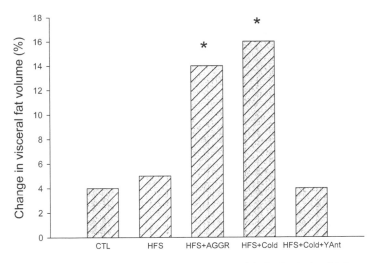

FIGURE 11.2.   Mice exposed to cold or aggression and fed a high-fat, high-sugar diet had large increases in their visceral fat. YAnt = NPY receptor antagonist. Data from Kuo et al., 2007.

stressor and a high-fat high-sugar diet. The combination of a stressor and a high-fat high-sugar diet increased visceral fat volume, while a high-fat high-sugar diet alone did not; in addition, injection of a Y2 receptor antagonist abolished the effect (Kuo et al., 2007; Figure 11.2).

The effects of NPY on adipokine secretion are uncertain; there are conflicting data. Kos and colleagues found that treatment with NPY reduced leptin secretion from adipocytes from subcutaneous abdominal fat in vitro, but had no effect on adiponectin or TNFα (Kos et al., 2007). However, Kuo and colleagues found that NPY was as potent as insulin in increasing both leptin and resistin secretion from preadipocytes from visceral fat in vitro (Kuo et al., 2007; Figure 11.3). This may be another example of tissue-specific effects of an information molecule and of the differences between subcutaneous and visceral fat.

Obesity and Inflammation

Obesity is associated with a chronic state of low-grade inflammation (Clement and Langin, 2007). Adipose tissue contains the cells involved in

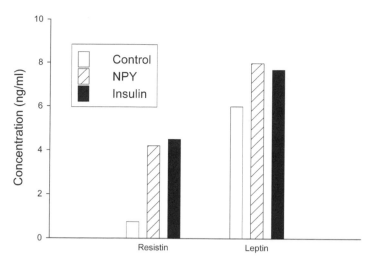

FIGURE 11.3. NPY is as potent as insulin in stimulating resistin and leptin secretion from preadipocytes from visceral fat in vitro. Data from Kuo et al., 2007.

immunological defense of the body (Fain, 2006). Adipose tissue secretes a number of peptides and cytokines associated with inflammation, and obesity generally results in higher circulating levels of these molecules. Although most adipokines appear to be inflammatory molecules, adiponectin is anti-inflammatory. Adiponectin is decreased in obesity.

Macrophages gather in adipose tissue, in clusters between adipocytes (Weisberg et al., 2003). Many of the active secretions from adipose tissue come from the macrophages (Roth et al., 2004a, b). The production of cytokines and other proinflammatory molecules from macrophages contributes to the pathologies associated with excess adipose tissue. Insulin resistance is particularly linked with the inflammation due to obesity (Xu et al., 2003; Roth et al., 2004a, b).

Hypoxia might play a role in the inflammation due to excess adipose tissue. As adipose mass increases, angiogenesis (possibly through the action of NPY; see above) assures that blood flow can reach the adipocytes. However, at large adipose tissue mass, some adipocytes and macrophages may not be well connected to the circulatory system and will suffer from hypoxia. In a normal, adaptive response, these cells will begin to secrete the appropriate inflammatory cytokines in response. This response may be locally beneficial but systemically harmful.

Central versus Peripheral Obesity

"Not all fat is alike" (Arner, 1998). Central or abdominal obesity, excess adipose tissue in the abdominal area, is associated with higher risks of comorbid disease states, such as type 2 diabetes, hypertension, dyslipidemia, and cardiovascular disease in both men and women (Karelis et al., 2004; Goodpaster et al., 2005; Racette et al., 2006; Van Pelt et al., 2006). For example, abdominal obesity was found to be the strongest predictor of insulin resistance among men and women over 50 (Racette et al., 2006). Lower body adiposity is associated with a less unhealthy metabolic profile. Overweight and obese women and obese men who had a higher proportion of fat in subcutaneous thigh adipose tissue were significantly less likely to display symptoms of the metabolic syndrome (Goodpaster et al., 2005). Obese individuals with mostly peripheral fat, distributed in subcutaneous depots in the glutealfemoral region, are at lower risk of the common comorbidities of obesity than are obese individuals with a large proportion of their fat in intra-abdominal depots (Van Pelt et al., 2006).

There are data to support a protective effect of peripheral fat (glutealfemoral adipose tissue) against disease risk. The adverse risks associated with a high waist-to-hips ratio can arise from either a large waist or a small hip or thigh circumference (Seidell et al., 2001; Snijder et al., 2003). Arterial stiffness, a risk factor for cardiovascular disease, is associated with increased trunkal fat, but increased peripheral fat conferred a small degree of protection (Ferreira et al., 2004).

Although the accumulation of subcutaneous fat in the lower body might represent a healthier regulation of fat stores compared with abdominal fat, excess adipose tissue is still associated with poor health outcomes. Metabolically healthy obese people may be less at risk than other obese people, but they still appear to be more at risk than the general population (Karelis et al., 2004).

Abdominal fat mainly consists of visceral and subcutaneous adipose tissue; the proportions of fat between these depots differ between men and women, and also differ among racial/ethnic groups. The metabolic and health consequences appear to differ as well. Visceral fat is associated with a greater likelihood of adverse health conditions (Karelis et al., 2004;

Racette et al., 2006), although excess subcutaneous abdominal fat has been implicated in poor glucose regulation (Garg, 2004; Jensen, 2006).

Visceral fat is found within the peritoneal cavity. Many authors have suggested that visceral adipose tissue differs from subcutaneous fat in ways that increase the health risks of obesity. In men, visceral fat is a significant mortality risk factor (Kuk et al., 2006). Excess visceral fat is a significant risk factor for the metabolic and health complications of obesity (Fujioka et al., 1987; Karelis et al., 2004; Racette et al., 2006). About 20% of obese men and women have metabolically healthy profiles. These individuals generally have significantly smaller proportion of adipose tissue as visceral fat (Karelis et al., 2004). There are also men and women who exhibit the opposite phenotype: normal in weight but exhibiting a metabolically "obese" profile. These individuals have a higher fat mass than would be predicted from their BMI (body mass index), but also a higher proportion of adipose tissue as visceral fat (Karelis et al., 2004). A higher proportion of fat as visceral adipose tissue was a significant risk factor for the metabolic syndrome (insulin resistance, dyslepidemia, and hypertension) in older men and women, even among those of normal weight (Goodpaster et al., 2005).

There are two main, nonexclusive hypotheses why visceral fat has more unhealthy consequences. One suggests that adipokine (e.g., leptin, interleukin-1, interleukin-6, tumor necrosis factor α, adiponectin) secretion by visceral fat differs from subcutaneous fat, and that these differences underlie the different risks to health (Karelis et al., 2004). Although the secretion of some adipokines has been shown to differ between visceral and subcutaneous fat (i.e., less leptin from visceral fat) few data are available to assess the health consequences. The other hypothesis is based on the fact that free fatty acids (FFA) released by much (but not all) visceral fat go directly into the portal vein. Thus large amounts of visceral fat will result in the liver being exposed to a greater concentration of FFA than would be predicted from systemic FFA availability. The contribution of visceral adipose tissue to hepatic FFA delivery increases with the amount of visceral fat in both men and women (Nielson et al., 2004). Liver fat has been shown to be associated with poor glucose control and higher concentrations of FFA (Seppälä-Lindroos et al., 2002). Visceral fat may play a significant role in hepatic insulin resistance (Bergman et al., 2006); however, some have questioned its importance for overall systemic insulin resistance, noting that visceral adipose tissue contributes a small proportion of total systemic FFA. Some authors point to abdominal subcutaneous fat as the major source of circulating FFA (Garg, 2004; Jensen, 2006; Koutsari and Jensen, 2006)

Visceral fat is associated with dysregulation of cortisol production and metabolism. Cushing's syndrome, in which there is adrenal hypersecretion of cortisol, is associated with increased visceral fat. Conversely, women with visceral obesity (but not suffering from Cushing's syndrome) are more sensitive to a CRH challenge than are normal-weight women or obese women with excess glutealfemoral fat as opposed to visceral obesity (Pasquali et al., 1993). Urinary excretion of cortisol and its metabolites is increased in women with excessive visceral adipose tissue (Pasquali et al., 1993).

There appear to be racial differences in the susceptibility to acquiring visceral fat. Asians have higher percent body fat for any given BMI than do Caucasians or people of sub-Sahara African descent (Deurenberg et al., 2002), with a greater proportion of fat in visceral adipose tissue (Park et al., 2001; Yajnik, 2004).

Obese postmenopausal African American women have less visceral fat for any given BMI than do postmenopausal Caucasian women, but a higher proportion of subcutaneous abdominal fat (Conway et al., 1995; Tittelbach et al., 2004). Young African American men and women have less visceral adipose tissue on average than do their Caucasian counterparts, despite African American women generally having higher total fat (Cossrow and Falkner, 2004). Interestingly, African Americans and Caucasians differ in their susceptibility to different aspects of the metabolic syndrome, with Caucasians more likely to express dyslipidemia (i.e., unfavorable cholesterol pattern and high triglycerides) while African Americans appear more susceptible to dysregulation of glucose metabolism (Cossrow and Falkner, 2004).

Summary

Adipose is a complex tissue; it contains adipocytes but also other cell types. Adipose tissue has multiple functions, of which storing fat is only one. Adipose tissue is important in the regulation of steroid hormones.

Factors released by adipose tissue have endocrine and immune functions. Many of those factors are proinflammatory; obesity is associated with a state of chronic low-grade inflammation, probably stimulated by factors released by adipose. The circulating levels of most adipokines are increased in obesity; circulating levels of the anti-inflammatory and insulin-action-enhancing adipokine adiponectin are decreased in obesity.

Some of the morbidity associated with obesity probably reflects allostatic load; that is, an exaggeration of normal function that over time can degrade and compromise organ systems. Obesity results in adipose mass being out of balance with the rest of the metabolic systems.

Not all fat presents the same risks; visceral fat is associated with a more unhealthy profile than is subcutaneous fat. Indeed, leg fat has been associated with an amelioration of symptoms of the metabolic syndrome. Some of the variation in health risks due to obesity among populations is likely due to the different propensities to accumulate fat in different depots.

CHAPTER TWELVE

Fat and Reproduction

··

M en and women differ in many ways, for good biological reasons. This is especially true for fat and fat metabolism. Although
both men and women are susceptible to obesity, they get fat in different
ways, and suffer different potential health consequences. Men and women
differ in the patterns of fat deposition, fat mobilization, utilization of fat
as a metabolic fuel, and the consequences of both excess and insufficient
fat stores. Many of these differences may reflect evolved adaptive differences that stem from the differences in male and female reproductive
costs. Reproduction is more nutritionally expensive for women than it is
for men. The costs of gestation and lactation dwarf male reproductive
effort. This asymmetry in reproductive cost is reflected in the asymmetry
in fat storage and in the utilization of fat as fuel.

In this chapter we examine differences in fat storage and metabolism
between men and women and the ways in which those differences might
underlie the differences in incidence and types of obesity experienced by
men and women. We approach these topics from the perspective of evolutionary biology. We hypothesize that many of the characteristics that
predispose people to weight gain derive from adaptive forces in our past.
Other characteristics may have been selectively neutral, due to the infre-
quency with which the obesity phenotype was expressed in the past, and
thus have accumulated in our lineage via genetic drift (e.g., Speakman,
2007). We propose that modern obesity can be explained as evolutionary
adaptive (or neutral) responses that in the modern environment result in
maladaptive physiological responses. We further propose that many of the
differences between men and women in the propensity to obesity and the

associated health consequences are reflections of the different adaptive pressures that have shaped male and female biology.

Fat, Leptin, and Reproduction

Reproduction is a key element in evolutionary fitness. Variation in reproductive success is a primary determinant of evolutionary success. It is possible to assist in passing on your genes without reproducing (e.g., inclusive fitness, increasing the reproductive fitness of your relatives), but producing offspring that survive to reproductive age is the direct way to succeed.

Reproduction is the most energy-intensive and nutritionally demanding aspect of life for female mammals. The nutritional costs of gestation and lactation are substantial, usually far greater than male nutritional costs associated with reproduction. This is certainly true for humans. The ability to regulate fertility, such that pregnancy and lactation occur at an appropriate time relative to female nutritional condition, would be adaptive. This would be especially true for avoiding pregnancy when the female is in poor condition. Expending energy and other nutrients on a fetus that has a poor chance of survival or when the expenditure has a high probability of seriously reducing maternal health in ways that would decrease her chances of future reproductive success (the extreme case being maternal death) would be maladaptive.

There is considerable evidence that female mammals, including humans, modulate their reproductive effort in relation to their nutritional status (Wade and Jones, 2004). How it is accomplished is not yet well understood. And of course there are a variety of ways to modulate reproductive output for a long-lived species such as ourselves. The number of opportunities for successful reproduction depends on many parameters that influence such factors as the length of the reproductive life span and the frequency of reproductive events in addition to the immediate reproductive functions (fertility, implantation, and maintenance of pregnancy).

Fat stores were probable crucial for reproductive success in early adult female *Homo*. Humans show little evidence of any seasonality in reproduction. It is reasonable to hypothesize that our early ancestors were responsive to food availability rather than any seasonal or annual parameters in regulating fertility. Human females have an enhanced ability to store fat on their lower extremities, as many these days lament. Adipose tissue, leptin, and fertility are linked in women. However, other factors and signaling molecules also appear to be involved. The associations between adiposity, leptin, and fertility may reflect indirect interactions via other metabolic or hormonal signals (Wade and Jones, 2004).

We are *not* proposing that obesity was adaptive in our past or that it is adaptive now. Obesity is associated with lower fertility in both men and women and in poor reproductive outcomes. Obesity per se would not have been likely to lead to increased reproductive success in our past. Instead, modern obesity may reflect the actions of adaptive characteristics that create a vulnerability to sustained weight gain under modern conditions. Obesity itself would not appear to be an adaptive strategy, but the metabolic adaptations that make many of us susceptible to obesity may have been.

Sex Differences in Adiposity

Women and men differ in the amount of fat they have, the proportion of body fat relative to lean mass, and how that fat is distributed. These differences, which begin early in life, even at birth, and are further strengthened during puberty, stem from metabolic and hormonal differences between the sexes and contribute to differences between women and men in health risks attributable to obesity.

Women have proportionately greater adipose stores than men, even after correcting for BMI (body mass index). This is true for all races and all cultures. Indeed, the mean percent of body fat for normal-weight women (BMI between 18 kg/m^2 and 25 kg/m^2) is similar to the percent of body fat of men who are classified as obese (BMI>30 kg/m^2) (Nielsen et al., 2004; Figure 12.1). This is partly explained by higher lean body mass (muscle mass) in men, but women will often have absolutely more fat than will men, even given male-female size differences.

Men and women differ in more than how much fat they have on their bodies; they also differ in where that fat tends to be located. Body fat is distributed differently between men and women (Figures 12.1 and 12.2). Women have greater adipose stores in thighs and buttocks (Williams, 2004); men tend to be more likely to have significant amounts of abdominal fat and to be more susceptible to abdominal adiposity (Nielson et al., 2004). Women have larger stores of subcutaneous fat; men are more likely to have visceral fat (Lemieux et al., 1993). This sex difference in adiposity is present at birth. Female babies have more subcutaneous fat than do male babies for all gestational ages (Rodríguez et al., 2005). Prepubertal girls have more fat in their legs and pelvis than do prepubertal boys (He et al., 2004). All of this is a matter of degree. Obese women will have large amounts of visceral fat; obese men will have large amounts of subcutaneous fat on their legs (Figures 12.1 and 12.2).

The difference in fat distribution between men and women affects other health risk factors associated with obesity. Waist circumference is a significant risk factor for the comorbidities of obesity. As would be expected, waist circumference in men and women is associated with abdominal subcutaneous and visceral fat; your waist size reflects your abdominal adipose tissue. However the relationships differ significantly between the sexes. The regression lines of waist circumference against subcutaneous abdominal fat for men and women are parallel but offset;

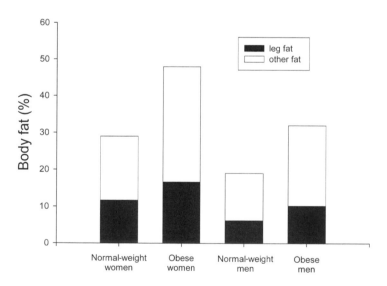

FIGURE 12.1. Women have higher total percent body fat and a greater proportion of fat in the leg than do men at all BMI values. Normal-weight men and women: BMI <25 kg/m^2. Obese men and women: BMI >30 kg/m^2. Data from Nielson et al., 2004.

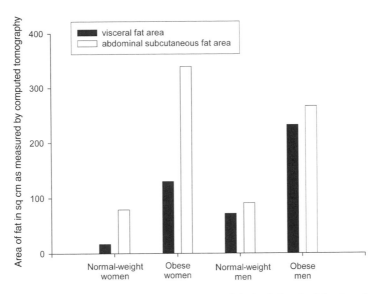

FIGURE 12.2. Women have a greater proportion of their abdominal fat in subcutaneous depots compared to men; men have significantly more visceral fat at all values of BMI. Obese men and women: BMI >30 kg/m². Data from Nielson et al., 2004.

women have on average 1.8 kg more subcutaneous abdominal fat than men for any given waist circumference (Kuk et al., 2005). In contrast, the slope of the regression line of waist circumference against visceral fat is significantly greater for men than for premenopausal women (Kuk et al., 2005).

Age and menopausal status also have significant effects on the relationships between waist circumference and visceral fat. Older men and women have significantly higher regression slopes than do their younger counterparts. The slopes of the regression lines for men are greater than for women standardized to any age, but the standardized slope for 40-year-old women is the same as the standardized slope for 25-year-old men. The slope for menopausal women is greater than the slope for premenopausal women and is parallel to older men (Kuk et al., 2005). The health consequences of obesity in menopausal women are also more similar to that of men. The above findings provide evidence that testosterone and estrogen affect fat distribution and consequences (see below).

Central versus Peripheral Obesity

Central obesity is associated with higher risks of comorbid disease states, such as type 2 diabetes, hypertension, dyslipidemia, and cardiovascular disease (see chapter 11). Lower-body adiposity is associated with a less unhealthy metabolic profile. Indeed, there is some evidence that increased leg adipose tissue (within reason) positively affects metabolic profiles (Ferreira et al., 2004).

Interestingly, not only do men on average have a greater proportion of fat as visceral fat, apparently turnover of visceral fat is higher in men than in women. Men have consistently been shown to have greater rates of fatty acid release (lipolysis) and fatty acid uptake (lipogenesis) in visceral fat compared with women (Williams, 2004). Adrenergic stimulation increases splanchnic fatty acid release in men but not in women (Jensen et al., 1996). Thus men are more susceptible to excess visceral fat, and the effects of visceral fat on health may differ between the sexes as well. Visceral fat has been shown to predict mortality risk in middle-aged white men (Kuk et al., 2006).

Visceral fat is associated with dysregulation of cortisol production and metabolism. Cushing's syndrome, in which there is adrenal hypersecretion of cortisol, is associated with obesity, especially increased visceral fat. Conversely, women with visceral obesity (but not suffering from Cushing's syndrome) are more sensitive to a corticotropin-releasing hormone challenge than are normal-weight women or obese women with excess glutealfemoral fat as opposed to visceral obesity (Pasquali et al., 1993). Urinary excretion of cortisol and its metabolites is increased in women with excessive visceral adipose tissue (Pasquali et al., 1993).

Effects of Sex Hormones on Fat Deposition and Metabolism

Not surprisingly, the sex hormones play a role in sex differences in the biology of fat. The gonadal hormones affect adipose tissue metabolism and appear to play significant roles in the resulting distribution and consequences of stored fat. Testosterone acts to increase lipolysis, inhibit lipoprotein lipase activity, and decrease triglyceride accumulation in adipose tissue. Lowering circulating testosterone levels in healthy young men increases total adipose tissue, with the largest percent increase occurring in subcutaneous adipose tissue; raising circulating testosterone decreases total adipose tissue (Woodhouse et al., 2004). Estrogens play multiple roles in the regulation of adipose tissue, in both men and women. Estradiol has direct effects on adipose tissue and also acts centrally to affect food intake and energy expenditure. Androgens appear to block proliferation and differentiation of preadipocytes (Singh et al., 2006). Estradiol enhances proliferation of preadipocytes from both men and women in vitro (Anderson et al., 2001). The effect was greater in preadipocytes from females compared to those from males.

Estradiol favors the deposition of subcutaneous fat; lack of estrogen in women leads to weight gain and to a larger proportion of fat gain in visceral fat. Menopausal women have higher visceral fat mass than do premenopausal women for the equivalent percent body fat (Tchernof et al., 2004) or waist circumference (Kuk et al., 2005). Estradiol-treated postmenopausal women have lower lipoprotein lipase (LPL) activity (Pedersen et al., 2004). Estrogens are important determinants of the female fat pattern.

Adipose tissues express both androgen and estrogen receptors. Visceral fat has higher levels of androgen and estrogen receptors than does subcutaneous fat, and this is true for both men and women (Rodriguez-Cuenca et al., 2005). The alpha and beta estrogen receptors are found in adipose tissue (Pedersen et al., 2004). In subcutaneous fat, estradiol acts through the α receptor to up-regulate α2A-adrenergic receptors, which results in decreased lipolysis. Estradiol does not appear to affect the concentration of α2A-adrenergic receptors in adipocytes from visceral fat (Pedersen et al., 2004). Subcutaneous adipocytes from premenopausal women have higher α2A-adrenergic receptor density and lower lipolytic activity in response to epinephrine than do subcutaneous adipocytes from men (Richelsen, 1986).

Leptin and Insulin

To date, the only circulating hormones that meet the criteria to be an adiposity signal are leptin and insulin. Basal circulating concentrations of both insulin and leptin are in proportion to fat mass. Both are transported across the blood-brain barrier and act centrally to regulate appetite, reduce food intake, and possibly increase energy metabolism (Woods et al., 2003). Although there are conceptual and empirical difficulties with the lipostatic models of regulation of both food intake and reproduction (Wade and Jones, 2004), the dynamic changes in adipose tissue certainly are associated with feeding and with reproduction. Insulin and leptin, among many others, appear to serve as important signaling molecules.

Leptin and insulin differ in important ways; circulating levels of leptin and insulin appear to reflect different fat depots. Leptin concentration is more reflective of subcutaneous fat; insulin is more reflective of visceral fat. Visceral fat also is more sensitive to insulin than is subcutaneous fat. Because of the differences between men and women in the proportion of visceral to subcutaneous fat, leptin is, in general, better correlated with total adipose mass in women, and insulin is more highly correlated to total adipose mass in men (Woods et al., 2003).

Insulin is not produced by adipose tissue but does have potent effects on adipose, and its function is affected by adipose. Increased fatness, whether measured by BMI, waist-to-hip ratio, waist circumference, or actual measures of body fat, is associated with a reduction in peripheral insulin sensitivity. Men and women differ in this regard. Despite women having a greater amount of body fat than do men, insulin sensitivity in women appears to be less affected by the amount of body fat. Increases in body fat among women are associated with smaller decreases in insulin sensitivity compared with men (Sierra-Johnson et al., 2004). Visceral fat and subcutaneous fat differ in their responses to insulin, both metabolically and in the synthesis and secretion of adipokines (Einstein et al., 2005). Excess visceral fat is associated with insulin resistance (Karelis et al., 2004; Racette et al., 2006). Thus the fat distribution differences between men and women have metabolic, endocrine, and health consequences.

Men and women differ in the responses to central insulin and leptin. Men are more sensitive to central insulin; women are more sensitive to central leptin. Intranasal administration of insulin led to weight loss, specifically fat loss, in men; it resulted in weight gain, primarily extracellular water, in women. Intranasal insulin reduced feelings of hunger in men but not in women (Hallschmid et al., 2004). The same results have been ob-

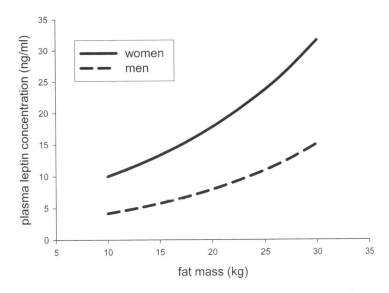

FIGURE 12.3. Plasma leptin concentration increases exponentially with fat mass; women have higher plasma leptin concentrations than do men for any fat mass. The equations for the curves are from Saad et al., 1997.

tained in rats. Male rats are more sensitive to central insulin, female rats to central leptin (Clegg et al., 2003).

These differences appear to stem from effects of the gonadal hormones. Male rats given exogenous estrogen are more sensitive to the effects of central leptin than are control males (Clegg et al., 2006). Estrogen appears to blunt the effects of central insulin; intact male and ovarectomized female rats reduced food intake after central administration of insulin. Intact female rats and male rats given exogenous estrogen did not. Interestingly, castrated male rats without exogenous estrogen showed no effect of central insulin on food intake (Clegg et al., 2006), implying that testosterone may directly affect central insulin signaling.

Serum leptin concentration shows persistent sex differences that begin before birth. Maternal circulating serum leptin is higher in pregnancies where the fetus is a girl (Al Atawi et al., 2005). Women have higher leptin levels than do men, even at birth, and this difference persists throughout life. These differences do not simply reflect differences in total adipose tissue between men and women (Figure 12.3); women have higher circulating leptin for any given amount of fat mass (Ostlund et al., 1996; Rosenbaum et al., 1996; Kennedy et al., 1997; Saad et al., 1997).

Fat Metabolism

Fat metabolism in women and men differs in a number of ways consistent with the sex differences in body fat percentage and adipose tissue distribution. Women appear to be metabolically inclined to store fat more so than are men. Interestingly, women also appear to utilize fat as an energy substrate during periods of sustained exertion more so than do men.

At rest, women shunt more circulating free fatty acids into reesterfication pathways than do men (Nielsen et al., 2003). Women have higher VLDL-triglyceride production rates but similar circulating concentrations (Mittendorfer, 2003). This is further evidence that women have higher rates of reesterfication and thus greater reuptake of free fatty acids into adipose tissue than do men. In the basal condition, women are more physiologically adapted to store fat.

The rates of fatty acid uptake and release depend on the type of adipose tissue as well as sex, and this is reflected in the differing patterns of fat deposition between men and women. Women have higher rates of fat uptake into leg fat depots (Votruba and Jensen, 2006). Rates of fatty acid release from abdominal adipose tissue are higher in women than men, but they are lower from gluteal or femoral adipose tissue (Williams, 2004). Fatty acid uptake is higher in abdominal adipose tissue relative to gluteal or femoral in both men and women after feeding. However, in women the majority of fatty acid uptake in abdominal adipose tissue is into subcutaneous fat, while in men a larger proportion goes into visceral fat (Williams, 2004). These findings are consistent with women being more likely to store fat subcutaneously and preferentially in the gluteal and femoral regions compared with men.

Women have higher rates of fat oxidation than men during sustained bouts of increased energy expenditure, such as endurance training. Men are more likely to up-regulate glucose and amino acid metabolism during sustained exercise bouts (Lamont et al., 2001; Lamont, 2005). The difference is associated with estrogen. Giving exogenous estrogen to males decreases carbohydrate and amino acid metabolism during exercise and increases fat oxidation (Hamadeh et al., 2005). Thus it would appear that women are more physiologically geared to use fat as a metabolic fuel under conditions of sustained increased demand, while men rely more on glucose and protein metabolism. This difference is mediated by the sex hormones.

Despite women having a greater increase in fat oxidation with sustained exercise, men are more likely to lose fat through a program of increased exercise than are women (Ross, 1997; Donnelly et al., 2003). This puzzling result is not well understood. It isn't known to what extent social and psychological factors might play a role. The studies to date have not shown that men are more motivated or dedicated when exercising to lose weight. A possibility is that the fat metabolism of women is more geared to expend fat during sustained exertion but then to store fat outside of those times. Thus women may burn more fat through exercise than do men, but afterward they may be more likely to replenish their adipose stores during recovery.

Certainly our ancestors exerted a great deal of energy and effort to survive and reproduce. Differences in fat metabolism during sustained exertion probably relate to differences in past evolutionary pressures between men and women. Perhaps in men exercise is a good model for selection on metabolism and the importance of different metabolic fuels. Male metabolism may reflect selective pressures primarily based on muscular activity, both intense and sustained. Our women ancestors also expended large amounts of energy in physical tasks; we are not suggesting that women did not physically work as hard as men, though there are and always have been differences between the sexes in physical capability. More importantly, however, women differed from men in that they expended large amounts of energy during pregnancy and lactation, perhaps the most energetically costly events in the lives of our female ancestors. The demands of reproduction were more likely to be the important selective pressures in women and thus had a larger influence on female metabolism.

Fat Advantages for Reproduction

There are several advantages to having both a greater capacity to store fat and a greater reliance on fat as a metabolic fuel during periods of sustained increased need, for example during pregnancy and lactation. Upregulating fat metabolism spares glucose; during pregnancy the glucose demands of the fetus and placenta must be balanced with the glucose need of the maternal brain. Increasing fat oxidation to provide fuel for maternal muscle and peripheral organs relieves some of this conflict.

After birth the infant receives nutrition via milk. Maternal transfer of nutrients is delivered through the infant's digestive tract rather than through the placenta. This again has a glucose sparing effect on women. It also allows energy from the past to be used to support current reproductive effort. Although energy requirements during the last trimester of gestation are certainly significant, even women in poor countries generally are able to gain maternal mass during early gestation. Appetite is usually increased during pregnancy, and often energy expenditure will be reduced. Excess energy intake during early pregnancy can be stored as fat and then utilized during lactation.

Many nonhuman mammals take this strategy to extremes. For example, female black bears gain large amounts of adipose tissue in the fall before getting pregnant. They den up to hibernate for the winter while pregnant. They give birth during hibernation. Thus most of gestation and all of early lactation takes place while the female is fasting.

The biology of bears is well adapted to this strategy. The cubs, which are born after a short gestation, are extremely small and altricial (undeveloped). Female black bears that weigh more than 200 pounds will produce cubs that weigh less than 1 pound each (Oftedal et al., 1993). The early birth of cubs provides the mother bear with significant metabolic advantages. Because she is fasting, she has no external source of glucose; during gestation the glucose needs of her brain (reduced by hibernation of course) compete with the glucose needs of the placenta and fetuses. By giving birth early she eliminates the placental-fetal drain on her glucose and switches her offspring to a diet of high-fat, high-protein milk (Oftedal et al., 1993) delivered through their intestinal tracts. She mobilizes energy ingested from before hibernation that was stored as fat to produce glucose for herself and to deliver energy in the form of milk fat directly to her cubs. The bigger puzzle is how can she manage her water balance if she is not eating or drinking, yet is producing milk that, while it is high-fat, contains more water than solids (Oftedal et al., 1993)? It turns out that even while hibernating the female is capable of ingesting all the excreta that her cubs produce. This is an efficient system that recycles a very large percentage of the water and nitrogen (from protein) that is delivered to the cubs (Oftedal et al., 1993). And it all depends on storing fat.

Bears that do not hibernate also utilize this reproductive strategy; cubs are born early and very small, and they are fed a high-fat, high-protein milk. This is even true for pandas (Figure 12.4a and b). The adaptation supporting gestation and lactation while fasting during hibernation has not disappeared in species that no longer hibernate. Pandas carry the legacy of their hibernating bear ancestors.

Humans aren't bears; women don't fast during pregnancy and lactation. They couldn't. But they are capable of storing significant nutritional resources that can be mobilized during reproduction, especially during lactation. We are large animals, and large animals are more able to use what is called a capital investment reproductive strategy as opposed to an income investment strategy. In other words, small mammals are generally incapable of storing sufficient energy to support reproduction. They must meet most of the nutrient demands of lactation through current food intake. Extremely large mammals, such as some whales, can support lactation while fasting; in effect they support lactation through past food intake (Oftedal et al., 1993). Bears manage with a combination of being large, storing fat, and hibernating. Humans support the costs of lactation by combining the use of nutrient stores with current intake and modulation of energy expenditure. Part of the energy cost of lactation is met by stored energy from past food intake.

This is true for nutrients besides energy. Mothers transfer a significant amount of calcium to their fetuses and then to their nursing infants to support bone growth and mineralization. During gestation calcium metabolism changes to increase the efficiency of absorption from the intestines, resulting in greater extraction from food and also greater retention by the kidney tubules, decreasing urinary excretion (Kovacs and Kronenberg, 1998; Power et al., 1999). The endocrine profile of gestation enhances calcium deposition onto bone.

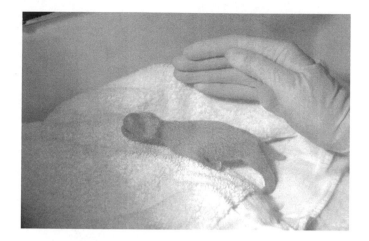

FIGURE 12.4. (a) Pandas produce small cubs at birth and produce high-fat, high-protein milk, retaining the adaptations of hibernating bears even though they do not hibernate. (b) Panda nursing a cub. Photo: Jessie Cohen, Smithsonian's National Zoo.

The endocrine profile of lactation does the reverse. During lactation, women lose bone mineral, mostly calcium; typically women lose 3 to 10% of bone mineral over the course of lactation (Prentice et al., 1995; Kalkwarf et al., 1997). This loss is partially buffered by the increase in bone mineral during gestation and is rapidly replenished after menses resumes (Kalkwarf et al., 1997). Interestingly, dietary calcium doesn't seem to affect the loss during lactation. A high-calcium diet may lead to increased bone calcium during gestation, but supplementing the diet with calcium during lactation has no measurable effects on calcium loss from bone; it merely increases urinary calcium excretion (Fairweather-Tait et al., 1995; Prentice et al., 1995; Kalkwarf et al., 1997).

Thus human female reproduction relies, to an extent, on past food intake stored on the mother's body to be mobilized during the time of greatest nutrient transfer to her infant—lactation. Since men do not experience this selective pressure, it is unsurprising that men and women should differ in aspects of metabolism related to storing and mobilizing nutrients. Men do not go through cyclical changes in bone mineral density; they also do not need to store and mobilize large amounts of fat in order to be reproductively successful.

Fat Babies

There is another human adaptation that likely influenced female fat metabolism during reproduction. Human neonates are among the fattest known baby mammals (Kuzawa, 1998; Figure 12.5). Only hooded seal pups have as great or greater a percentage of body fat when born. Hooded seals represent an extreme case of short but intense maternal effort during lactation; lactation lasts only four days, but the milk is 56 to 60% fat, and the pups literally double their body weight during that four-day period, with almost all of the tissue deposition being lipid (Oftedal et al., 1993). The mother leaves after the four-day lactation period and the pup then grows while fasting, metabolizing the stored fat. The high percentage of fat at birth in hooded seals is part of this extreme reproductive adaptation.

Why are human infants born so fat? Human lactation is long; in hunter-gatherer societies it routinely continues for three or more years. Nursing is fairly frequent; human babies are not left on their own for long periods of time without food. Human milk is quite dilute, containing only 3 to 4% fat and less than 1% protein; the daily transfer of nutrients from mother to child is fairly low, at least compared to many other mammals. This is the general anthropoid primate pattern: a long lactation (usually longer than gestation), with frequent nursing that transfers a large volume of dilute milk to the infant (Oftedal, 1984; Power et al., 2002). There

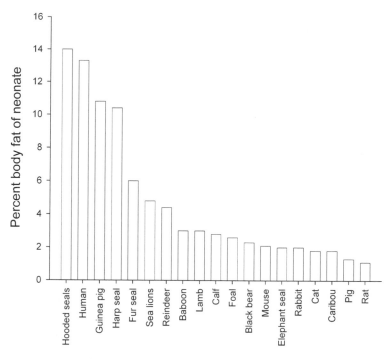

FIGURE 12.5. Human babies are among the fattest mammalian neonates—almost as fat as seal pups. Data from Kuzawa, 1998, except hooded seal value from Oftedal, 1993.

would not appear to be any adaptive advantage for a primate infant to be born fat, and most are not. Human babies are the exception.

There has been extensive speculation that the larger brain characteristic of the genus *Homo* would have required metabolic adaptations to support its energetic cost to grow and maintain (Aiello and Wheeler, 1995). Brain metabolism accounts for a considerable percentage of human energy requirement. This is especially true for babies, where over 50% of energy expenditure is due to brain metabolism (Kuzawa, 1998). The energetic expense of a human baby's brain is estimated to be three or more times that of a chimpanzee infant's brain. The higher neonatal fat in human babies may represent an appropriate increase in energy reserves in response to this energy demand (Kuzawa, 1998). The large amount of fetal adipose deposition that occurs in the third trimester in humans is coincident with the start of rapid brain growth. An exceptionally fat neonate, especially when compared to our nonhuman primate relatives, would appear to be a key adaptation for human beings probably related to supporting the extensive postnatal brain growth and maintenance energy requirement necessary for our large brains (Kuzawa, 1998).

There are good, evolutionary reasons why most of our brain growth occurs after birth. The constraints on the pelvis and the birth canal that enable adult females to walk, run, and otherwise locomote in an efficient and adaptive manner would not allow such an extremely large-headed infant to be born. If human babies achieved most of their brain growth in utero our species probably would have gone extinct either because all the women would have died in childbirth or because they would never have been able to make it to adulthood if they grew a pelvis that had a birth canal of sufficient size. They would have become easy prey.

There are other possible advantages to being a fat baby. Neonates are generally at high risk of infectious disease; their immune systems are immature. Many diseases disrupt feeding, digestion, or both (Kuzawa, 1998). For example, diarrhea is a leading cause of infant mortality in the world. Large amounts of fat stored in adipose tissue is protective during many diseases, for both infants and adults. Of course that would be true for chimpanzee babies as well as human babies; so is there any reason to suspect that human babies had higher disease risks in the past?

There are immune-function molecules in milk that assist an infant's immune system (e.g., secretory immunoglobulin A or SIgA), as well as many other molecules that appear to perform an antimicrobial function. Because of their immature immune systems infants are susceptible to disease; one function of milk is to boost infant immunity and reduce pathogen risk. Indeed, during the first few days of lactation, mammal mothers generally produce milk called colostrum, which is high in SIgA and other immunoglobulins. There are other antimicrobial molecules in milk. For example, indigestible oligosaccharides in milk serve a decoy function. Bacteria invade cells by attaching to oligosaccharide residues on the cell membrane. Bacteria in milk attach to similar oligosaccharides secreted into the milk and are effectively carried through the infant's digestive tract and eliminated in the feces.

Human milk is not different nutritionally from chimpanzee or gorilla milks (Milligan, 2008). However, human milk is different from other milks in its immune and antimicrobial function molecules. Human milk contains the highest known concentrations of SIgA; mature human milk (e.g., milk produced after the first week, not colostrum) contains a higher concentration of SIgA than does rhesus macaque colostrum milk (Milligan, 2005, 2008). Human milk also contains higher concentrations and higher diversity of oligosaccharides compared to milks of all other primates so far examined. Human milk has high concentrations of lactoferrin, a molecule that binds iron and serves to effectively "starve" bacteria in the milk. All the evidence indicates that human milk has greatly enhanced antimicrobial and immune function.

For much of our past humans lived in an environment with a high density of pathogens. Although a chimpanzee or gorilla wandering through a tropical forest comes into contact with billions of microbes, very few of those microbes are relevant to the animal's health. Indeed, most of the billions of microbes we are exposed to each day are just as irrelevant to our health and well-being. Only a small fraction of microbes are actually pathogenic for any given species. Humans increase both the concentration of and exposure to pathogens by some of the very characteristics of our life that have made us so successful. We live in large concentrations of our fellow humans, and we occupy that living space more or less continuously. We are thus exposed to numerous potential disease vectors (each other) every day. We also end up concentrating our pathogens in our environment just by our natural biological functions. The rules of sanitation and the invention of sewer systems were key events in improving human health, but they are relatively modern occurrences, especially in the evolutionary sense. Indeed, at the beginning of agriculture it is very likely that night soil, human feces, was one of the first fertilizers. It also increases pathogen exposure and risk. As our ancestors increased in number, spread out across the world (and thus encountered novel diseases), began to settle for long periods of time in the same place (and thus concentrated the microbes, good and bad, that make a living off of us), and domesticated animals (another source of novel pathogens; think bird flu), our disease risk increased (Barrett et al., 1998). Our milk evolved to respond; perhaps this increased disease risk was another selective pressure that favored fat babies.

Maternal adiposity is associated with infant body fat content (Catalano et al., 2007). There would appear to be adaptive advantages to producing a fat baby, or at least there was one in the past. Part of the selective advantage for the human female pattern of fat deposition may have arisen to enable fat babies.

Fat and Female Reproduction

Fat is intimately tied to female reproduction. The association between fatness and reproductive success in women may start at birth. A high ponderal index at birth (birth weight divided by the cube of birth length) in female infants is associated with both higher estradiol levels (Jasienska et al., 2005) and resistance to estradiol suppression by physical activity as adults (Jasienska et al., 2006). Thus a measure of fatness at birth is associated with ovarian function as an adult. These data suggest that in our past women producing lean female babies may have been selectively at a disadvantage due to the potentially lower fertility of their daughters. Maternal adiposity is associated with infant adiposity (Catalano et al., 2007), so a potential connection exists between maternal fatness and the reproductive potential of daughters.

Adiposity is associated with the age at menarche (Matkovic et al., 1997), fertility (Gesink Law et al., 2006), and pregnancy outcome (Pasquali et al., 2003). Fat and leptin play important roles in fertility and are implicated in nutritional infertility, though the exact mechanisms are still largely unknown. Certainly leptin is required for both male and female fertility. A lack of leptin signaling, either through leptin deficiency or faulty receptor function, causes infertility. Exogenous leptin administered to leptin-deficient animals with competent receptor signaling restores fertility (Chehab et al., 1996). A minimal level of leptin signaling is required for reproduction. Leptin is thus certainly permissive of reproduction, although it isn't clear if leptin is a primary signal or acts indirectly through other metabolic and endocrine pathways (Wade and Jones, 2004).

Female reproduction is linked with fat in several ways, partly through leptin. At all life stages, leanness in women is potentially associated with lower fertility. Leanness at birth is associated with reduced ovarian function (Jasienska et al., 2005, 2006). Leanness in adolescents is associated with delayed menarche (Lee et al., 2007). Leanness in adult women is associated with irregular or absent ovulatory cycles. Fat is important to female reproductive fitness.

Fat, Leptin, and Puberty

Adolescent girls with higher fat mass attain puberty at an earlier age (Matkovic et al., 1997; Tam et al., 2006). The link between adiposity and age at menarche in adolescent girls is not simple, however. There does not appear to be a threshold level of fatness that women must attain in order to be fertile as proposed by Frisch and Revelle (1971). Indeed, energy balance may be as important in some circumstances (Wade and Jones, 2004); in other words, the trajectory (gaining or losing fat) plays a role. However, adiposity and leptin certainly have effects on puberty.

The idea of an association of leptin with puberty and fertility was a logical extension of the fact that leptin-deficient mice, both male and female, were not only obese but also infertile. Exogenous leptin produced a reduction in weight and also reversed the infertility (Chehab et al., 1996). Leptin is important in regulating the transition through puberty. For example, giving exogenous leptin to mice resulted in their attaining sexual maturity at a significantly earlier age (Chehab et al., 1996). Female rats under food restriction have delayed onset of sexual maturation; giving exogenous leptin reversed this effect (Cheung et al., 1997). An increase in nocturnal leptin, as well as insulin-like growth factor-I (IGF-I) just before the onset of puberty in male rhesus monkeys gave further weight to the theory that leptin, IGF-I, or both were involved in the pubertal transformation (Suter et al., 2000).

The links between fat, leptin, and female reproduction may be manifested in the historically recent decline in the age of menarche in the United States and other wealthy nations. The age of menarche has shown a consistent decline over time in the United States (McDowell et al., 2007), paralleling the increase in overweight and obesity among adolescent girls. Girls with higher BMIs from early life on average begin menstruating at an earlier age (Lee et al., 2007). Circulating leptin is correlated with BMI in adolescent girls (Matkovic et al., 1997). It is reasonable to hypothesize that the higher on-average BMIs of today's young girls are associated with higher on-average levels of circulating leptin, and that this is one possible mechanism behind the decrease in the age at menarche. Indeed, there is an inverse relationship between leptin levels and the age at menarche in adolescent girls (Matkovic et al., 1997). Those data were supportive of a threshold leptin level being important for puberty. There is a great deal of variation in leptin levels for all menarchal categories (early, normal, and late), however, and not all studies have found a significant relationship between leptin and age at menarche (e.g., Tam et al., 2006). At present the evidence suggests that leptin is permissive of puberty but is not the only factor involved.

Indeed, there are heritable components to the changes in circulating leptin and IGF-I in pubertal girls. Based on twin studies, estimates of heritability of circulating IGF-I (54 to 77%) are similar though slightly higher than those for leptin (38 to 73%). The greater range for leptin and higher intratwin differences in circulating leptin support the hypothesis that leptin may be more affected by environmental effects (Li et al., 2005).

Circulating leptin levels consistently rise with age in adolescent girls; circulating leptin declines during puberty in adolescent boys (Ahmed et al., 1999; Kratzsch et al., 2002). In both sexes leptin is strongly correlated with fat mass (Ahmed et al., 1999). The relationship between fat and leptin begins to differ between males and females at puberty.

A soluble short form of the leptin receptor (sOB-R) acts as a binding protein for circulating leptin (Lammert et al., 2001; Kratzsch et al., 2002). In the first years of life sOB-R circulates at high concentrations; circulating sOB-R steadily declines until puberty. Levels of sOB-R appear to plateau after the first stages of puberty (Kratzsch et al., 2002). The free leptin index (FLI), the ratio of leptin to sOB-R, is more strongly associated with growth and sexual maturity than is leptin by itself (Kratzsch et al., 2002; Li et al., 2005). This is another example of the complexity and flexibility of the biology driven by these information molecules.

The timing of menarche appears to be a complex interaction between genetics and environment. There are strong heritable components, but birth weight and early life also have significant effects. For example, Tam and colleagues (2006) demonstrated that girls who were long and light at birth were generally more likely to begin menstruating at an early age, but that result was modified by the BMI at age 8. Girls with high BMI had an earlier average age at menarche. Thus the group with the earliest age at menarche was composed of girls who were long and light at birth and had a high BMI at age 8. The girls with the latest age at menarche were short and heavy at birth but had a low BMI at age 8. BMI at age 8 was the strongest predictor of age at menarche, with birth body form second. But both those factors explained only 12% of the variation in age at menarche (Tam et al., 2006).

Obesity and Fertility

Although a minimal level of fat (and leptin) appears to be required for fertility, obesity does not enhance fertility; in fact it does just the opposite. Obese people, both male and female, have lower fertility. Obesity in both men and women increases the time to pregnancy. In one study the probability of conceiving in a given cycle was on average 18% less in obese women (Gesink Law et al., 2006). Obesity increases the risks of a large number of reproductive disorders in women, including effects of hyperandrogenism, which can lead to anovulation (reviewed in Pasquali et al., 2003).

Obesity also has significant effects on many aspects of male sexual and reproductive performance. Obesity is associated with erectile dysfunction, possibly through the same means by which obesity plays a role in hypertensive disease and cardiovascular disease (reviewed in Hammoud et al., 2006). Sperm density and total sperm per ejaculate was lower in young men with BMI above 25 kg/m^2 (Jensen, 2004). Obesity also negatively affected sperm motility (Kort et al., 2005). The altered spermatogenesis in obesity is due to hypoandrogenism and to the effects of increased estrogens from aromatase conversion of androgen to estrogens in adipose tissue (reviewed in Hammoud et al., 2006).

Obesity, Pregnancy, and Birth Outcome

Obese women experience a higher risk for a number of pregnancy complications and increased risks of morbidity for both mother and infant (Pasquali et al., 2003). Obese pregnant women are more at risk for hypertension, toxemia, gestational diabetes, and urinary infections. They are at greater risk for spontaneous abortion after infertility treatment (Wang et al., 2002). They are more likely to have complicated labor and delivery and to have a cesarean section (Pasquali et al., 2003; Catalano and Ehrenberg, 2006). Maternal obesity complicates anesthesia (Catalano, 2007).

TABLE 12.1   *Maternal obesity and risk for 7 birth defects*

	Odds ratio for maternal BMI* of 30 kg/m² or more	95% confidence interval
Spina bifida	2.10	1.63–2.71
Heart defects	1.40	1.24–1.59
Anorectal atresia	1.46	1.10–1.95
Second- or third-degree hypospadias	1.33	1.03–1.72
Limb reduction defects	1.36	1.05–1.77
Diaphragmatic hernia	1.42	1.03–1.98
Omphalocele	1.63	1.07–2.47
Gastroschisis	0.19	0.10–0.34

Source: Waller et al., 2007.
Note: Odds ratio >1 means increased risk; <1 means decreased risk.
*BMI = body mass index

Obese women suffer from more complications after cesarean birth, such as greater incidence of infection and poor wound healing (Catalano, 2007).

There are increased risks for many poor birth outcomes for obese women. These include stillbirth and various congenital anomalies, including neural tube defects. In an ongoing study that examined over 10,000 births with at least 1 identified birth defect and over 4,000 control births without any congenital anomalies, maternal obesity (BMI≥30 kg/m²) was identified as a significant risk factor for 7 birth defects and as protective for 1 birth defect, gastroschisis (Waller et al., 2007; Table 12.1). Maternal overweight was a significant risk factor for 3 of the birth defects and protective of gastroschisis; maternal thinness (BMI <18.5 kg/m²) was a risk factor for cleft lip and palate. The results of this study are consistent with a number of other studies (e.g., Werler et al., 1996; Shaw et al., 1996; Anderson et al., 2005) that indicate that infants of obese women have approximately a twofold increase in the risk of spina bifida. The increased risk for the other birth defects was more modest. Gastroschisis, or the development of the intestines and sometimes other organs outside of the fetal abdomen, is a birth defect related to young maternal age and low birth weight, among other risk factors (Feldkamp et al., 2007).

The functional link between maternal obesity and increased risk for this spectrum of congenital anomalies in offspring is unknown at this time. Poor glycemic control is associated with many structural birth defects in both humans and animal models (Eriksson et al., 2003). Glucose is a mild teratogenic substance at high concentrations. Thus the mechanisms by which diabetes and obesity increase the risks of birth defects may have common underpinnings. However, controlling for gestational diabetes reduced but did not eliminate the increased risk for birth defects associated with maternal obesity, suggesting that other metabolic aspects of obesity may play a role as well (Waller et al., 2007).

Perhaps most relevant to this book, maternal obesity is associated with macrosomia. In the United States there has been a steady increase in both maternal body weight and birth weight (Catalano et al., 2007; Figures 12.6 and 12.7). Maternal obesity is a strong risk factor for macrosomia; fat mothers are more likely to produce excessively fat babies. There are a number of mechanisms for this effect; maternal obesity is associated with poor glucose control, which at its extreme becomes gestational diabetes. Gestational diabetes is associated with large, overly fat babies. Indeed, the excess weight gain of infants of diabetic mothers is disproportionate in the fat compartment.

Macrosomia is associated with a number of increased risks for both

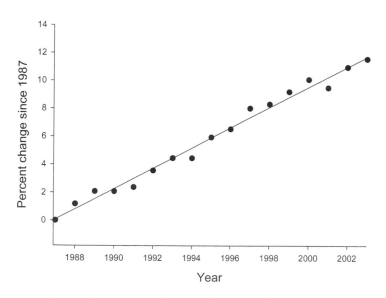

FIGURE 12.6. Maternal weight at delivery has been steadily increasing at Metro-Health Medical Center, Cleveland, Ohio. Between 1987 and 2003 mean maternal weight increased by almost 12%. Data from Catalano et al., 2007.

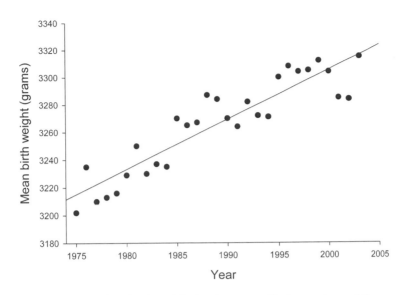

FIGURE 12.7. Mean infant birth weight has been steadily increasing at Metro-Health Medical Center, Cleveland, Ohio. Data from Catalano et al., 2007.

mother and child. The risk of a cesarean section certainly increases with fetal size (Ehrenberg et al., 2004; Catalano, et al., 2007). Macrosomic babies are at enhanced risk of obesity and diabetes in later life. For all the proposed evolutionary advantages to fat babies that have been discussed previously, once again it appears that fat is a double-edged sword, and too much of a good thing may lessen rather than enhance health.

Summary

Many authors have suggested that obesity results from a mismatch between our evolved, adaptive responses to past conditions in which obtaining food required extensive physical effort and food scarcity was common and the modern condition of easy access to plentiful, energy-dense foods. Many of the arguments have focused on survival during famines as an evolutionary force behind what are now obesity-prone traits among humans (reviewed in Speakman, 2007).

We don't dispute these arguments, though in many cases they appear somewhat weak and have been criticized (e.g., Speakman, 2007). On close examination, famine in our past may not have provided a sufficiently strong selective force to favor an obesity-prone genotype (Speakman, 2006, 2007). However, evolutionary success depends on reproductive success, which includes more than survival. We argue that the effects of even milder (and probably quite common) conditions of food insecurity in our past would have had significant consequences on female fertility and reproductive success and led to an adaptive advantage for genes that enabled females to store body fat in readily metabolizable depots.

Fat and reproduction are intimately linked in women. Leptin, the molecule of "fat homeostasis," has direct effects on female fertility and fetal growth and development. There appears to be an asymmetry between the effects of low leptin (strong effects that decrease reproduction) and high leptin (no known reproductive consequences, though maternal obesity can have negative metabolic consequences for both mother and child, and likely reduces reproductive success), which implies that evolution would have favored (or at least not selected against) genes that led to a propensity to positive energy balance when environmental conditions permitted.

The female pattern of adiposity, with predominantly lower-body, subcutaneous adipose stores, appears to be a healthier pattern than the male pattern of more visceral fat. It is associated with fewer comorbidities and is less likely to adversely affect health. The liver will be exposed to lower concentrations of FFA than for equivalent amounts of visceral fat. The metabolic costs of storing fat are lower and the advantages for reproduction, at least in the past, significant. The costs of female reproduction would provide a potent adaptive force driving adipose tissue metabolism in women. In addition, the ability to produce "fatter" female infants and thus fatter children (within the context of our past, not at the level of fatness today) may have had reproductive benefits in terms of earlier age at menarche and more resilient ovarian function for these offspring as adults, which would result in an increased total reproductive life span (see chapter 13 for a more complete discussion). The advent of our larger brain may have selected for fatter neonates to support the higher energy costs of that larger brain. Fat babies also may have been more resistant to mortality from infectious disease. Fat appears to be very important to our reproductive success.

Fat affects reproduction. A certain level of fatness appears necessary for successful female reproduction and infant health. We are not arguing that obesity in women was adaptive. Indeed, maternal obesity is associated with a number of reproductive problems. The effects of obesity on reproduction suggest that obesity would lower reproductive success in both men and women. Obesity is not adaptive; however, it may result from an exaggeration of formerly adaptive characteristics that are over- or inappropriately expressed in the modern environment. Our evolutionary history appears to have selected for both higher than usual (among primates) maternal and infant fatness. A sexual dimorphism in adiposity is understandable given the potential benefits to sustaining reproduction (e.g., fertility, lactation, age at menarche) in women and a lack of such adaptive pressures in men. Indeed, excess body fat in men is associated with erectile dysfunction, lower fertility, and decreased fatty acid availability and oxidation during endurance exercise. Storing large amounts of adipose tissue would not have been a reproductive or survival advantage to our hunter-gatherer forefathers. It may have been critical for our foremothers and their babies. Unfortunately modern conditions allow an exaggeration of that adaptation, and too much of that formerly good thing results in morbidity and mortality.

Men are more susceptible to central adiposity. Central adipose tissue deposits are more resistant to mobilization. There would appear to be little adaptive advantage to storing visceral fat. We suggest that the pattern of central obesity, more commonly seen in men, and associated with greater comorbidity, reflects the genetic drift hypothesis of human susceptibility to obesity. Under conditions common in our past, few individuals would have been able to remain in positive energy balance for long enough for significant visceral adipose tissue to accumulate.

The differing fat storage patterns between men and women and the metabolic differences in how they meet sustained energy demands reflect their asymmetrical costs of reproduction. In the past, fat was more important to the reproductive success of women. The female pattern of excess adiposity in the lower extremities reflects an exaggeration of an adaptation for female reproductive success. The modern environment allows the adaptive pattern to go beyond its evolved function and into pathology.

CHAPTER THIRTEEN

Genetic and Epigenetic Correlates of Obesity

....

The epidemic of obesity has occurred too rapidly for it to represent genetic change in the population. However, the fact that under the same environmental and social conditions there are lean as well as obese people implies that there are intrinsic, and thus likely genetic, differences among people that either protect them from or predispose them to obesity. A concerted, ongoing effort to identify genetic correlates of obesity has produced interesting, if sometimes confusing, results.

Adoption studies (e.g., Stunkard et al., 1986) and twin studies (e.g., Stunkard et al., 1990) have demonstrated that obesity does have a genetic underpinning; studies generally estimate the heritability of BMI (body mass index), adiposity, and fat distribution at about 60%. Adopted children are more like their biological parents and not their adoptive parents in adiposity (Stunkard et al., 1986). Dizygotic twins are no more similar to each other than are nontwin siblings; monozygotic twins are more likely to have similar body weight (Stunkard et al., 1990).

Although obesity has a significant heritable component, the genetic basis is poorly defined. Single-gene polymorphisms linked with obesity are known in people and in animal models. The prevalence of the known single-gene mutations that cause obesity in humans is very low. There certainly are mutations that cause leptin deficiency, and these not surprisingly are associated with a tendency to overeat (Roth et al., 2004a, b). Monogenic mutations that cause malfunction of the melanocortin 4 receptor are associated with binge eating (Farooqi et al., 2003; Branson et al., 2003; List and Habener, 2003). The study of these mutations increases our understanding of the ways in which people might become obese and is of help in treating the rare individuals who suffer from these disorders;

however, most human obesity does not appear to fit a monogenic model. In general, human obesity arises from the interaction of multiple genes with the environment over a significant amount of time (Roth et al., 2004a, b).

In addition, the consequences of obesity are not the same for all people in all places. There are cultural and social differences in the perception and acceptance of obesity. People vary in their susceptibility to the diseases associated with obesity as well. There are significant racial (and ethnic) differences in the associations between disease risks and obesity. Fat distribution and BMI have different associations among racial groups. Europeans appear to differ from Asians and sub-Saharan Africans in the risk of metabolic syndrome conditions by BMI (Abate and Chandalia, 2003; Yajnik, 2004). The consequences of being obese depend on a complex set of environmental and genetic factors.

Although a genetic propensity to obesity could be thought maladaptive in the modern, developed world with easy and reliable access to plentiful food, it is unclear what, if any, adaptive consequences polymorphisms that affected the development, regulation, and metabolism of fat stores would have had in our past. There appears to be substantial variation among people in the propensity to deposit fat and where that fat gets deposited. Some of this variation is associated with geographic regions of origin, and may reflect genetic differences. For example, people from the Indian subcontinent appear to have higher adiposity at any particular body mass index than do Caucasians or sub-Saharan Africans, and a greater propensity to central adiposity (Yajnik, 2004). We hypothesize that many polymorphisms among human beings that make them susceptible to obesity and the negative health consequences of excess weight in the modern milieu may have been selectively invisible, or even possibly favored, in our past. The advantages of storing energy obtained from episodic conditions of plentiful food outweighed the long-term health consequences associated with the rare possibility of becoming obese. As a species, we favored a fat-storing phenotype in our past. Under the modern milieu, these thrifty genes result in less than optimal health.

In this chapter we examine aspects of the relationship between genetics and obesity. We examine racial differences in fat deposition and metabolism; the case of Pima Indians is an example of genetic and environment interactions in obesity and diabetes. There is some support for the hypothesis that people tend to choose people with body builds like their own as spouses; assortative mating may be affecting the distribution of fat and lean people. Finally, we suggest a speculative hypothesis regarding a latitudinal cline in the ability to up-regulate fat oxidation in response to dietary fat.

The Old Genetics

Genetics and evolution are the guiding principles for understanding living organisms. The basic discoveries of genetic material occurred in the mid- to late 1800s, but their importance wasn't recognized at that time. In 1865 Mendel published his work on the inheritance of characteristics in pea plants. In 1869 Johann Miescher, a Swiss biologist, isolated phosphate-rich molecules from the nuclei of white blood cells; he called them nuclein. Phoebus Levene, born in that same year, spent most of his science career studying the structure of sugars; he was the first to discover that Miescher's "nuclein" were composed of adenine, guanine, thymine, cytosine, deoxyribose, and a phosphate group. Nuclein became deoxyribonucleic acid (DNA). Modern genetics began in the early 1900s with the rediscovery of Gregor Mendel's work. At that time proteins were thought to be the material that carried genetic inheritance. In 1944 Oswald Avery, with colleagues Colin MacLeod and Maclyn McCarty, demonstrated that DNA, not proteins, was the material of genetic inheritance. This was confirmed in 1952 by Alfred Hershey and Martha Chase in the Hershey-Chase experiments that used radioactive labeling to demonstrate that DNA was the genetic material for bacterial phages (viruses that infect bacteria). The following year the structure of DNA was determined by James Watson and Francis Crick (Watson and Crick, 1953) based on the x-ray diffraction photography of Rosalind Franklin (Franklin and Gosling, 1953) and Maurice Wilkins (Wilkins et al., 1953).

It took roughly 100 years from the discovery of the basic laws of genetic inheritance and the fortuitous discovery of the genetic material to finally determine the biochemical structure of that material: DNA. But before that happened the mathematics of genetics had already been greatly advanced by Ronald Fisher, Sewall Wright, and J. B. S. Haldane. These three essentially invented population genetics and set the stage for the modern synthesis of evolution and genetics. The concept of a gene didn't need an actual structure and mechanism to be useful. The key importance of the mathematical work of these statistician-biologists was that formal models could be constructed to show what natural selection could and could not do, and in the case of Sewall Wright, additionally what random variation and genetic drift could and could not do. The work of these and other early population geneticists reconciled the theory of Darwin with Mendel's laws of inheritance. The discrete inheritance of characteristics and the Darwinian notion of gradual change were not incompatible.

In the modern synthesis a gene is a segment of DNA that codes for a gene product. DNA is transcribed to RNA (ribonucleic acid), and the RNA is transcribed into a sequence of amino acids, forming a peptide. There are regulatory elements in the DNA that determine the beginning and end of a gene; there are even elements that determine when a gene is active and when it is turned off. It is an elegant system; perhaps too simple and elegant to truly describe how the incredibly complex and often messy biological processes of life worked. But it was an incredibly important start.

The New Genetics

Fast-forward 50 years and the science of genetics has become marvelously full of additional complexity. The notion of a gene being a region of DNA that codes for a particular product is now known not to be strictly true. Of course in many instances the original concept still holds. However, we now know that a gene may produce more than one product; the same region of DNA may be part of more than one gene; a stop codon (a DNA sequence that stops RNA transcription) in the middle of a gene may not always stop transcription; and post-translation effects are often key for the function of many gene products. As an example of the last phenomenon, the gene products OB-R and sOB-R, the long and short form of the leptin receptor, are produced by the same gene, but a post-translational cleavage of OB-R produces sOB-R. The two molecules have very different functions: OB-R is the signaling receptor for leptin while sOB-R acts as a binding protein.

The human genome appears to contain 20,000 to 25,000 genes; far fewer than was previously estimated. However, the post-translational modification of gene products increases the number of functional molecules produced by the genome. In many cases a gene does not create a functional protein but rather a preproprotein that can, depending on the post-translational cleavage, become one of several proteins. In addition, other post-translational events may regulate the function of the peptide.

For example, ghrelin, the important appetite-enhancing peptide, is the product of the preproghrelin gene that codes for a 117-amino-acid peptide. Ghrelin itself has both 27- and 28-amino-acid-length forms. In humans the 28-amino-acid-length form predominates; in rats the 27-amino-acid-length form is more common. Both forms have biological activity (reviewed in Gil-Campos et al., 2006). In addition, an alternative cleavage of the preproghrelin gene produces the peptide obestatin, which may have opposite effects on appetite than does ghrelin (Zhang et al., 2005), though that is not certain (Gourcerol et al., 2007). Finally, for ghrelin to exert its appetite-increasing effects via NPY neurons in the arcuate nucleus (and possibly other neurons and brain regions as well), it must be acylated with a medium chain fatty acid at the serine-3 position. Nonacylated ghrelin appears to have peripheral physiological effects, but it does not affect appetite. Thus an up-regulation of the preproghrelin gene can result in a variety of physiological and metabolic consequences, of which an increase in appetite is only one possibility. And it is possible that increased synthesis of acylated ghrelin can occur without any changes in preproghrelin transcription, via regulation of the cleavage of preproghrelin or the acylation of ghrelin.

There are other fascinating mechanisms by which the creation and function of gene products are regulated. Some genes have a stop codon within the DNA sequence. Under certain conditions, that halts transcription of the mRNA, the mRNA is then degraded, and its components reused. Thus no peptide will be formed. However, under other conditions the stop codon is ignored, transcription continues, and a peptide is produced.

There are also micro RNAs (miRNAs), 20- to 22-nucleotide RNAs that associate with certain proteins (e.g., the Argonaut proteins, Ago1 to Ago4) and bind to mRNA and regulate mRNA translation and degradation (Buchan and Parker, 2007). A particular miRNA-Ago complex will bind to a specific mRNA at complementary sequences within the 3'-

untranslated region (3'UTR). The 3'UTRs are sites where complexes important to mRNA translation and degradation are assembled. In proliferating cells the effect of these miRNA-protein complexes appears generally to be to inhibit mRNA translation and to promote mRNA degradation (Buchan and Parker, 2007). However, during cell cycle arrest these complexes can act to promote mRNA translation (Vasudevan et al., 2007). Thus the regulatory effect of miRNA-protein complexes may change from inhibition to activation during the cell cycle (Vasudevan et al., 2007).

There are enhancer regions of DNA; so called because they function to up- or down-regulate gene transcription. These regions are often found close to, even within, noncoding segments of a gene, but they can also be thousands of nucleotides away from the gene they regulate. Transcription factors bind to these noncoding segments of DNA and regulate the extent of gene transcription.

The concepts underlying genetics and metabolism are in effect converging. Regulation has become a key word in genetics. The transcription of DNA to mRNA is regulated, as is the translation of the mRNA into an amino acid sequence. In many cases gene products are not functional until they have been metabolized. Peptides often require acylation, methylation, glycosylation, or phosphorylation before they become functional. Some peptides function only in complexes; for example, adiponectin forms high-molecular-weight complexes with itself. It is these complexes that have the highest biological activity. There is a great deal of genetic metabolism.

Scientists have known for quite a while that there are genes that regulate other genes. It has become clear that these may be the most important. The amount of DNA that actually codes for peptides that are then used in cellular or whole-body metabolism is a small fraction of the total. Only a few percent of human DNA actually codes for a protein. Many of these genes are highly conserved as well. For example, mice and humans are quite different animals; however, 99% of human protein-producing genes have mouse homologs. In essence mice and humans are built out of the same basic material.

Evolution appears to have acted more on regulatory DNA sequences than on those that encode a product. It has long been known that most of the genome is noncoding. Much of this DNA has been termed *junk DNA*, and it has been assumed to have little or no function. That view is now changing. For example, the length of repetitive DNA sequences often appears to have effects on gene transcription even though the repetitive sequences are not transcribed.

Selection can act on any level of gene function. It can act on the structure of a produced peptide; the regulatory elements that determine when and how frequently a gene is transcribed; and the post-translational metabolism a gene product undergoes. We are learning that, in some ways, the genetic machinery is as complex, interactive, and regulatory as the resulting organism. Another way of looking at it is that there is metabolism at all levels of biological organization, from the whole organism, to the cell, and finally down to the genome.

Single Nucleotide Polymorphisms

Many kinds of mutations can occur to change the DNA; one of the simplest to conceive of is a single change in a nucleotide base pair. In other words, in the codon sequence one of the base pairs is replaced with another to form a single nucleotide polymorphism (SNP, pronounced "snip") in the population. For example, in a gene, the triplet CTC becomes CAC. Sometimes the change in codon sequence will change the gene product. In the example above, CTC codes for the amino acid glutamic acid while CAC codes for valine. Thus, if this SNP exists in a transcribed DNA sequence and that mRNA is translated into a peptide, there will be a protein polymorphism in the population with two different amino acid sequences. In the hemoglobin molecule this SNP is part of the mutation that causes sickle-cell anemia.

Of course there are other SNPs that would not change the amino acid. For example, CTC and CTT both code for glutamic acid. Previously these SNPs were considered to be silent mutations; they did not affect the gene product because they did not change the amino acid sequence. New evidence has shown, however, that even though two triplets code for the same amino acid they may still affect protein synthesis. They may affect the rate at which the protein is assembled, which may have effects on protein folding. Protein folding can be very important to protein function. We now know that even so-called silent mutations may have functional and adaptive significance.

There is a concerted effort under way to locate SNPs that are associated with obesity. The results have been interesting, though often confusing and unclear. For example, a variety of SNPs have been found for the preproghrelin gene, but none of them have been reliably shown to be associated with obesity or the metabolic syndrome (Gil-Campos et al., 2006). Again, although certain rare single-gene mutations can cause a vulnerability to obesity, most human obesity probably stems from the interactions of multiple genes with the environment.

The more we know, the more biological complexity we uncover. And that complexity enables this elegant genetic system to produce the amazing intricacy and diversity of living creatures, including variability in how vulnerable people are to obesity and its associated diseases.

In Utero Programming of Physiology

We know that we pass on traits and characteristics to our children. We do this through genetics, passing on biological attributes, and through socialization, passing on our culture and ideals. There is now convincing evidence that we influence our children's adult metabolic characteristics by their early nutrition and that it starts before birth. The fetal and early-life environment, especially the nutritional environment, appears to have profound effects on the future risks of obesity, type 2 diabetes, and cardiovascular disease.

The theory is that physiology and metabolism become programmed at some point in life in response to the nutritional state of that time. The concept of critical time periods for physiological development is well established. Many aspects of morphology, physiology, and even behavior can be modified by interventions at critical times but are unaffected if the interventions occur outside of that critical period. For example, the sex steroid hormones are known to induce long-term changes in brain structure during critical developmental stages that are necessary for competent sexual behavior (Goy and McEwen, 1980). Male rats deprived of testosterone during this critical period of development (by gonadectomy) fail to respond to testosterone as adults; however, if the gonadectomy occurs after the critical time period the rats respond to exogenous testosterone. Female rats exposed to testosterone during this critical window of development become masculinized (reviewed in Goy and McEwen, 1980). Early-life experience can have fundamental influences on long-term physiology and behavior.

Poverty, Nutrition, and Heart Disease

In the 1970s Anders Forsdahl investigated the wide variation in mortality rates due to heart disease throughout counties in Norway. This variation could not be explained by variation in current living standard, but it did appear to track differences in living standards when the adults were born (Forsdahl, 1977). Specifically, heart disease rates in adults, both men and women, were correlated with infant mortality rates from the adults' childhood (Forsdahl, 1977). Forsdahl hypothesized that poverty in early life led to a vulnerability to heart disease in later life, but only if living conditions improved. It was the mismatch between early life and later life that created increased risk. In 1986 Barker and Osmond published data from England and Wales that confirmed Forsdahl's earlier observations: infant mortality from 1921 to 1925 was strongly, positively correlated with ischemic heart disease mortality rates from 1968 to 1978 (Barker and Osmond, 1986).

Barker went on to show that birth weight and weight at 1 year of age were highly predictive of heart disease risk. Individuals at the lower end of weight at these time points were much more likely to die from cardiovascular disease (Barker, 1997). The hypothesis was now refined from poverty during early life to poor nutrition during early life increasing the risk of heart disease, but only under conditions of improved nutrition in later life. Populations in economically disadvantaged countries where nutrition was poor all through life did not have high rates of mortality from heart disease; mortality was high from many other diseases and causes perhaps, but heart disease appears to be a disease of plenty, not want. The evidence of Forsdahl and then Barker indicated that a change from a low-nutrition state when young to a high-nutrition state when older conferred the highest risk.

The original Barker hypothesis links poor early-life nutrition with growth and development of tissue and organs that in essence program a vulnerability to heart disease (Barker, 1994). Barker has continued to expand on the concept of early-life programming of physiology and metabolism to influence future disease risk. Barker and colleagues have shown a link between pubertal growth in girls and later breast cancer risk in their daughters (Barker et al., 2007). Many aspects of metabolism and disease risk appear to have their origins, or at least susceptibility, in fetal and early life.

Epigenetic Factors

There are key developmental periods in any organism's life when environmental factors can exert profound effects on eventual morphology, physiology, and metabolism. For instance, the sex of a hatchling crocodile is not determined by its genetics but rather by the temperature at which the egg was incubated during the key period of sexual development. There may be epigenetic and genetic imprinting correlates with obesity. Both in utero and in early life the regulation of genes in critical time periods can have lifelong consequences. Some proportion of the variation among people may reflect differential regulation of genes during development caused by both genetic and environmental factors. There is evidence that variation in the expression of developmental genes among individuals is associated with the variation in their adiposity, both total and the distribution on the body (Gesta et al., 2006).

The concept of epigenetics precedes our modern understanding of genetic mechanisms. C. H. Waddington (1942) introduced the term *epigenetics,* combining embryology with genetics, to describe how genotype interacting with environment gives rise to phenotype. At that time genes were theoretical constructs without actual physical identity. The modern, narrower definition of epigenetics refers primarily to mechanisms that silence genes (e.g., DNA methylation) without changing the underlying DNA structure (Crews and McLachlan, 2006). Holliday (1990) defines epigenetics as "the study of the mechanisms of temporal and spatial control of gene activity during the development of complex organisms." Development depends on communication between cells; the environment of a cell is critical to its properties in the developing organism (Holliday, 2006).

The concept is perhaps most easily understood in the silencing or activation of genes in different tissues. All cells in our body have the same inherent genetic potential. Heart cells do not differ from kidney, brain, or gonad cells in terms of DNA gene sequencing. Instead, gene activation differs between cell types. Some genes that are expressed in heart are silenced in gonads and vice versa. One mechanism is via DNA methylation, which generally acts to silence a gene. Genes with methylated promoter regions are inactivated (Holliday, 2006).

Epigenetic change is stable over cell division and can even be passed from generation to generation. For example, an epigenetic mutation in the plant *Linaria vulgaris* changes the symmetry of the flowers from bilateral to radial. This mutant was first described by Linnaeus more than 250 years ago. It fascinated biologists, including Darwin. The mutation is passed on to future generations in a recessive manner. Darwin actually showed this in pyloric snapdragons, but being unaware of Mendel's work, he did not arrive at the correct conclusion. It has now been shown that the mutation results from methylation of the *Lcyc* gene, thus silencing it (Cubas et al., 1999). The original genetic potential has not been lost in these mutants; flowers will occasionally revert to wild type through re-methylation of *Lcyc*.

Information molecules act to channel physiology within the limits of phylogeny. Human beings will never lay down blubber layers like marine mammals; despite the aquatic ape theory there is no evidence that our adipose tissue provides an effective insulation function (Kuzawa, 1998). The extent and distribution of fat stores is significantly variable among people, however. In addition to genetic factors, risk factors for obesity, especially central obesity, include birth weight, maternal BMI, and weight gain during pregnancy. Both small- and large-for-gestational-age babies are at higher risk of becoming obese, and more importantly, of developing obesity-related diseases such as type 2 diabetes (e.g., Yajnik, 2004).

Maternal characteristics that affect the uterine environment, such as maternal nutrition, BMI, weight gain, and glucose regulation, will affect physiology and metabolism of the mother's offspring. This provides a mechanism for a form of "inheritance of acquired characteristics" from mother to child. The mother's weight disorders or her dysregulation of glucose metabolism will affect the development of her offspring's physiology in utero. The infant will be born with a propensity for particular metabolic and physiologic responses to food intake. The interaction of this in utero programmed physiology with the childhood environment will then further determine the eventual physiology, metabolism, and health of the individual. A mother may pass on her own weight disorders to her offspring, leading to the specter of obese mothers passing on to their daughters characteristics that will increase the likelihood that those daughters will be obese and will subsequently have offspring who also are at higher risk of obesity and its associated diseases.

The original conception of in utero programming of physiology involved low-birth-weight and small-for-gestational-age infants; these infants were at higher risk for obesity, dyslipidemia, poor glucose control, and cardiovascular disease, the so-called metabolic syndrome (Barker 1991, 1998; Barker et al., 1993). There was an even higher risk for those small-for-gestational-age babies that experienced more rapid growth during childhood, so-called catch-up growth. The concept of the "thrifty" phenotype was proposed. Simply put, the maternal-placental-fetal axis provides a conduit of information about the external environment to the developing fetus. Conditions that lead to restricted growth in the fetus, such as maternal malnutrition, result in channeling the physiology of the fetus to a state appropriate for conditions of low energy intake. This is proposed as an evolved, adaptive response. At the simplest, it allows the fetus to survive an adverse uterine environment until birth. In addition, it was proposed that the infant has been physiologically adapted to survive in an environment of food scarcity. However, after birth many of these infants instead are exposed to relative food abundance. There is a mismatch between the programmed physiology and later energy intake.

The epidemiological evidence now shows that either end of the birth weight distribution has increased risks of excessive weight gain, eventual obesity, and the metabolic syndrome. After a certain point, bigger is not better for either moms or babies. The proposed mechanisms for the programming of fetal physiology associated with maternal obesity include poor maternal glucose regulation and maternal dysregulation of cortisol metabolism. This condition is more one of pathology as opposed to an inappropriate adaptive strategy. For example, high levels of glucose can have toxic effects. At the least, the premature and substantial up-regulation of insulin secretion by the fetal pancreas appears to lead in many cases to an eventual exhaustion of this organ later in life, an example of a phenomenon termed *allostatic overload* (McEwen, 2005).

Another epigenetic mechanism that may be influencing humanity's vulnerability to weight gain concerns environmental toxins. Humans are now producing and releasing into the environment a large number of chemicals, many of which have potential biological activity. Many of these molecules have been termed *endocrine-disrupting chemicals* because of their ability to either mimic or block the actions of naturally occurring hormones; often these chemicals have estrogen-like effects (Crews and McLachlan, 2006). Many of these chemicals may influence metabolism, appetite, and other factors relevant to obesity. There are many examples of toxicity tests of chemicals in animals in which nontoxic doses of the chemical (and thus exposures deemed safe) result in weight gain (Baillie-Hamilton, 2002).

The Thrifty Phenotype

A refinement of the basic Barker hypothesis links the programmed changes in physiology to predictive adaptation. The thrifty phenotype hypothesis posits that a fetus has many possible metabolic and physiological paths it can follow. The end result is not genetically fixed; it is guided by genetics but influenced by environment. The hypothesis proposes that under most circumstances (at least in our past) current conditions were the best predictor of future conditions. Thus the early-life nutritional environment was the best indicator of the adult nutritional environment. Clues about the early-life nutritional environment come from the mother, through the placenta and the uterine environment before birth, and through milk after birth. Poor early-life nutrition results in a programmed physiology and metabolism that is geared to poor nutrition. When the nutritional environment dramatically improves, the programmed metabolism may not properly match the new nutritional plane. The thrifty phenotype will be susceptible to sustained weight gain and eventual obesity if living in circumstances of plenty.

The Barker hypothesis of an infant with a thrifty phenotype, programmed by a poor fetal environment due to maternal undernutrition or other causes of intrauterine growth restriction (IUGR), born into a world of plenty provides a reasonable hypothesis to explain the increasing prevalence of obesity in developing countries. As hunger and poor nutrition become a thing of the past and the nutrition transition exposes the population to higher-calorie foods, the mismatch between the fetal and early-life nutritional experience and the adult experience would lead to a vulnerability to sustained weight gain.

However, the thrifty phenotype hypothesis does not explain why obesity rates in economically advantaged countries continue to climb. The concept of in utero programming of physiology still appears to be operative, however, as the vulnerability to obesity appears to increase at both ends of the birth weight spectrum. Large as well as small babies appear vulnerable to obesity. Indeed, in general both large- and small-for-gestational-age infants share an increased fat mass relative to lean body mass at birth (Kunz and King, 2007). This is also true of babies born of diabetic mothers, even if birth weight is within the normal range (Catalano et al., 2003). A risk factor for adult obesity and the associated diseases is a high proportion of fat mass relative to lean mass at birth, regardless of birth weight (Kunz and King, 2007).

Mechanisms for in Utero Programming

There are many possible mechanisms by which events in utero could alter development of the fetus in ways that could have lifelong consequences. Maternal condition and nutrition can affect placentation and early embryo development. During organogenesis, environmental conditions might affect the number and function of cell types in the heart, kidney, pancreas, and other organs. Later in pregnancy, the uterine environment could influence various regulatory set points of fetal metabolism. Even after birth, early nutrition and other circumstances are likely to influence the growth and development of organ systems and metabolic systems. The effects of the environment on the fetal and infant physiology could be transduced through alteration of gene function (epigenetics) or by altering the molecular and metabolic environment in ways that change cell growth and differentiation.

Nutrition can affect gene expression. Recent studies have shown that food is more than a source of metabolic fuel and raw materials for tissue synthesis. Food components can act as signaling molecules. Food can have genetic effects. Dietary supplementation of pregnant rats with folic acid, vitamin B12, choline, and betaine changed coat color in the offspring via a flipping of a transposon in the *agouti* gene (Waterland and Jirtle, 2003). Presumably this effect was related to DNA methylation, as these nutrients are significant methyl donors.

Critique of the Thrifty Phenotype Hypothesis

Some researchers have criticized the concept of environmental programming as a mechanism, though they accept that different in utero and early-life conditions result in different physiological expressions. There are certainly some weaknesses in the theory as proposed; however, the fundamental concept that circumstances during gestation can strongly influence metabolism and physiology later in life has now been well documented.

Whether there really are thrifty genes or thrifty phenotypes is still a question (Speakman, 2006). No good candidates have yet been found. Speakman (2006) questions whether episodic famines would actually provide sufficient and appropriate selective pressure to favor fat deposition. The focus of his analysis is largely on adult adiposity. In the past, subtle variation in the propensity to store fat in different depots may have had little if any adaptive significance.

The hypothesis that in utero conditions serve as a predictor of postnatal conditions and that the thrifty genotype and phenotype were selected via this predictive mechanism does push the adaptationist perspective to an extreme. Conditions in most environments are unlikely to remain stable for such an extended period of time. Of course it is possible that the selective pressure was episodic sustained periods of food scarcity. However, the hypothesis that the in utero programming of physiology serves to prepare the neonate and eventual adult for conditions of scarcity is not required for the basic concept that in utero conditions can set physiological and metabolic systems on a particular path. All that is required is that conditions in utero (e.g., conditions conducive to intrauterine growth restriction) are met by changes in anatomy and physiology (e.g., nephron number, pancreatic β cell function) that will later determine the metabolic response to whatever conditions the animal meets in later life.

It is a truism that before an adult can be reproductively successful that adult must have successfully passed through all the nonreproductive stages of growth and development that come before. You must have been a successful subadult, juvenile, infant, neonate, and initially a successful fetus in order to become a successful reproductive adult. If the metabolic adjustments necessary for a fetus in a nutrient-poor, growth-restricted uterine environment to be a successful neonate result in a propensity to obesity or other disease later in life, that is still adaptive compared with failing to be born.

Human neonates are remarkably fat. This high level of body fat at birth may have served an important adaptive function relative to supporting the dramatic and unique postnatal brain growth in humans. It is interesting to note that IUGR babies are still relatively fat compared with nonhuman primate infants. This suggests that defense of neonatal adiposity was adaptive. Human fetal metabolism appears to strive to attain a level of adiposity by birth. Thus the metabolic changes that encourage fat deposition do not have to be predictive of scarcity after birth; they are just the response needed to produce an appropriately fat neonate. Unfortunately it appears that those metabolic adaptations cannot be undone. When the child grows up under circumstances of plenty there is a mismatch between the physiology necessary to produce a fat baby and the nutrient conditions of later life, resulting in a vulnerability to excess adipose tissue accumulation.

Under conditions of excessive fetal nutrition (e.g., gestational diabetes), human infants tend to be large (macrosomia). The increase in neonatal mass is disproportionately skewed to the fat compartment. Macrosomic infants do have greater fat-free mass as well, but most of the weight difference can be accounted for by greater fat mass. Again, our evolved fetal physiology appears to favor fat deposition. Under conditions of poor nutrient flow to the fetus a bias toward fat deposition probably has adaptive function. It is likely that low-birth-weight babies were far more common than macrosomic babies for most of our past. When fetal nutrition exceeds the requirements for normal growth, much of that energy is shuttled into fat deposition. This appears to have long-lasting effects on physiology and metabolism. We propose that one aspect of our biological propensity to gain weight in the modern environment stems from the selective pressure on fetal and maternal metabolism to favor fat babies.

Not all people react to the modern obesogenic environment in the same way. Nor do all people suffer the same health effects of obesity. There is considerable variation in human response to our obesogenic environment and to obesity itself. Much of that variation is not randomly distributed; there are familial and group associations. Epidemiological evidence indicates there are racial and ethnic differences in the vulnerability to obesity and to its associated diseases. For example, Native Americans, Hispanics, and African Americans appear to be more vulnerable to obesity and to type 2 diabetes than are European Americans (Abate and Chandalia, 2003).

Complicating a discussion of race, ethnicity, and genetics is the fact that race and ethnicity do not have universally agreed-upon definitions (Collins, 2004; Keita et al., 2004). The concept of race can be biologically or socially constructed; there is often little agreement between those who hold one view versus the other. And the categories that are used in most epidemiological research are a mixture of the two concepts.

The evidence from the Human Genome Project and other recent research on human genetic variation demonstrates both the uniqueness of the individual and the commonalities between humans (Royal and Dunston, 2004). Even with more than 6 billion humans alive today, humanity is currently expressing only a fraction of the possible genetic variation inherent in our genome. Thus any newborn is likely to be unique in at least some genetic aspects. At the same time the average genetic difference between randomly selected human beings is fairly low compared with many other species with much lower population numbers. This reflects the fact that modern humans appear to derive from a fairly recent (within the last 100,000 years) small founding population (reviewed in Jorde and Wooding, 2004).

Groups of people can certainly be defined based on geographic origin; until fairly recently (on an evolutionary time scale) most people never traveled very far from their place of birth. People within a geographic region are more likely to have similar genomes, but the genetic variation within these geographic groups generally is larger than the differences between groups (Mountain and Risch, 2004). If a large number of polymorphic markers are used, and we now know of thousands of potential candidates thanks to the staggering amount of new data produced by modern genetic research, the patterns of genetic variation among people from different continents show consistent patterns. African populations are the most genetically diverse; the largest genetic distance is between African and non-African populations; and the root of the genetic tree is closest to the African populations (Jorde and Wooding, 2004). All the evidence is consistent with the hypothesis that the ancestral human population originated in Africa and that a small subset of that African population spread out to occupy the rest of the world approximately 50,000 to 60,000 years ago. Our species is relatively young.

Race and ancestry certainly are related; however ancestry is a more subtle and complex description of an individual (Jorde and Wooding, 2004). For example, when 100 polymorphisms were used to sort 107 sub-Saharan Africans, 67 East Asians, and 81 Europeans, the result was three clusters with 100% accuracy. However, most individuals did not map to their respective cluster with 100% probability. They shared most of their ancestry with their geographic group but had some probability of shared ancestry with the other clusters as well. In addition, when a sample of 263 South Asians was added to the analysis, they did not map into a single cluster, but were spread between the East Asian and European clusters. Some mapped primarily to the East Asian cluster; others, to the European cluster; and many, to spots between the two clusters (Bamshad et al., 2003; Jorde and Wooding, 2004). This is quite reasonable given the known history of migration (and thus gene flow) through the Indian subcontinent from both East Asia and Europe.

Genetic similarities and differences among individuals derive from shared or different ancestry. Race and ethnicity are imperfect markers for ancestry. However, they can still be useful (Jorde and Wooding, 2004; Mountain and Risch, 2004). If, for a particular trait under consideration, a significant amount of variation among people assorts by racial and ethnic categories, then this variation provides useful information to be studied further. To leap to the conclusion that the differences between the groups derives from genetic differences is usually unwarranted (Mountain and Risch, 2004). At present, race and ethnicity are at best weak surrogates for genetic, cultural, environmental, and socioeconomic factors deriving from ancestry.

Fat Deposition and Metabolism

Of special importance is the greater tendency of men to store fat in the visceral compartment. This appears to be true for all human racial groups; within race a greater proportion of total fat in men is stored in visceral adipose tissue (e.g., Park et al., 2001; Sumner et al., 2002). However there are also difference between the races in the distribution of fat for any given percent body fat. African American men have lower amounts of visceral fat than do European American men; African American women have higher amounts of subcutaneous fat than women of other races (Hoffman et al., 2005). A study that examined fat distribution in Asian American men and women compared to European Americans found that the Asian Americans were shorter, weighed less, and had lower mean BMI. Despite the lower BMI the percent body fat was identical between the Asian Americans and the European Americans (Park et al., 2001). In that study the percentage of fat in the visceral depot was not different in the men at about 9 to 10% of total adipose tissue; the Asian American women had a higher percentage of their total fat in the visceral depot than did European American women (5.1 versus 3.4%).

In a sample of older women in the United States (mean age approximately 65 years), Filipino women had more visceral fat despite having lower BMI than the African American women and no difference in BMI between them and the white women (Araneta and Barrett-Conner, 2005). The Filipino women also had the highest rates of type 2 diabetes. The African American women had the greatest amount of subcutaneous fat and the lowest ratio of visceral to subcutaneous fat.

People of Asian descent appear to be more vulnerable to type 2 diabetes under a Western diet (Abate and Chandalia, 2003; Yajnik, 2004). The prevalence of diabetes for people of Asian descent is higher in the United States than in the countries of origin. This is also true for Hispanics and for Native American people living in the United States compared to Mexico. Although Hispanic is an ethnic designation, and can include all races, a large proportion of Hispanics in the United States have Native American ancestry, and Native Americans are descendants of Asian populations that crossed into North America approximately 10,000 years ago. Thus the vulnerability to obesity and type 2 diabetes that Asians, Hispanics, and Native Americans share may have some common genetic underpinnings.

Pima Indians

The prevalence of obesity and type 2 diabetes among Pima Indians living in the United States is extraordinarily high, among the highest in the world (Schulz et al., 2006). The high rates of obesity and type 2 diabetes among Pima Indians predate the general increase in the prevalence of these conditions in the United States (Bennett et al., 1971; Knowler et al., 1990).

Pima Indians appear to be susceptible to obesity and type 2 diabetes, however, that susceptibility only becomes evident under environmental conditions of high energy-density food and low physical activity. A change of lifestyle to a Westernized diet and significantly lower physical activity is implicated in the high rates of obesity and type 2 diabetes in Pima Indians in the United States. Pima Indians living in Mexico were significantly leaner and had negligible rates of type 2 diabetes (Schulz et al., 2006; Figure 13.1). Pima Indians in Mexico, both men and women, were also significantly more physical active than the U.S. Pima Indians (Schulz et al., 2006; Figure 13.2). The Pima Indians appear to be unusually vulnerable to obesity in the modern, Westernized environment.

The pancreatic polypeptide (PP) family may influence vulnerability to obesity and diabetes in Pima Indians. Variation in both fasting and postprandial circulating pancreatic polypeptide are associated with vulnerability to adiposity, but the two measures were opposite in direction. Fasting PP was positively associated with an increase in waist circumference, and the postprandial PP response was negatively associated (Koska et al., 2004). Single nucleotide polymorphisms (SNPs) in peptide YY (PYY) and the Y2 receptor (Y2R) were associated with obesity, though only in male Pima Indians (Ma et al., 2005).

Polymorphisms in other genes have been implicated in variation in the vulnerability to obesity among Pima Indians, including a missense substitution in the fatty acid synthase gene that lowers carbohydrate oxidation and is protective against obesity (Kovacs et al., 2004) and several

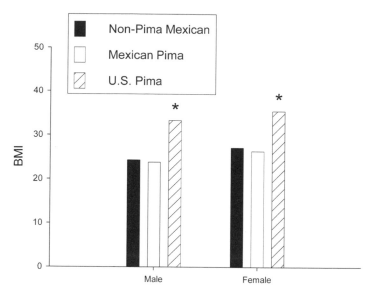

FIGURE 13.1.   Men and women of Pima and non-Pima ancestry living in Mexico had essentially the same BMI. All had lower mean BMI than Pima Indians living in the United States (* = p < .05). Data from Schulz et al., 2006.

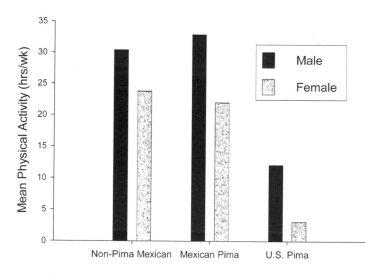

FIGURE 13.2.   Physical activity for Pima Indians living in Mexico was about the same as for non-Pima Mexicans living in the same community. Both groups were far more physically active than were Pima Indians living in the United States. Data from Schulz et al., 2006.

polymorphisms in the melanocortin 4 receptor gene that appear to contribute to obesity among a small percentage of the Pima Indian population (Ma et al., 2004). In addition, an allele of the *ARHGEF11* gene, with a frequency of 10% in Pima Indians, is associated with type 2 diabetes, and if present in Pima Indians without type 2 diabetes is associated with a nominal insulin resistance (Ma et al., 2007).

Of course the genetic studies only identify certain vulnerabilities to obesity; the obesity seen in the U.S. Pima Indian population stems from the interaction of these genetic factors with the recent shift away from their ancestral lifestyle. The vulnerability to obesity could also represent the result of a positive feedback loop wherein in utero programming of physiology produces offspring with physiology that is geared toward storing fat.

Assortative Mating and the Obesity Epidemic

There is evidence that genetic propensities to obesity in the modern environment can be self-reinforcing due to both genetic and epigenetic processes. The fetal programming of metabolism by maternal condition, an epigenetic mechanism, is in essence a Lamarckian transmission of characteristics. There is evidence for a straightforward genetic transmission of obesity risk as well. BMI has been shown to have a heritable component (Stunkard et al., 1986, 1990). It also appears that there is assortative mating with respect to BMI among people (Hebebrand et al., 2000; Jacobson et al., 2007; Speakman et al., 2007). In other words, people are more likely to marry and have children with individuals who are similar to them in BMI (Hebebrand et al., 2000; Jacobson et al., 2007; Speakman et al., 2007).

It is possible, even likely, that BMI is not the most appropriate measure to investigate this phenomenon. Body shape or form may be a better parameter to examine. Certainly people vary in both parameters, and body morph type is probably highly correlated with BMI. The advantage to hypothesizing body morph type as opposed to BMI as a parameter influencing mating and thus eventual obesity vulnerability is that body morph type is less dependent on socioeconomic status and modern environmental conditions that affect its expression. There is some evidence that this may be true; both men and women with heavy arms, either due to a greater fat mass or a greater lean mass, were more likely to be married to a partner with the same trait (Speakman et al., 2007). These individuals were also more likely to be married to a person with lower-than-average leg fat.

A propensity for people to prefer mates with body morphs of their own type would lead to a bimodal or multimodal distribution to obesity risk under modern conditions. It would, in part, explain both the existence of obesity-prone and obesity-resistant individuals. Speakman and colleagues (2007) developed a model that demonstrated that assortative mating could, in theory, cause a doubling of obesity prevalence in a few generations. The effect of assortative mating of course will depend strongly on the genetic characteristics that underlie human vulnerability to obesity.

Latitude and Dietary Fat

The ability to use fat as a metabolic fuel differs among people; a significant proportion of that variation appears to be explained by racial and ethnic differences. Generally, Europeans are more able to up-regulate fat oxidation than are other racial groups. For example, African American women have lower metabolic flexibility compared to white American women (Berk et al., 2006). The white women were able to up-regulate fat oxidation in response to a high-fat meal to a greater extent than were the African American women. A very interesting complexity to this general association is that Inuits and other inhabitants of the far northern latitudes differ from both Europeans and Asians on aspects of fat oxidation and the consequences of excess adipose tissue. The incidence of obesity among Inuits has become quite high, based on the standard BMI criteria. However, at each level of BMI, Inuits had lower blood pressure and circulating lipid levels compared to Europeans (Young et al., 2007). This is opposite to the general pattern of people of Asian descent having higher risk factors for the metabolic syndrome at lower BMI than do people of European descent. Thus there appear to be racial aspects to the variation, but perhaps also adaptation to conditions found in high latitudes away from the equator.

This raises a very interesting though highly speculative hypothesis: as you move farther from the equator the prey animal species that are available are more and more likely to carry significant amounts of fat on their bodies. In other words, sub-Saharan African prey species are generally lean. Prey species in northern latitudes, in northern Europe or Asia and especially near the Arctic circle, are higher in fat content. This is certainly true on a seasonal basis. Many species far from the equator put on significant amounts of fat before winter, either in order to hibernate, or simply to buffer the losses of body energy during the winter months when food availability is insufficient to satisfy energy needs. Both mammalian and avian prey would be high-fat just before winter, into early winter, and perhaps beyond. Plant resources would have been lowest in availability to our ancestors living far from the equator during winter months as well. As you move farther from the equator, at the very least seasonally there would be periods with low availability of carbohydrate and relatively high abundance of fat in the food environment. During the summer months the availability of carbohydrate would increase relative to fat.

Thus the ability to be metabolically flexible and switch between predominantly fat versus predominantly carbohydrate metabolism may have been more adaptive to our ancestors that lived in far northern latitudes compared to those that lived nearer the equator. Although fat is important in the diet of all people, perhaps it was more important, and more prevalent, in the portion of humanity that lived at high latitudes. There may be a latitudinal cline in the ability to up-regulate fat oxidation and thus the ability to metabolize a high-fat diet. This might be one aspect of the differences between African American and European American people that contributes to the greater prevalence of obesity among African Americans eating a Western diet; high-fat foods may contribute more to fat accumulation in African Americans.

Summary

The knowledge base for biology is expanding at an incredible rate. Scientists are studying phenomena that hadn't even been identified not too long ago. Nowhere is this more evident then in genetics. We have now sequenced the entire genomes of multiple species, ranging from humans to zebra fish to plant root microbes (e.g., Paulsen et al., 2005). Among nonhuman primates, sequencing of the genomes of the chimpanzee, orangutan, rhesus macaque, and common marmoset have been completed. The genome sequences for baboons, squirrel monkeys, bush babies, and tarsiers are scheduled to be completed within a few years. One outcome from this expanded knowledge base is that our perception of how genetics works is far more complex, and realistic, than the genetics either of the authors learned in graduate school.

Human beings are variable for a wide range of characteristics; this is true for the propensity to accumulate fat, both in quantity and in distribution. In the modern environment many of us, unfortunately, are vulnerable to obesity. At the same time many people remain lean. Genetic variation is certainly one aspect that underlies the variability in human fatness today; however, the genetics is rarely simple. The search for obesity-related SNPs has value but is also likely to turn up many hundreds or even thousands of potential candidates, each of which on its own contributes little to obesity in the general population.

Modern genetic research has shown both an underlying simplicity and a complexity in our genetic heritage. The basic building blocks of life, represented by the approximately 20,000 to 30,000 genes that code for proteins, are remarkably similar across species. The regulation of those genes can be very complex, and it is the regulatory functions of DNA that account for the amazing diversity of form and function found in living things. Epigenetic mechanisms, in which gene expression is changed without changing the structure of the gene, contribute to the diversity of biological expression seen among species, within species, and within different cells in our bodies. The early environment, from maternal nutritional state to concentrations of man-made chemicals, can influence adult physiology and metabolism.

Pima Indians provide an excellent example of genetic vulnerability and of the importance of environmental factors in determining the likelihood of that vulnerability being expressed. Pima Indians that follow a more traditional lifestyle, with extensive physical activity, look very different from their inactive brethren.

Surviving the Perils of Modern Life

··

These are very exciting times for biology; our knowledge is increasing at an exponential rate. New scientific techniques allow glimpses of the very core of life, the intricate molecular mechanisms that enable animals to function. The more we learn, the more we are amazed by the diversity, flexibility, and adaptability of life.

In this book we examine a broad range of biology relevant to human obesity. Our approach is comparative and based on the foundation of evolution. We try to integrate multiple levels of analysis, from the molecular to the organismal and even to the very structure of our communities. We approach the topic from an evolutionary perspective, from a broad viewpoint, such as an examination of diverse functions of ancient information molecules, to specific features of human evolution. We examine biology that is likely common to most mammals or even most vertebrates and some biological features that appear to be unique to humans.

We emphasize a systems approach because we believe it is the most productive path for understanding the biology and behavior of organisms. Molecular techniques provide amazing information, key insights into how life works. But in the end, none of the gut-brain peptides can function without a gut and a brain; indeed they are irrelevant without a liver, kidney, adipose tissue, and all the other organs, structures, and tissues that make up a whole organism. And at the end it is the organism that acts.

We value molecular techniques; we use them in our research. We value in-depth investigations of organs and neural circuits. We value most of all putting these different levels together toward an understanding of whole-body physiology. Our ultimate goal is to put that understanding of human whole-body physiology into a comparative and evolutionary context to understand not only how it works but also how it came into being.

Information Molecules and Evolution

All life has a common beginning; that is evident in the common molecular basis of life. Life requires the ability to transmit, distribute, and manage information, from outside and from inside the organism. We examine the functions of different molecules and how they regulate the physiology, metabolism, and behavior of animals. There are far more of those information molecules than we could possibly discuss in this book, with more being discovered every year. Still, there are a finite number of them, and most appear ancient, or derived from ancient molecules that were probably present in the first multicellular organisms, if not before. Evolution acts on what already exists. It does not design for the future; it is a synthesis of the past.

In addition to learning about the functions of the specific molecules we discuss in this book, the reader, we hope, has integrated what we consider to be a fundamental concept regarding the approach to understanding these molecules. The evolutionary process has resulted in these molecules having diverse functions that can vary with tissue, with developmental stage, and always with the specific circumstances the organism finds itself under. Evolution has, in effect, co-opted these molecules to perform multiple functions beyond their original function, knowledge of which may be lost forever in the changes over hundreds of millions of years. And of course these molecules interact and affect each other's functions. They need to be viewed from a regulatory perspective, as opposed to a structural perspective—a dynamic model with multiple inputs and outputs.

That doesn't mean that investigating a single aspect of a molecule's function is not a productive means of increasing knowledge. Much of science advances in just that way: narrowing the focus and limiting the parameters to gain a clearer look at a phenomenon. However, no molecule or system exists in isolation, and a full understanding of both form and function comes from integrating the reductionist methodology into a more all-inclusive approach.

We give many examples of this in the book; leptin is perhaps the best example relevant to the overall theme of the book. Yes, leptin functions as a satiety signal. It thus has direct relevance to the consideration of obesity. But leptin also functions as a developmental hormone and a reproductive hormone. A key function may be to link body condition to reproductive function. From an evolutionary perspective, leptin's function may have been more to shut off reproduction during lean times than it was to stop our ancestors from eating when food was plentiful. A narrow view of leptin gives some insight but misses the more complete understanding that comes from considering adaptation and evolution.

Obesity and Evolution

A person's size and his vulnerability to sustained weight gain leading to obesity certainly have genetic components. This fact has been shown in countless studies (e.g., Stunkard et al., 1986, 1990). At the same time the rapid and sustained increase in the number of human beings who are overweight and obese does not represent a global genetic shift; our gene pool has not suddenly and dramatically changed. As a species, humanity probably was always this vulnerable to obesity. In the past the obese phenotype was rare; the most likely reason was that external, environmental factors acted to suppress the expression of obesity, even though the susceptibility was there. The modern environment has relaxed many of those constraints and may even have added factors that further predispose many of us to weight gain. The underlying biology that interacts with our modern obesogenic environment contains both adaptive and what were probably selectively neutral traits. One of the conclusions we have arrived at is that there are a large number of paths to obesity in the modern world. Interestingly, many individuals remain lean in this same environment. Human beings appear to be very diverse.

There are three key aspects of human biology that we believe represent adaptive evolutionary changes in our lineage that have led to physiology and metabolism that increase humanity's susceptibility to obesity under modern conditions. These are our large (and metabolically expensive) brains; an increase in neonatal fatness; and an increase in maternal fatness. Both of the latter two adaptations may have arisen due to the selective pressures of producing offspring who will have an extended period of substantial postnatal brain growth (Kuzawa, 1998; Cunnane and Crawford, 2003). In other words, our larger brains have been a selective pressure for increased fatness, at least for certain age and sex categories of humanity, and enabled us to produce an environment where that ability for enhanced fatness can be expressed beyond the possibilities of our past.

The importance of fat, both in our diet and on our bodies, appears to have increased in human beings compared to our nonhuman primate relatives. We suggest that this change in nutritional biology was linked to the seminal evolutionary event in our lineage: our larger brain. Our larger brains provide us with greatly enhanced cognitive abilities with which we take on the world. A larger brain was certainly a successful adaptation; however, it also comes with substantial metabolic and nutrient costs. The larger brains of our ancestors allowed them to increase their foraging efficiency and obtain higher energy foods and higher net total energy, but it might have been what required that increase as well. The brain must be fed. In the past, external constraints limited our ability to consume fat and to metabolically produce fat from excess energy intake. Calories were scarce and they required significant energy expenditure to obtain. Positive energy balance was not easy to achieve, and only occurred episodically. We propose that these circumstances acted to select for an enhanced ability to store fat in our ancestors, at least in the females, and as fetuses and infants.

Although maternal adiposity is a reproductive strategy found in many mammals, the fatness of our babies is unusual (Kuzawa, 1998). We produce babies who rival marine mammals in fatness. This fat does not serve the same purpose as it does in marine mammals; our fat distribution does not appear to act as effective insulation against cold (Kuzawa, 1998). The amount of brown fat on adult humans apparently does vary with cold exposure, with higher levels after prolonged cold exposure (Nedergaard et al., 2007). Brown fat enables nonshivering thermogenesis, an effective way to generate body heat. Still, the amount of brown fat on adult humans is quite limited. Human fatness is not an adaptation to cold.

Producing fat infants is a significant change from the pattern seen in the rest of our nonhuman primate relatives. Although the adaptive function of the increased adipose tissue deposited before birth is uncertain, it is a reasonable hypothesis that it is related in some fashion to the increased postnatal brain growth (e.g., Kuzawa, 1998; Cunnane and Crawford, 2003). The high metabolic expense of the larger brain may have selected for greater body energy stores. In addition, the brain has a need for particular fatty acids, largely supplied by diet, or in the case of neonates, by fat stores laid down in utero from maternal sources. The nutritional value of our milk did not change (Milligan, 2008), so any added needs for brain development were not met by changes in milk. At least part of the extra energetic expense to the infant because of its larger brain apparently was met by the mother during gestation and stored on the infant in the form of fat (Cunnane and Crawford, 2003).

There are other possible adaptive advantages to a fat baby. Extra adiposity can confer an advantage during illness. Diseases often disrupt feeding and digestion (Kuzawa, 1998). Diarrhea and stomach flu come to mind. Large body size increases starvation time. This may have been one advantage to our ancestors to become larger as a species; however, our babies are much smaller than us. Their starvation times would be correspondingly shorter than an adult's. A greater amount of fat on a newborn's body would buffer the baby from short- to medium-term starvation, for example from an intestinal ailment.

Human milk did not change nutritionally to become different from the apes, but it does appear that it changed in terms of its antimicrobial function. Human milk contains the highest concentrations of immunoglobulins, oligosaccharides, and other antimicrobial molecules of all the other milks that have been examined (Goldman et al., 1982; Milligan, 2005, 2008). The inference is that human infants were under high pathogen risk for an extended period of time in our evolution. This provides independent support for the hypothesis that disease risk to infants was a selective factor for the increase in neonatal fat mass in our lineage.

This higher pathogen risk can be seen as a logical outcome of our increase in population density (more disease vectors and exposure); expansion into novel environments as we spread across the globe (novel pathogens); domestication of other animals (more novel pathogens in close proximity and contact); and the inevitable concentration of human waste (with its pathogens and parasites) as we began to occupy the same places for extended periods of time. Indeed, once agriculture was developed, our pathogen load probably skyrocketed. The first fertilizers were probably human and animal wastes (night soil), linking pathogens directly to the food supply. Our successes as a species have resulted in an increased pathogen-risk environment, at least until the relatively recent inventions of sewers and other sanitary techniques.

In many ways we can blame the obesity epidemic on our brains; at least in an evolutionary sense. The biology of building and maintaining that valuable yet expensive organ may explain many of our preferences that put us at risk of obesity now that we have used our technology and culturally transmitted knowledge to increase our access to food and decrease our required exertion to live. Our large brains, which arguably have made us the most successful species of our time, have also allowed us to construct an obesogenic environment. Many of the aspects of that environment represent understandable preferences on our part for energy-dense foods and relief from extreme and sustained exertion. These preferences are probably part of our evolved survival strategies dating to times when food was often scarce, was fraught with risks (such as the risk of becoming food for other creatures), and required substantial energy expenditure to obtain. In addition, our success as a species and the social and cultural systems that arose due to our increased intelligence appear to have put our ancestors under an increased pathogen load. Being able to carry "a little extra" may have been a successful adaptive strategy in a world of uncertain food and increased chances of debilitating intestinal disease. In effect, the adaptation of a larger brain appears to have been accompanied by selection for physiology and behavior that make many of us susceptible to obesity. Our brains have also given us the capability to construct an environment where that susceptibility can be expressed.

Of course our brains give us the capability to understand this about ourselves. We can evaluate the danger that this vulnerability to obesity raises and work to alleviate the problem.

Framing the Debate

Why should we be worried about the increase in human obesity? The legitimate concern about human obesity is due to its effects on health. If there were no health consequences, then human adiposity would be a social, cultural, and esthetic concern, and people could arrive at diverse opinions with few significant consequences. Indeed, the cultural opinions regarding weight and fatness have changed over recent history within the United States; prior to the 1900s being above average in weight was associated with prosperity and was thought to provide a healthy reserve in the event of disease (Cassell, 1995). In the early 1900s moral judgments began to become associated with obesity; obesity was thought by many to represent gluttony and a lack of self-control (Cassell, 1995). The word *stout* went from being a compliment to being mildly derogative. There was social pressure to be thin and for fat people to lose weight.

These opinions were reinforced by two events in the early 1900s. In 1912 a study of insurance policyholders provided evidence that there was a relationship between body weight and health, with higher-than-average weight associated with excess mortality (Cassell, 1995). Fatness had now been linked with causing poor health rather than serving as a healthy reserve. The other event was World War I; rationing efforts in the United States were implemented to ensure adequate war rations for the troops overseas. Posters and slogans implied that to be thin was evidence of patriotism; to be fat implied selfishness (Cassell, 1995).

The motion picture industry and television have contributed to shaping the popular concept of attractive body form: Movie stars, especially the women, are usually thin. Indeed, there is legitimate concern that many women are trying to achieve an unhealthy state of thinness in the pursuit of fashionable beauty. This has led to a serious tension between those who are concerned that obesity is one of the most significant threats to public health today and those who perceive that much of the negative connotations regarding obesity reflect bias and discrimination based on social norms of beauty. Saguy and Riley (2005) have identified two broad camps of thought, labeled "antiobesity" and "fat acceptance," that are engaged in a public struggle to frame the question of obesity. Are the antiobesity efforts a legitimate matter of public health or part of a weight obsession based on fashion and perception?

We are scientists. We put a premium on evidence, large bodies of data gathered through studies based on sound, scientific principles. All evidence can be viewed from different perspectives, of course, which will influence the conclusions that are drawn. There are always assumptions underlying any logical inferences. It is important, especially for scientists, to continually review, challenge, and test our assumptions.

Humanity is changing in body size and shape. On balance, human beings as a species are the fattest they have ever been, and the trend appears likely to continue, at least for awhile. Some data indicate that the prevalence of obesity is plateauing or possibly even declining in the United States. This is true for both adults and children. The reasons for this change are not clear. It may be due to changes in behavior such as increases in physical activity and decreases in food consumption. It is possible that it merely represents a saturation of the vulnerable population.

The primary goal of this book is to explore the possible origins of the modern increase in human fatness. We focus more on understanding human obesity than on to what extent it represents a health problem versus a social and cultural concern. The evidence supports the assertion that it is both. Obese people do face barriers and challenges that affect their day-to-day living; many of those problems are amenable to structural accommodations and do not require weight loss. Health consequences are also directly attributable to obesity, and these need to be studied and addressed. The focus of this book is to try and understand where our vulnerability to obesity came from. We believe that achieving an understanding of the evolutionary biology that underlies human obesity can help frame societal and individual responses.

Obesity, Health, and Lifestyle

It is important to remember, of course, that there are many aspects of our environment that pose health risks; obesity is only one of many. Cigarette smoking, substance abuse, poor nutrition, and lack of exercise can all have serious health consequences independent of adiposity. And there are people with high BMI (body mass index) and high percent body fat who are metabolically healthy.

Can there be healthy obese people? Certainly, at least in the short to medium term; possibly some obese individuals can have good health over a normal, long life. Adipose tissue is metabolically active, however, and a substantial increase in adipose will have endocrine and immune-function consequences. Just as there would be great concern if a person's adrenal or parathyroid gland doubled in size, so too a large increase in adipose tissue may have long-term consequences that will decrease health; it will certainly affect the individual's physiology. It will affect the circulating concentrations of the sex hormones and other steroids, bioactive peptides, cytokines, and immune-function molecules (see chapter 11). This dys-regulation of the internal information molecule milieu may produce a physiological strain (allostatic load) on many organ systems. Obesity has metabolic consequences.

Lifestyle factors can influence the association among BMI and risk factors for elements of the metabolic syndrome. Inuits who migrated to Denmark differed from Inuits living in Greenland in the association of obesity with cardiovascular disease risk; the Inuits living in Denmark more closely resembled Danes of European descent (Jørgensen et al., 2006).

Physical activity is an important covariate for the association between BMI and disease risk. In a very large prospective study of older men and women (ages 50 to 71) both moderate and vigorous physical activity were associated with decreased all-cause mortality (Lietzmann et al., 2007). Physical fitness is an important factor in decreasing mortality in men at most ages (e.g., Lee et al., 1999). A 12-year study in the United States conducted on 2,603 people over the age of 60 showed that physical activity prolongs life regardless of body fat (Sui et al., 2007). Cardiovascular fitness helps adults over 60 live longer, regardless of their body fat. Indeed, obese but fit individuals were at lower risk of death than normal-weight individuals who were not physically fit. Among individuals with low cardiovascular fitness, high BMI (>35 kg/m^2) conferred the highest mortality risk (Sui et al., 2007). Other studies have also consistently found that moderately overweight individuals are not at increased risk of death provided they are physically active (e.g., Gale et al., 2007). The exercise need not be especially strenuous; walking for at least half an hour or engaging in physical but recreational activities like golf, dance classes, or swimming confer health benefits (Sui et al., 2007).

Physical activity appears to be associated with aging on the cellular level, indeed at the level of the chromosome. In a twin study, leukocyte telomere length was found to be positively associated with physical activity during leisure time (Cherkas et al., 2008). The difference was such that active subjects had telomere lengths equal to inactive subjects 10 years younger.

Physical activity is good; cardiovascular fitness is an important determinant of overall health. A lack of physical activity is associated with many metabolic problems that lessen health, including obesity and its sequelae. The data from Pima Indians presented in chapter 13 drive this message home. Pima Indians who are physically active generally are not obese; those with low levels of physical activity are obese (Figures 13.1 and 13.2).

Physical Activity and Evolution

There are many perspectives within which to view modern human obesity. This book takes an evolutionary biology perspective. We are exploring the how and why of humanity's susceptibility to become obese, how the biology predisposes and how it interacts with the modern environment we have created. The evolutionary perspective shows us why we have created many of the obesogenic features of our modern-day world. It makes intuitive and evolutionary sense that we as a species value energy-dense foods and inventions that reduce our required exertion. For millions of years perhaps the most significant constraint on our ancestors was the difficulty in balancing the exertion and energy expenditure required by the environment with the amount of food that exertion could obtain. For most of our past that balance hovered, often precariously, very close to zero. It is little wonder that our preference is to arrange things so that food intake easily meets or exceeds exertion. That was rare in our past. Our technology has allowed those inherent urges to be met, and now we have to deal with the consequences. Because there are always consequences; no matter how clever or technologically advanced we become as a species, our biology still carries the past.

Is obesity pathology? Does it represent a failure of physiological systems to correctly perform their function? In some specific types of obesity that is demonstrably true, but it doesn't appear to explain the general case of obese people in today's world. Many obesity researchers frame the obesity question as a failure of energy homeostasis. That is certainly true; however, it is important to add the qualifier "in the modern environment." To say that energy homeostasis should be maintained regardless of environment denies the reality of evolution. Evolved systems don't generally work globally.

We evolved as an active species; we are not sessile creatures. Physical activity was a part of our ancestors' lifestyle for millions of years. The modern era has afforded us the luxury of being exceptionally lazy, if we so choose; we don't have to exert ourselves to live. But our biology is geared toward an active lifestyle. It is perhaps naive to assume that our energy intake regulation system will work effectively at extremely low levels of energy expenditure.

The evolutionary perspective suggests that we should be building environments that require more physical effort. Not intense, acute, short-term efforts; those can be enjoyable, in the context of sports, dance, and other physically challenging human endeavors, but they also have dangers and by their very nature are not sustainable for extended periods of time or as we age. Both the medical literature and the evolutionary perspective suggest that moderate but sustained exertion is something our bodies are very well adapted to undertake and indeed may be the healthiest option. Perhaps if we were once again required to spend more time and energy in things physical, our food intake would be less of a concern.

Homeostasis, Allostasis, and Anticipatory Regulation

The regulation of appetite is often framed as a homeostatic process. There probably really isn't any such thing as weight homeostasis, except as a consequence of body energy homeostasis. However, for most people, changes in weight are usually changes in adipose tissue, and thus weight homeostasis generally equals energy balance. And there are convincing data that under many conditions adult nonreproductive animals regulate their feeding and activity to achieve energy balance over moderate time scales. The epidemiological data also say that over the last few decades this has not been the case for many humans. Is this failure of homeostasis pathology? Or is it within the norm of evolved physiology and behavior but now inappropriate due to the rather dramatic differences between our modern world and the one in which we evolved? Is there underlying pathology that can be treated to return obese people to leanness? Or is the vulnerability to obesity a result of a mismatch between evolved adaptive biology and the modern circumstance?

The homeostatic perspective is a powerful tool for understanding regulatory physiology. It does not represent all physiology, however. There are many nonhomeostatic physiological processes. Animals do not achieve evolutionary success by maintaining stability; they do so by maintaining viability. For many aspects of biology that means maintaining a level of stability; for example, serum ionized calcium is maintained within a fairly narrow range; too high or too low results in pathology and even death. But for many other aspects of living, animals must abandon homeostasis, at least temporarily. Homeostasis is achieving viability through resistance to change (stability); allostasis is achieving viability through change. Allostasis provides an alternative framework within which to view physiological regulation.

Adaptive physiological changes are necessary for life; they also can impose a cost. This concept has been termed *allostatic load*. Normal, adaptive physiological mechanisms can lead to pathology if they are extended beyond their usual time scales or exceed limits that were generally found in nature prior to the modern environment. Adipose tissue is involved in endocrine regulation of physiology and immune function. It produces biologically active molecules that affect other organ systems and whole-body physiology. The excessive amount of adipose tissue that defines obesity also results in many dysregulations of these normal regulatory functions.

Pharmacology and the Evolutionary Perspective

We end this book with a cautionary note. It is tempting, and very human, to hope for a simple, pharmacological fix to obesity. We hope that the readers of this book will come away with a healthy skepticism of that approach. Not that it is impossible for such an approach to work, but the complexity of an evolved biological system suggests that most simple molecular interventions will have multiple unintended consequences and may trigger compensatory metabolic systems. Biological systems are inherently messy; that is part of the power of evolved systems.

There are dangers to studying metabolic pathways without accounting for the whole organism physiology. For example, attractin was first discovered as a circulating secreted molecule expressed by activated T lymphocytes. The discovery that a transmembrane form affects basal metabolism suggested that attractin might be a potential extracellular target for the pharmacological treatment of obesity. However, mutations of the attractin locus also are associated with juvenile onset neurodegeneration (Duke-Cohan et al., 2004). Our knowledge of the functional activity of attractin continues to evolve, but systemic alterations of attractin function are likely to have many and varied effects beyond that of changing metabolic rate.

The fact that leptin is an ancient information molecule with pleiotropic actions should give pause to those who would eagerly ascribe potential pharmacological benefits to leptin. Leptin is well documented to have many different functions; more are likely to be discovered. Therapeutic interventions based on the leptin system will likely have many and varied effects on physiology and metabolism that vary by tissue, in context with other signaling pathways, and with age. There may well be many unintended consequences, some that we may not yet have even considered possible. This caution is likely to be relevant for most other pharmacological manipulations of metabolic signaling targeting information molecules.

There are also dangers in not taking an organismal view of health conditions such as those associated with obesity. We argue that in many, but not all instances, obesity may represent natural, adaptive responses to the conditions of the modern world—responses that were adaptive in the past when external factors constrained both the upper limit of food intake and the lower limit of energy expenditure. Many of the diseases associated with obesity are also due, at least in part, to the body's physiological adaptations to increased adipose tissue and its effects on metabolism, the immune system, and other end-organs. To reverse or ameliorate these health consequences probably will require a whole-body approach, but many therapies these days are targeted toward changing the values of certain, indisputably important parameters as if merely returning the level of, for example, systolic blood pressure or fasting blood glucose to a "normal" level will return the whole body to health. This approach can be fraught with unintended, and unpredictable, consequences. For example, a recent drug trial that sought to treat type 2 diabetics by dramatically lowering fasting glucose levels to the "healthy" range had to be stopped because of an increase in heart attacks and strokes (NHLBI press release, 2008).

The body adjusts and adapts to its circumstances; if a parameter is significantly outside of the normal healthy range, then it is likely that other aspects of physiology and metabolism are adjusting to that change, influencing it, and possibly driving that change. To make an intervention to return the parameter to the normal range without taking into account the metabolic and physiologic conditions of the patient risks perturbing a long chain of connected metabolic parameters with important health consequences. Although high fasting blood glucose is dangerous and debilitating over the long term, it may represent a response to other dysregulations in organ systems and actually be the body's attempt to maintain function in the face of multiple pathologies. Or the dramatic lowering of circulating glucose may key other, normal physiological responses that, under the circumstances, challenge weakened or vulnerable organ systems. It is quite possible that a person who has experienced poor glucose control over an extended period is not in an appropriate physiological state to have a low fasting blood glucose level that would be appropriate and healthful in an individual whose glucose metabolism has not been compromised.

Addressing obesity in our society likely will require multilayered, integrated interventions. People can be vulnerable to obesity for various reasons. There are many paths to obesity, and probably many paths to avoiding excessive fat gain. One-size-fits-all solutions have not worked in the past, and we argue they are unlikely to ever work. Understanding the complex biology that underlies our behavior relevant to feeding, food preferences, and the importance of physical activity, not just in function but also as evolutionary history, is one way to guide the necessary integrated approaches to modifying our modern world and our behavior to reduce the health and economic burden of obesity.

References

Abate N, Chandalia M. 2003. The impact of ethnicity on type 2 diabetes. J Diabetes Complications 17: 39–58.

Abdallah L, Chabert M, and Lois-Sylvestre J. 1997. Cephalic phase responses to sweet taste. Am J Clin Nutr 65: 737–743.

Ahima RS, Osei SY. 2004. Leptin signaling. Physiol and Behav 81: 223–241.

Ahmed ML, Ong KKL, Morrell DJ, Cox L, Drayer N, Perry L, Preece MA, Dunger DB. 1999. Longitudinal study of leptin concentration during puberty: sex differences and relationship to changes in body composition. JCEM 84: 899–905.

Ahren B, Holst JJ. 2001. The cephalic insulin response to meal ingestion in humans is dependent on both cholinergic and noncholinergic mechanisms and is important for postprandial glycemia. Diabetes. 50 (5): 1030–1038.

Aiello LC, Bates N, and Joffe T. 2001. In defense of the expensive tissue hypothesis. In Evolutionary Anatomy of the Primate Cerebral Cortex. Falk D, Gibson KR (eds.), 57–78. Cambridge: Cambridge University Press.

Aiello LC, Wheeler P. 1995. The expensive-tissue hypothesis: the brain and the digestive system in human and primate evolution. Curr Anthropol 46: 126–170.

Al Atawi F, Warsy A, Babay Z, Addar M. 2005. Fetal sex and leptin concentrations in pregnant females. Ann Saudi Med 25: 124–128.

Alexe D-M, Syridou G, Petridou ET. 2006. Determinants of early life leptin levels and later life degenerative outcomes. Clin Med Res 4: 326–335.

Allison KC, Ahima RS, O'Reardon JP, Dinges DF, Sharma V, Cummings DE, Heo M, Martino NS, Stunkard AJ. 2005. Neuroendocrine profiles associated with energy intake, sleep, and stress in the night eating syndrome. JCEM 90: 6214–6217.

Anderson LA, McTernan PG, Barnett AH, Kumar S. 2001. The effects of androgens and estrogens on preadipocyte proliferation in human adipose tissue: influence of gender and site. JCEM 86: 5045–5051.

Anderson RE, Crespo CJ, Bartlett SJ, Cheskin LJ, Pratt M. 1998. Relationship of physical activity and television with body weight and level of fatness among children. JAMA 279: 938–942.

Andrew R, Phillips DIW, Walker BR. 1998. Obesity and gender influence cortisol secretion and metabolism in man. JCEM 83: 1806–1809.

Aquila S, Gentile M, Middea E, Catalano S, Morelli C, Pezzi V, Andò S. 2005. Leptin secretion by human ejaculated spermatozoa. JCEM 90: 4753–4761.

Arai M, Assil IQ, Abou-Samra AB. 2001. Characterization of three corticotropin-releasing factor receptors in catfish: a novel third receptor is predominantly expressed in pituitary and urophysis. Endocrinology 142: 446–454.

Araneta MRG, Barrett-Conner E. 2005. Ethnic differences in visceral adipose tissue and type 2 diabetes: Filipino, African-American, and white women. Obesity Res 13: 1458–1465.

Araneta MRG, Wingard DL, Barrett-Connor E. 2002. Type 2 diabetes and metabolic syndrome in Filipina-American women: a high-risk nonobese population. Diabetes Care 25: 494–499.

Arita Y, Kihara S, Ouchi N, et al. 1999. Paradoxical decrease of an adipose-specific protein, adiponectin, in obesity. Biochem Biophys Res Commun 257: 79–83.

Ariyasu H, Takaya K, Tagami T, Ogawa Y, Hosoda K, Akamizu T, Suda M, Koh T, Natusi K, Toyooka S, Shirakami G, Usui T, Shimatsu A, Doi K, Hosoda H, Kojima M, Kanagawa K, Nakao K. 2001. Stomach is major source of circulating ghrelin and feeding state determines plasma ghrelin-like immunoreactivity levels in humans. JCEM 86: 4753–4758..

Arner P. 1998. Not all fat is alike. Lancet 351: 1301–1302.

Arosio M, Ronchi CL, Beck-Peccoz P, Gebbia C, Giavoli C, Cappiello V, Conte D, Peracchi M. 2004. Effects of modified sham feeding on ghrelin levels in healthy human subjects. JCEM 89: 5101–5104.

Arunagh S, Pollack S, Yeh J, Aloia JF. 2003. Body fat and 25-hydroxyvitamin D levels in healthy women. JCEM 88: 157–161.

Arvat E, et al. Endocrine activities of ghrelin, a natural growth hormone secretagogue (GHS), in humans: comparison and interactions with hexarelin, a nonnatural peptidyl GHS, and GH-releasing hormone. JCEM 86: 1169–1174.

Aschoff J, Pohl H 1970. Rhythmic variations in energy metabolism. Federation Proceedings 29: 1541–1552.

Ashwell CM, et al. 1999. Hormonal regulation of leptin expression in broiler chickens. AJP-Regul, Integrative, and Comp Physiol 276 (1): R226–R232.

Ashworth CJ, Hoggard N, Thomas L, Mercer JG, Wallace JM, Lea RG. 2000. Placental leptin. Rev Reprod 5: 18–24.

Ategbo JM, Grissa O, Yessoufou A, Hichami A, Dramane KL, Moutairou K, et al. 2006. Modulation of adipokines and cytokines in gestational diabetes and macrosomia. JCEM 91: 4137–4143.

Avery OT, MacLeod CM, McCarty M. Studies on the chemical nature of the substance inducing transformation of Pneumococcal types. 1944. J Experimental Med 79: 137–158.

Ayas NT, White DP, Al-Delaimy WK, Manson JE, Stampfer MJ, Speizer FE, Patel, S, Hu FB. 2003. A prospective study of self-reported sleep duration and incident diabetes in women. Diabetes Care 26: 380–384.

Babey SH, Hastert TA, Yu H, Brown ER. 2008. Physical activity among adolescents: when do parks matter? Am J Prev Med 34: 345–348.

Bado A, Levasseur S, Attoub S, Kermorgant S, Laigneau JP, Bortoluzzi MN, Moizo L, Lehy T, Guerre-Millo M, Le Marchand-Brustel Y, Lewin MJ. 1998. The stomach is a source of leptin. Nature 394: 790–793.

Baillie-Hamilton PF. 2002. Chemical toxins: a hypothesis to explain the obesity epidemic. J Alt Comp Med 8: 185–192.

Bajari TM, Nimpf J, Schneider WJ. Role of leptin in reproduction. 2004. Curr Opin Lipidol 15: 315–319.

Bamshad M, Wooding SP. 2003. Signatures of natural selection in the human genome. Nat Rev Genet 4 (2): 99–111.

Banks WA, Clever CM, Farrell CL. 2000. Partial saturation and regional variation in the blood to brain transport of leptin in normal weight mice. Am J Physiol 278: E1158–E1165.

Banks WA, Phillips-Conroy JE, Jolly CJ, Morley JE. 2001. Serum leptin levels in wild and captive populations of baboons (*Papio*): implications for the ancestral role of leptin. JCEM 86: 4315–4320.

Barker DJP. 1991. Fetal and Infant Origins of Adult Disease. London: BMJ.

Barker DJP. 1997. The fetal origins of coronary heart disease. Eur Heart J 18: 883–884.

Barker DJP. 1998. Mothers, Babies, and Health in Later Life. Edinburgh: Churchill Livingstone.

Barker DJP, Gluckman PD, Godfrey KM, et al. 1993. Fetal nutrition and cardiovascular disease in adult life. Lancet 341: 938–941.

Barker DJP, Osmond C. 1986. Infant mortality, childhood nutrition, and ischaemic heart disease in England and Wales. Lancet 8489: 1077–1081.

Barker DJP, Osmond C, Thornburg KL, Kajantie E, Forsen TJ, Eriksson JG. 2008. A possible link between the pubertal growth of girls and breast cancer in their daughters. Am J Hum Biol 20 (2): 127–131.

Barrachina MD, Martinez V, Wang L, Wei JY, Tache Y. 1997. Synergistic interaction between leptin and cholecystokinin to reduce short-term food intake in lean mice. PNAS USA 94: 10455–10460.

Barrenetxe J, Villaro AC, Guembe L, Pascual I, Munoz-Navas M, Barber A, Lostao MP. 2002. Distribution of the long leptin receptor isoform in brush border, basolateral membrane, and cytoplasm of enterocytes. Gut 50: 797–802.

Barrett R, Kuzawa CW, McDade T, Armelagos GJ. 1998. Emerging and re-emerging infectious disease: the third epidemiologic transition. Ann Rev Anthro 27: 247–271.

Batterham RL, et al. 2002. Gut hormone PYY3-36 physiologically inhibits food intake. Nature 418: 650–654.

Bauman DE, Currie WB. 1980. Partitioning of nutrients during pregnancy and lactation: a review of mechanisms involving homeostasis and homeorrhesis. J Dairy Sci 1514–1529.

Beall MH, Haddad ME, Gayle D, Desai M, Ross MG. 2004. Adult obesity as a consequence of in utero programming. Clin Obstet Gynecol 47: 957–966.

Beck BB. 1980. Animal Tool Behavior: The Use and Manufacture of Tools. New York: Garland Press.

Bennett PH, Burch TA, Miller M. 1971. Diabetes mellitus in American (Pima) Indians. Lancet 2 (7716): 125–128.

Berglund MM, et al. 2003. the use of bioluminescence resonance energy transfer 2 to study neuropeptide y receptor agonist: induced b-arrestin 2 interaction. J of Pharmacol and Experim Therapeutics 306: 147–156.

Bergman RN, Kim SP, Catalano KJ, Hsu IR, Chiu JD, Kabir M, Hucking K, Ader M. 2006. Why visceral fat is bad: mechanisms of the metabolic syndrome. Obesity 14 (suppl): 16S–19S.

Berk ES, Kovera AJ, Boozer CN, Pi-Sunyer FX, Albu JB. 2006. Metabolic inflexibility in substrate use is present in African-American but not Caucasian healthy, premenopausal, nondiabetic women. JCEM 91: 4099–4106.

Bernard C. 1865. An Introduction to the Study of Experimental Medicine. Trans. Henry Cooper Greene. New York: Dover Publications, 1957.

Berridge KC. 1996. Food reward: brain substrates of wanting and liking. Neurosci Biobehav Rev 20: 1–25.

Berridge KC. 2004. Motivation concepts in behavioral neuroscience. Physiol Behav 81: 179–209.

Berridge KC, Grill HJ, Norgren R. 1981. Relation of consummatory responses and preabsorptive insulin response to palatability and taste aversions. J Comp Physiol Psychol 95: 363–382.

Berthoud HR, Morrison C. The brain, appetite, and obesity. 2008. Ann Rev Psychol 59: 55–92.

Berthoud HR, Trimble ER, Siegal EG, Bereiter DA, Jeanrenaud B. 1980. Cephalic-phase insulin secretion in normal and pancreatic islet-transplanted rats. Am J Physiol 238: E336–E340.

Birketvedt GS, Sundsfjord J, Florholmen JR. 1999. Hypothalamic-pituitary-adrenal axis in the night eating syndrome. Am J Endocrinol Metab 282: E366–E369.

Blaxter K. 1989. Energy Metabolism in Animals and Man. Cambridge: Cambridge University Press.

Blouin K, Richard C, Belanger C, Dupont P, Daris M, Laberge P, Luu-The V, Tchernof A. 2003. Local androgen inactivation in abdominal visceral adipose tissue. JCEM 88: 5944–5950.

Boden G, Chen X, Mozzoli M, Ryan I. 1996. Effect of fasting on serum leptin in normal human subjects. JCEM 81: 3419–3423.

Bondeson J. 2000. The Two-Headed Boy, and Other Medical Marvels. Ithaca: Cornell University Press.

Booth DA. 1972. Conditioned satiety in the rat. J Comp Physiol Psychol 81: 457–471.

Boswell T, et al. 2006. Identification of a non-mammalian leptin-like gene: characterization and expression in the tiger salamander (*Ambystoma tigrinum*). Gen and Comp Endocrinol 146 (2): 157–166.

Boulus Z, Rossenwasser AM. 2004. A chronobiological perspective on allostasis and its application to shift work. In Allostasis, Homeostasis, and the Costs of Adaptation, Schulkin J (ed.). Cambridge: Cambridge University Press.

Bouret SG, Draper SJ, Simerly RB. 2004. Trophic action of leptin on hypothalamic neurons that regulate feeding. Science. 304: 108–110.

Bouret SG, Simerly RB. 2004. Minireview: leptin and development of hypothalamic feeding circuits. Endocrinol 145: 2621–2626.

Bouret SG, Simerly RB. 2006. Developmental programming of hypothalamic feeding circuits. Clin Genet 70: 295–301.

Bowman SA. 2007. Low economic status is associated with suboptional intakes of nutritious foods by adults in the national health and nutrition examination survey 1999–2002. Nutr Res 27: 515–523.

Bowman, SA, Gortmaker, SL, Ebbeling, CB, Pereira, MA, Ludwig, DS. 2004. Effects of fast food consumption on energy intake and diet quality among children in a national household survey. J Pediatrics 113: 112–118.

Bowman SA, Vinyard BT. 2004. Fast food consumers vs. non-fast food consumers: a comparison of their energy intakes, diet quality, and overweight status. J Am College Nutr 23: 163–168.

Brady LS, et al. 1990. Altered expression of hypothalamic neuropeptide mRNAs in food-restricted and food-deprived rats. Neuroendocrinology 52: 441–447.

Brand-Miller JC, Holt SHA, Pawlak DB, McMillan J. 2002. Glycemic index and obesity. Am J Clin Nutr 76 (suppl): 281S–285S.

Branson R, Potoczna N, Kral JG, Lentes KL, Hoehe MR, Horber FF. 2003. Binge eating as a major phenotype of melanocortin 4 receptor gene mutations. NE J Med 348 (12): 1096–1103.

Bray GA. 2004. Medical consequences of obesity. JCEM 89: 2583–2589.

Bray GA, Gray D. 1988. Obesity. Part 1. Pathogenesis. West J Med 149: 429–441.

Bray GA, Jablonski KA, Fujimot, WY, Barrett-Connor E, Haffner S, Hanson RL, Hill JO, Hubbard V, Kriska A, Stamm E, Pi-Sunyer FX. 2008. Relation of central adiposity and body mass index to the development of diabetes in the Diabetes Prevention Program. Am J Clin Nutr 87: 1212–1218.

Bray GA, Nielsen SJ, Popkin BM. 2004. Consumption of high-fructose corn syrup in beverages may play a role in the epidemic of obesity. Am J Clin Nutr 79: 537–543.

Brenna JT. 2002. Efficiency of conversion of α -linolenic acid to long chain n-3 fatty acids in man. Curr Opinion in Clin Nutr and Metabolic Care 5: 127–132.

Bribiescas RG. 2005. Serum leptin levels in Ache Amerindian females with normal adiposity are not significantly different from American anorexia nervosa patients. Am J Hum Biol 17: 207–210.

Brody S. 1945. Bioenergetics and Growth. New York: Hafner.

Brotanek JM, Gosz J, Weitzman M, Flores G. 2007. Iron deficiency in early childhood in the United States: risk factors and racial/ethnic disparities. Pediatr 120: 568–575.

Brotanek JM, Halterman J, Auinger P, Flores G, Weitzman M. 2005. Iron deficiency, prolonged bottle-feeding, and racial/ethnic disparities in young children. Arch Pediatr Adolesc Med 159: 1038–1042.

Brown PJ, Condit-Bentley VK. 1998. Culture, evolution, and obesity. In Handbook of Obesity, Bray G, Bouchard C, James WPT (eds.), 143–155. New York: Marcel Dekker.

Brownson RC, Baker EA, Housemann RA, Brennan LK, Bacak SJ. 2001. Environmental and policy determinants of physical activity in the United States. Am J Pub Health 91: 1995–2003.

Bruce DG, Storlien LH, Furler SM, Chisolm DJ. 1987. Cephalic phase metabolic responses in normal weight adults. Metabolism 36: 721–725.

Brunet M, Guy F, Pilbeam D, Mackaye HT, Likius A, Ahounta D, Beauvilain A,

Blondel C, Bocherens H, Boisserie J-R, De Bonis L, Coppens Y, Dejax J, Denys C, Duringer P, Eisenmann V, Fanone G, Fronty P, Geraads D, et al. 2002. A new hominid from the upper Miocene of Chad, Central Africa. Nature 418: 145–151.

Bruttomesso D, Pianta A, Mari A, Valerio A, Marescotti MC, Avogaro A, Tiengo A, Del Prato S. 1999. Restoration of early rise in plasma insulin levels improves the glucose tolerance of type 2 diabetic patients. Diabetes 48: 99–105.

Bucham JR, Parker R. 2007. The two faces of miRNA. Science 318: 1877–1878.

Bungum TH, Satterwhite M, Jackson AW, Morrow JR, Jr. 2003. The relationship of body mass index, health costs and job absenteeism. Am J Health Behav 27: 456–462.

Bunn HT. 1981. Archeological evidence for meat-eating by Plio-Pleistocene hominids from Koobi-Fora and Olduvai Gorge. Nature 291: 574–577.

Bunn HT. 2001. Hunting, power scavenging, and butchering by Hadza foragers and by Plio-Pleistocene *Homo*. In Meat-Eating and Human Evolution, Stanford CB and Bunn HT (eds.), 199–218. New York: Oxford University Press.

Butte NF, Hopkinson JM, Nicolson MA. 1997. Leptin in human reproduction: serum leptin levels in pregnant and lactating women. JCEM 82: 585–589.

Cammisotto PG, Gingras D, Renaud C, Levy E, Bendayan M. 2006. Secretion of soluble leptin receptors by exocrine and endocrine cells of the gastric mucosa. Am J Physiol--Gastrointestinal and Liver Physiol 290: G242–249.

Cammisotto PG, Renaud C, Gingras D, Delvin E, Levy E, Bendayan M. 2005. Endocrine and exocrine secretion of leptin by the gastric mucosa. J Histochemistry Cytochemistry 53: 851–860.

Campfield LA Smith FJ, Guisez Y, et al. 1995. Recombinant mouse OB protein: evidence for a peripheral signal linking adiposity and central neural networks. Science 269: 546–549.

Campos P, Saguy A, Ernsberger P, Oliver E, Gaesser G. 2006. The epidemiology of overweight and obesity: public health crisis or moral panic? Int J Epidemiol 35: 55–59.

Cannon WB. 1932. The Wisdom of the Body. New York: Norton.

Cannon WB. 1935. Stresses and strains of homeostasis. Am J Med Sci 189: 1–14.

Casabiell X, Piñeiro V, Peino R, Lage M, Camiña J, Gallego R, Vallejo LG, Dieguez C, Casanueva FF. 1998. Gender differences in both spontaneous and stimulated leptin secretion by human omental adipose tissue in vitro: dexamethasone and estradiol stimulate leptin release in women, but not in men. JCEM 83: 2149–2155.

Casabiell X, Piñeiro V, Tome MA, Peino R, Dieguez C, Casanueva FF. 1997. Presence of leptin in colostrum and/or breast milk from lactating mothers: a potential role in the regulation of neonatal food intake. JCEM 82: 4270–4273.

Cassell JA. 1995. Social anthropology and nutrition: a different look at obesity in America. J Am Dietetic Assoc 95: 424–427.

Catalano PM. 2007. Management of obesity in pregnancy. Obstet Gynecol 109: 419–433.

Catalano PM, Ehrenberg HM. 2006. The short- and long-term implications of maternal obesity on the mother and her offspring. BJOG 113: 1126–1133.

Catalano PM, Hoegh M, Minium J, Huston-Presley L, Bernard S, Kalhan S, Hauguel-De Mouzon S. 2006. Adiponectin in human pregnancy: implications for regulation of glucose and lipid metabolism. Diabetologia 49: 1677–1685.

Catalano PM, Thomas A, Huston-Presley L, Amini SB. 2003. Increased fetal adiposity: a very sensitive marker of abnormal in utero development. Am J Obstet Gynecol 189: 1698–1704.

Catalano PM, Thomas A, Huston-Presley L, Amini SB. 2007. Phenotype of infants of mothers with gestational diabetes. Diabetes Care 30 (suppl 2): S156–S160.

CDC. 2007. Prevalence of regular physical activity among adults—United States, 2001 and 2005. MMWR [Morbidity and Mortality Weekly Report] 56 (Nov. 23): 1209–1212. Accessed at www.cdc.gov/mmwr/preview/mmwrhtml/mm5646a1.htm.

Cerda-Reverter JM, et al. 2000. cNeuropeptide Y family of peptides: structure, anatomical expression, function, and molecular evolution. Biochem and Cell Biol 78 (3): 371–392.

Chakravarthy MV, Booth FW. 2004. Eating, exercise, and "thrifty" genotypes: connecting the dots toward an evolutionary understanding of modern chronic diseases. J Appl Physiol 96: 3–10.

Chan JL, Heist K, DePaoli AM, Veldhuis JD, Mantzoros CS. 2003. The role of falling leptin levels in the neuroendocrine and metabolic adaptation to short-term starvation in healthy men. J Clin Invest 111: 1409–1421.

Chan VO, Colville J, Persaud T, Buckley O, Hamilton S, Torreggiani WC. 2006. Intramuscular injections into the buttocks: are they truly intramuscular? Eur J Radiol 58: 480–484.

Chehab FF, Lim ME, Lu R. 1996. Correction of the sterility defect in homozygous obese female mice by treatment with human recombinant leptin. Nat Genet 12: 318–320.

Cherkas LF, Hunkin JL, Kato BS, Richards B, Gardner JP, Surdulescu GL, Kimura M, Lu X, Spector TD, Aviv A. 2008. Arch Intern Med 168: 154–158.

Christakis NA, Fowler JH. 2007. The spread of obesity in a large social network over 32 years. N Eng J Med 357: 370–379.

Clegg DJ, Brown LM, Woods SC, Benoit SC. 2006. Gonadal hormones determine sensitivity to central leptin and insulin. Diabetes 55: 978–987.

Clegg DJ, Riedy CA, Smith KA, Benoit SC, Woods SC. 2003. Differential sensitivity to central leptin and insulin in male and female rats. Diabetes 52: 682–687.

Clegg DJ, Woods SC. 2004. The physiology of obesity. Clin Obstet Gynecol 47: 967–979.

Cleland VJ, Schmidt MD, Dwyer T, Venn AJ. 2008. Television viewing and abnormal obesity in young adults: is the association mediated by food and beverage consumption during viewing time or reduced leisure-time physical activity? Am J Clin Nutr 87: 1148–1155.

Clement K, Langin D. 2007. Regulation of inflammation-related genes in human adipose tissue. J Int Med 262: 422–430.

Cohade C, Mourtzikos KA, Wahl RL. 2003. "USA-fat": prevalence is related to ambient outdoor temperature—evaluation with 18F-FDG PET/CT. J Nucl Med 44: 1267–1270.

Cohen P, Zhao C, Cai X, Montez JM, Rohani SC, Feinstein P, Mombaerts P, Friedman JM. 2001. Selective deletion of leptin receptor in neurons leads to obesity. J Clin Invest 108: 1113–1121.

Coimbra-Filho AF, Mittermeier RA. 1977. Tree-gouging, exudate-eating, and the "short-tusked" condition in *Callithrix* and *Cebuella*. In The Biology and Conservation of the Callitrichidae, D. G. Kleiman (ed.), 105–115. Washington, DC: Smithsonian Institution Press.

Coleman DL. 1973. Effects of parabiosis of obese with diabetic and normal mice. Diabetologia 9: 294–298.

Collins FS. 2004. What we do and don't know about "race," "ethnicity," genetics, and health in the dawn of the genome era. Nat Genet Suppl 36 (11): S13–S15.

Combs TP, Berg AH, Rajala MW, Klebanov S, Lyengar P, Jimenez-Chillaron JC, Patti ME, Klein SL, Weinstein RS. 2003. Sexual differentiation, pregnancy, calorie restriction, and aging affect the adipocytes-specific secretory protein adiponectin. Diabetes 52: 268–276.

Combs TP, Scherer PE. 2003. The significance of elevated adiponectin in the treatment of type 2 diabetes. Canadian Journal of Diabetes 27: 433–438.

Conway JM, Yanovski SZ, Avila NA, Hubbard VS. 1995. Visceral adipose tissue differences in black and white women. Am J Clin Nutr 61: 765–771.

Cordain L, Eaton SB, Brand-Miller J, Mann N, Hill K. 2002. The paradoxical nature of hunter-gatherer diets: meat-based, yet non-atherogenic. Eur J Clin Nutr 56: S42–S52.

Cordain L, Watkins BA, Florant GL, Kelher M, Rogers L, Li Y. 2002. Fatty acid analysis of wild ruminant tissues: evolutionary implications for reducing diet-related chronic disease. Eur J Clin Nutr 56: 181–191.

Cordain L, Watkins BA, Mann NJ. 2001. Fatty acid composition and energy density of foods available to African hominids. World Rev Nutr Diet 90: 144–161.

Cossrow N, Falkner B. 2004. Race/ethnic issues in obesity and obesity-related comorbidities. JCEM 89: 2590–2594.

Coursey DG. 1973. Hominid evolution and hypogeous plant foods. Man 8: 634–635.

Craig WC. 1918. Appetites and aversions as constituents of instincts. Biol Bull 34: 91–107.

Crespi EJ. Denver RJ. 2006. Leptin (*ob* gene) of the South African clawed frog *Xenopus laevis*. PNAS 103: 10092–10097.

Crews D, McLachlan JA. 2006. Epigenetics, evolution, endocrine disruption, health, and disease. Endocrinology 147 (suppl): S4–S10.

Crystal SR, Teff KL. 2006. Tasting fat: cephalic-phase hormonal responses and food intake in restrained and unrestrained eaters. Physiol Behav 89: 213–220.

Cubas P, Vincent C, Coen E. 1999. An epigenetic mutation responsible for natural variation in floral symmetry. Nature 401: 157–161.

Cummings DE, Overduin J. 2007. Gastrointestinal regulation of food intake. J Clin Invest 117: 13–23.

Cummings DE, Purnell JQ, Frayo RS, Schmidova K, Wisse BE. 2001. A preprandial rise in plasma ghrelin levels suggests a role in meal initiation in humans. Diabetes 50: 1714–1719.

Cunnane SC, Crawford MA. 2003. Survival of the fattest: fat babies were the key to evolution of the large human brain. Comp Biochem and Physiol Part A 136: 17–26.

Dallman MF, Akana SF, Strack AM, Hanson ES, Sebastian RJ. 1995. The neural network that regulates energy balance is responsive to glucocorticoids and insulin and also regulates HPA axis responsivity at a site proximal to CRF neurons. Ann NY Acad Sci 771: 730–742.

Dallman MF, Pecoraro N, Akana SF, la Fleur SE, Gomez F, Houshyar H, Bell ME, Bhatnagar S, Laugero KD, Manalo S. 2003. Chronic stress and obesity: a new view of "comfort food." PNAS 100: 11696–11701.

Dallman MF, Pecoraro NC, la Fleur SE. 2005. Chronic stress and comfort foods:

self-medication and abdominal obesity. Brain, Behavior, and Immunity 19: 275–280.

Dallman MF, Strack AM, Akana SF, Bradbury MJ, Hanson ES, Scribner KA, Smith M. 1993. Feast and famine: critical role of glucocorticoids with insulin in daily energy flow. Front Neuroendocrinol 14: 303–347.

D'Amour DE, Hohmann G, Fruth B. 2006. Evidence of leopard predation on bonobos (*Pan paniscus*). Folia Primatologica 77: 212–217.

Dannenberg AL, Burton DC, Jackson RJ. 2004. Economic and environmental costs of obesity: the impact on airlines. Am J Prev Med 27: 264.

Darmon, N, Drewnowski, A. 2008. Does social class predict diet quality? Am J Clin Nutr 87: 1107–1117.

Dart RA. 1925. *Australopithecus africanus*: the man-ape of South Africa. Nature 115: 195–199.

Dean WRJ, MacDonald IAW. 1981. A review of African birds feeding in association with mammals. Ostrich 52: 135–155.

Deaner RO, Isler K, Burkart J, van Schaik C. 2007. Overall brain size, and not encephalization quotient, best predicts cognitive ability across non-human primates. Brain Behav Evol 70: 115–124.

Decsi T, Koletzko B. 1994. Polyunsaturated fatty acids in infant nutrition. Acta Paediatr Suppl 83 (395): 31–37.

Degen L, Oesch S, Casanova M, Graf S, Ketterer S, Drewe J, Beglinger C. 2005. Effect of peptide YY3-36 on food intake in humans. Gastroenterology 129: 1430–1436.

DeLuca HF. 1988. The vitamin D story: a collaborative effort of basic science and clinical medicine. FASEB J 2: 224–236.

Demment MW, Van Soest PJ. 1985. A nutritional explanation for body-size patterns of ruminant and nonruminant herbivores. Am Nat 125: 641–772.

Denbow DM, et al. 2000. Leptin-induced decrease in food intake in chickens. Physiol and Behav 69 (3): 359–362.

Denton DA. 1982. The Hunger for Salt. New York: Springer-Verlag.

Denver RJ. 1999. Evolution of the corticotropin-releasing hormone signaling system and its role in stress-induced phenotypic plasticity. Ann NY Acad Sci 897: 46–53.

Department of Health. 2006. Forecasting obesity to 2010. Accessed at www.dh.gov.uk/en/Publicationsandstatistics/Publications/PublicationsStatistics/DH_4138630.

De Smet B, Thijs T, Peeters TL, Depoortere I. 2007. Effect of peripheral obestatin on gastric emptying and intestinal contractility in rodents. Neurogastroenterol and Motil 19: 211–7.

Dethier VG. 1976. The Hungry Fly: A Physiological Study of the Behavior Associated with Feeding. Cambridge: Harvard University Press.

Deurenberg P, Deurenberg-Yap M, Guricci S. 2002. Asians are different from Caucasians and from each other in their body mass index/body fat per cent relationship. Obesity Rev 3: 141–146.

de Waal FBM, Lanting F. 1997. Bonobo: The Forgotten Ape. Berkeley: University of California Press.

Dhurandhar NV, Israel BA, Kolesar JM, Mayhew GF, Cook ME, Atkinson RL. 2000. Adiposity in animals due to a human virus. Int J Obes 24: 989–996.

Dhurandhar NV, Whigham LD, Abbott DH, Schultz-Darken NJ, Israel BA, Bradley SM, Kemnitz JW, Allison DB, Atkinson RL. 2002. Human adenovirus Ad-36 promotes weight gain in male rhesus and marmoset monkeys. J Nutr 132: 3155–3160.

Diamond P, LeBlanc J. 1988. A role for insulin in cephalic phase of postprandial thermogenesis in dogs. Am J Physiol 254 (5 Pt 1): E625–32.

Dibaise JK, Zhang H, Crowell MD, Krajmalnik-Brown R, Decker GA, Rittmann BE. 2008. Gut microbiota and its possible relationship with obesity. Mayo Clin Proc 83: 460–469.

Dickinson S, Hancock DP, Petocz P, Ceriello A, Brand-Miller J. 2008. High-glycemic index carbohydrate increases nuclear factor-kB activation in mononuclear cells of young, lean healthy subjects. Am J Clin Nutr 87: 1188–1193.

Dierenfeld ES, Hintz HF, Robertson JG, Van Soest PJ, Oftedal OT. 1982. Utilization of bamboo by the giant panda. J Nutr 12: 636–641.

Donnelly JE, Hill, JO, Jacobsen DJ, Potteiger J, Sullivan DK, Johnson SL, Heelan K, Hise M, Fennessey PV, Sonko B, Sharp T, Jakicic JM, Blair SN, Tran ZV, Mayo M, Gibson C, Washburn RA. 2003. Effects of a 16-month randomized controlled exercise trial on body weight and composition in young, overweight men and women. Arch Intern Med 163: 1343–1350.

Doyon C, et al. 2001. Molecular evolution of leptin. Gen and Comp Endocrinol 124 (2): 188–189.

Drazen DL, Vahl TP, D'Alessio DA, Seeley RJ, Woods SC. 2006. Effects of a fixed meal pattern on ghrelin secretion: evidence for a learned response independent of nutrient status. Endocrinology 147: 23–30.

Drenick EJ, Bale GS, Seltzer F, Johnson DG. 1988. Excessive mortality and causes of death in morbidly obese men. JAMA 243: 443–445.

Drewnowski A. 2000. Nutrition transition and global dietary trends. Nutr 16: 486–487.

Drewnowski A. 2007. The real contribution of added sugars and fats to obesity. Epidemiol Rev 29: 160–171.

Drewnowski A, Darmon N. 2005. The economics of obesity: dietary energy density and energy cost. Am J Clin Nutr 82: 265S–273S.

Du S, Lu B, Zhai F, Popkin BM. 2002. A new stage of the nutrition transition in China. Pub Health Nutr 5: 169–174.

Duke-Cohan JS, Kim JH, Azouz A. 2004. Attractin: cautionary tales for therapeutic intervention in molecules with pleiotropic functionality. J Environ Pathol Toxicol Oncol 23: 1–11.

Dunbar RIM. 1998. The social brain hypothesis. Evol Anthropol 6: 178–190.

Eaton SB, Eaton SB. 2003. An evolutionary perspective on human physical activity: implications for health. Comp Biochem Physiol Pt A Mol Integr Physiol 136: 153–159.

Eaton SB, Konner, M. 1985. A consideration of its nature and current implications. N Eng J Med 312: 283–289.

Eaton SB, Nelson DA. Calcium in evolutionary perspective. 1991. Am J Clin Nutr 54 Suppl 1: 281S–287S.

Ehrenberg, HM, Durnwald CP, Catalano P, Mercer BM. 2004. The influence of obesity and diabetes on the risk of cesarean delivery. Am J of Obstetrics and Gynecology 191: 969–974.

Einstein A. 1905. Does the inertia of a body depend upon its energy content? Ann D Phys 17: 891.

Einstein F, Atzmon G, Yang X-M, Ma X-H, Rincon M, Rudin E, Muzumdar R, Barzilai N. 2005. Differential responses of visceral and subcutaneous fat depots to nutrients. Diabetes 54: 672–678.

Ellison PT. 2003. Energetics and reproductive effort. Am J Hum Biol 15 (3): 342–351.

Epstein, A. N. 1982. Mineralcorticoids and cerebral angiotensin may act to produce sodium appetite. Peptides 3: 493–494.

Erickson JC, Hollopeter G, Palmiter RD. 1996. Attenuation of the obesity syndrome of ob/ob mice by the loss of neuropeptide Y. Science 274: 1704–1707.

Erlanson-Albertsson C. 2005. How palatable food disrupts appetite regulation. Basic Clin Pharmacol Toxicol 97: 61–73.

Ezzati M, Martin H, Skjold S, Vander Hoorn S, Murray CJL. 2006. Trends in national and state-level obesity in the USA after correction for self-report bias: analysis of health surveys. J R Soc Med 99: 250–257.

Fain JN. 2006. Release of interleukins and other inflammatory cytokines by human adipose tissue is enhanced in obesity and primarily due to the nonfat cells. Vitamins and Hormones 74: 443–477.

Fain JN, Bahouth SW, Madan AK. 2004. TNFα release by the nonfat cells of human adipose tissue. Int J Obesity 28: 616–622.

Fain JN, Reed N, Saperstein R. 1967. The isolation and metabolism of brown fat cells. J Biol Chem 8: 1887–1894.

Fairweather-Tait S, Prentice A, Heumann KG, Jarjou LMA, Stirling DM, Wharf SG, Turnland JR. 1995. Effect of calcium supplements and stage of lactation on the calcium absorption efficiency of lactating women accustomed to low calcium intakes. Am J Clin Nutr 62: 1188–1192.

Farooqi IS, Keogh JM, Yeo GS, Lank EJ, Cheetham T, O'Rahilly S. 2003. Clinical spectrum of obesity and mutations in the melanocortin 4 receptor gene. N Eng J Med 348: 1085–1095.

Farquharson J, Cockburn F, Patrick WA, Jamieson EC, Logan RW. 1992. Infant cerebral cortex phospholipid fatty acid composition and diet. Lancet 340: 810–813.

Farrell JI. 1928. Contributions to the physiology of gastric secretion. Am J Physiol 85: 672–687.

Fei H, et al. 1997. Anatomic localization of alternatively spliced leptin receptors (Ob-R) in mouse brain and other tissues. PNAS 94: 7001–7005.

Feldkamp ML, Carey JC, Sadler TW. 2007. Development of gastroschisis: review of hypotheses, a novel hypothesis, and implications for research. Am J Med Genet Pt A 143: 639–652.

Feldman M, Richardson CT. 1986. Role of thought, sight, smell, and taste of food in the cephalic phase of gastric acid secretion in humans. Gastroenterology 90: 428–33.

Ferreira I, Snijder MB, Twisk JWR, Van Mechelen W, Kemper HCG, Seidell JC, Stehouwer CDA. 2004. Central fat mass versus peripheral fat and lean mass: opposite (adverse versus favorable) associations with arterial stiffness? The Amsterdam growth and health longitudinal study. JCEM 89: 2632–2639.

Feynman RP. 1964. Feynman lectures on physics. Vol. 1. Reading, MA: Addison-Wesley.

Fitzsimons JT. 1998. Angiotensin, thirst, and sodium appetite. Physiol Rev 78: 583–686.

Flatt JP. 2007. Differences in basal energy expenditure and obesity. Obesity 15: 2546–2548.

Flegal KM. 2006 Commentary: the epidemic of obesity—what's in a name? Int J Epidemiol 35: 72–74.

Flegal KM, Graubard BI, Williamson DF, Gail MH. 2007. Cause-specific excess deaths associated with underweight, overweight, and obesity. JAMA 298: 2028–2037.

Flint A, Moller BK, Raben A, Sloth B, Pedersen D, Tetens I, Holst JJ, Astrup A. 2006. Glycemic and insulinemic responses as determinants of appetite in humans. Am J Clin Nutr 84: 1365–1373.

Flynn FW, Berridge KC, Grill HJ. 1986. Pre- and postabsorptive insulin secretion in chronic decerebrate rats. Am J Physiol 250: R539–R548.

Foley RA. 2001. The evolutionary consequences of increased carnivory in hominids. In Meat-Eating and Human Evolution, Stanford CB, Bunn HT (eds.). New York: Oxford University Press.

Foley RA, Lee PC. 1991. Ecology and energetics of encephalization in hominid evolution. Phil Trans Royal Soc Ser B 334: 223–232.

Forsdahl A. 1977. Are poor living conditions in childhood and adolescence important risk factors for arteiosclerotic heart disease? Br J Prev Soc Med 31: 91–95.

Forssmann WG, Hock D, Lottspeich F, Henschen A, Kreye V, Christmann M, Reinecke M, Metz J, Catlquist M, Mutt V. 1983. The right auricle of the heart is an endocrine organ: cardiodilatin as a peptide hormone candidate. Anat Embryol 168: 307–313.

Fowler SP, Williams K, Hunt KJ, Resendez RG, Hazuda HP, Stern MP. 2005. Diet soft drink consumption is associated with increased incidence of overweight and obesity in the San Antonio heart study. ADA Annual Meeting 1058-P.

Frank BH, Willet WC, Li T, et al. 2004. Adiposity as compared with physical activity in predicting mortality among women. N Eng J Med 351: 2694–2703.

Franklin RE, Gosling RG. 1953. The structure of sodium thymonucleate fibres. I. The influence of water content. Acta Crystallographica 6: 673–677.

Freedman DS, Khan LK, Serdula MK, Galuska DA, Dietz WH. 2002. Trends and correlates of class 3 obesity in the United States from 1990 through 2000. JAMA 288: 1758–1761.

Friedmann H. 1955. The Honeyguides. United States National Museum, Bulletin 208. Washington, DC: Smithsonian Institution.

Friedman MI, Stricker EM. 1976. Evidence for hepatic involvement in control of ad libitum food intake in rats. Psychol Rev 83: 409–431.

Frisch RE, Revelle R. 1971. Height and weight at menarche and a hypothesis of menarche. Arch Dis Child 46: 695–701.

Fujioka S, Matsuzawa Y, Tokunaga K, Tarui S. 1987. Contribution of intra-abdominal fat accumulation to the impairment of glucose and lipid metabolism in human obesity. Metabolism 36: 54–59.

Gale CR, Javaid MK, Robinson SM, Law CM, Godfrey KM, Cooper C. 2007. Maternal size in pregnancy and body composition in children. JCEM 92: 3904–3911.

Gallagher D, Heymsfield SB, Moonseong H, Jebb SA, Murgatroyd PR, Sakamoto Y. 2000. Healthy percentage body fat ranges: an approach for developing guidelines based on body mass index. Am J Clin Nutr 72: 694–70.

Gallistel CR. 1980. The Organization of Action: A New Synthesis. Hillsdale, NJ: Erlbaum.

Gallup Organization (eds.). 1995. Sleep in America. Accessed at www.stanford.edu/~dement/95poll.html.

Garcia J, Hankins WG, Rusiniak KW. 1974. Behavioral regulation of the internal milieu in man and rat. Science 185: 824–831.

Garg A. 2004. Regional adiposity and insulin resistance. JCEM 89: 4206–4210.

Geier AB, Foster GD, Womble LG, McLaughlin J, Borradaile KE, Nachmani J, Sherman S, Kumanyika S, Shults J. 2007. The relationship between relative weight and school attendance among elementary schoolchildren. Obesity 15: 2157–2161.

Gentile NT, Seftchick MW, Huynh T, Kruus LK, Gaughan J. 2006. Decreased mortality by normalizing blood glucose after acute ischemic stroke. Acad Emerg Med 13: 174–180.

Gerloff U, Hartung B, Fruth B, Hohmann G, Tautz D. 1999. Intracommunity relationships, dispersal patterns, and paternity success in a wild living community of bonobos (*Pan paniscus*) determined from DNA analysis of faecal samples. Proc Biol Soc 266: 1189–1195.

German J, Dillard C. 2006. Composition, structure and absorption of milk lipids: a source of energy, fat-soluble nutrients and bioactive molecules. Critical Rev Food Sci Nutr 46: 57–92.

Gesink Law DC, Maclehose RF, Longnecker MP. 2006. Obesity and time to pregnancy. Hum Reprod 22: 414–420.

Gesta S, Blüher M, Yamamoto Y, Norris AW, Berndt J, Kralisch S, Boucher J, Lewis C, Kahn CR. 2006. Evidence for a role of developmental genes in the origin of obesity and body fat distribution. PNAS 103: 6676–6681.

Gibbs J, Smith GP. 1977. Cholecystokinin and satiety in rats and rhesus monkeys. Am J Clin Nutr 30: 758–761.

Gibbs J, Smith GP, Greenberg D. 1993. Cholecystokinin: a neuroendocrine key to feeding behavior. In Hormonally Induced Changes in Mind and Brain, Schulkin J (ed.). San Diego: Academic Press.

Gibbs J, Young RC, Smith GP. 1973. Cholecystokinin decreases food intake in rats. J Comp Physiol Psychol 84: 488–495.

Gilby IC. 2006. Meat sharing among the Gombe chimpanzees: harassment and reciprocal exchange. Anim Behav 71: 953–963.

Gilby IC, Eberly LE, Pintea L, Pussey, AE. 2006. Ecological and social influences on the hunting behavior of wild chimpanzees, *Pan troglodytes schweinfurthii*. Anim Behav 72: 169–180.

Gilby IC, Eberly LE, Wrangham WR. 2007. Economic profitability of social predation among wild chimpanzees: individual variation promotes cooperation. Anim Behav 4–10.

Gilby IC, Wrangham RW. 2007. Risk-prone hunting by chimpanzees (*Pan troglodytes schweinfurthii*) increases during periods of high diet quality. Behav Ecol and Sociobiol 61: 1771–1779.

Gil-Campos M, Aguilera CM, Cañete R, Gil A. 2006. Ghrelin: a hormone regulating food intake and energy homeostasis. Br J Nutr 96: 201–226.

Gingerich PD, et al. 2001. Origin of whales from early artiodactyls: hands and feet of Eocene Protocetidae from Pakistan. Science 293: 2239–2242.

Giovannini M, Radaelli G, Banderali G, Riva E. 2007. Low prepregnant body mass index and breastfeeding practices. J Hum Lac 23: 44–51.

Glazko GV, Koonin EV, Rogozin IB. 2005. Molecular dating: ape bones agree with chicken entrails. Trends Gen 21: 89–92.

Glazko GV, Nei M. 2003. Estimation of divergence times for major lineages of primate species. Mol Biol Evol 20: 424–434.

Gluckman P, Hanson M. 2006. Mismatch: Why Our World No Longer Fits Our Bodies. New York: Oxford University Press.

Goldman L, Cook EF, Mitchell N, Flatley M, Sherman H, Rosati R, Harrel, F, Lee K, Cohn PF. 1982. Incremental value of the exercise test for diagnosing the presence or absence of coronary artery disease. Circulation 66: 945–953.

Goldschmidt M, Redfern JS, Feldman M. 1990. Food coloring and monosodium glutamate: effects on the cephalic phase of gastric acid secretion and gastrin release in humans. Am J Clin Nutr 51: 794–797.

Goodall J. 1986. The Chimpanzees of Gombe: Patterns of Behavior. Cambridge: Harvard University Press.

Goodpaster BH, Krishnaswami S, Harris TB, Katsiaras A, Kritchevsky SB, Simonsick EM, Nevitt M, Holvoet P, Newman AB. 2005. Obesity, regional body fat distribution, and the metabolic syndrome in older men and women. Arch Intern Med 165: 777–783.

Gordon-Larsen P, Nelson MC, Page P, Popkin BM. 2006. Inequality in the built environment underlies key health disparities in physical activity and obesity. Pediatr 117: 417–424.

Gosman, GG, Katcher, HI, Legro, RS. 2006. Obesity and the role of gut and adipose hormones in female reproduction. Hum Repro Update 12: 585–601.

Gourcerol G, Coskun T, Craft LS, Mayer JP, Heiman ML, Wang L, Million M, St. Pierre DH, Taché Y. 2007. Preproghrelin-delivered peptide, obestatin, fails to influence food intake in lean or obese rodents. Obesity 15: 2643–2652.

Gourcerol G, St-Pierre DH, Taché Y. 2007. Lack of obestatin effects on food intake: should obestatin be renamed ghrelin-associated peptide (GAP)? Regulatory Peptides 141: 1–7.

Goy RW, McEwen BS. 1980. Sexual differentiation of the brain. Cambridge: MIT Press.

Grill, HJ. 2006. Distributed neural control of energy balance: contributions from hindbrain and hypothalamus. Obesity 14: 216S–221S.

Grill HJ, Kaplan JM. 2002. The neuroanatomical axis for control of energy balance. Frontiers in Neuroscience 23: 2–40.

Grill HJ, Norgren R. 1978. The taste reactivity test. II. Mimetic responses to gustatory stimuli in chronic thalamic and chronic decerebrate rats. Brain Res 143: 263–279.

Grill HJ, Smith GB. 1988. Cholecystokinin decreases sucrose intake in chronic decerebrate rats. Am J Physiol 254: R853–856.

Guilmeau S, Buyse M, Tsocas A, Laigneau JP, Bado A. 2003. Duodenal leptin stimulates cholecystokinin secretion: evidence of a positive leptin-cholecystokinin feedback loop. Diabetes 52: 1664–1672.

Gunderson EP, Rifas-Shiman SL, Oken E, Rich-Edwards JW, Kleinman KP, Taveras EM, Gillman MW. 2008. Association of fewer hours of sleep at 6 months postpartum with substantial weight retention at 1 year postpartum. Am J Epidemiol 167: 178–187.

Halaas JL, Gajiwala KS, Maffei M, Cohen SL, Chait BT, Rabinowitz D, Lallone RL, Burley SK, Friedman JM. 1995. Weight-reducing effects of the plasma protein encoded by the obese gene. Science 269: 855–856.

Hales CH, Barker DJP. 2001. The thrifty phenotype hypothesis. Brit Med Bull 60: 51–67.

Hall KRL, Schaller GB. 1064. Tool-using behavior of the California sea otter. J Mammalogy 45: 287–298.

Hallschmid M, Benedict C, Schultes B, Fem H-L, Born J, Kern W. 2004. Intranasal insulin reduces body fat in men but not in women. Diabetes 53: 3024–3029.

Hamadeh MJ, Devries MC, Tarnopolsky MA. 2005. Estrogen supplementation reduces whole body leucine and carbohydrate oxidation and increases lipid oxidation in men during endurance exercise. JCEM 90: 3592–3599.

Hambly C, Speakman JR. 2005. Contribution of different mechanisms to compensation for energy restriction in the mouse. Obesity Res 13: 1548–1557.

Hammoud AO, Gibson M, Peterson CM, Hamilton BD, Carrell DT. 2006. Obesity and male reproductive potential. J Androl 27: 619–626.

Hanover LM, White JS. 1993. Manufacturing, composition, and applications of fructose. Am J Clin Nutr 58 (suppl): 724S–732S.

Hany TF, Gharehpapagh E, Kamel EM, et al. 2002. Brown adipose tissue: a factor to consider in symmetrical tracer uptake in the neck and upper chest region. Eur J Nucl Med Mol Imaging 29: 1393–1398.

Hare B, Melis A, Woods V, Hastings S, Wrangham R. 2007. Tolerance Allows Bonobos to Outperform Chimpanzees on a Cooperative Task. Current Biol 17: 619–623.

Hart D, Sussman RW. 2005. Man the Hunted: Primates, Predators, and Human Evolution. New York: Basic Books.

Havel PJ. 2001. Peripheral signals conveying metabolic information to the brain: short-term and long-term regulation of food intake and energy homeostasis. Experimental Biol and Med 226: 963–977.

Havel PJ. 2005. Dietary fructose: implications for dysregulation of energy homeostasis and lipid/carbohydrate metabolism. Nutr Rev 63: 133–157.

Havel PJ, Kasim Karakas S, Mueller W, Johnson PR, Gingerich RL, Stern JS. 1996. Relationship of plasma leptin to insulin and adiposity in normal weight and overweight women: effects of dietary fat content and sustained weight loss. JCEM 81: 4406–4413.

Hay RL, Leakey MD. 1982. Fossil footprints of Laetoli. Sci Am Feb.: 50–57.

He Q, Horlick M, Thornton J, Wang J, Pierson RN, Jr., Heshka S, Gallagher D. 2004. Sex-specific fat distribution is not linear across pubertal groups in a multiethnic study. Obesity Res 12: 725–733.

Heaney RP, Davies KM, Barger-Lux MJ. 2002. Calcium and weight: clinical studies. J Am Coll Nutr 21: 152S–155S.

Hebebrand J, Wulftange H, Goerg T, Ziegler A, Hinney A, Barth N, Mayer H, and Remschmidt H. 2000. Epidemic obesity: are genetic factors involved via increased rates of assortative mating? Int J Obesity 24: 345–353.

Hedley AA, Ogden CL, Johnson CL, Carroll MD, Curtin LR, Flegal KM. 2004. Prevalence of overweight and obesity among US Children, Adolescents, and Adults, 1999–2002. JAMA 291: 2847–50.

Heekeren HR, Marrett S, Bandettini PA, Ungerleider LG. 2004. A general mechanism for perceptual decision-making in the human brain. Nature 431: 859–862.

Heekeren HR, Marrett S, Ruff DA, Bandettini PA, Ungerleider LG. 2006. Involvement of human left dorsolateral prefrontal cortex in perceptual decision making is independent of response modality. PNAS 103: 10023–100288.

Heindel JJ. 2003. Endocrine disruptors and the obesity epidemic. Toxicology Sci 76: 247–249.

Helmholtz, H von. 1847. Über die Erhaltung der Kraft, eine physikalische Abhandlung. Berlin: G. Reimer, 1847.

Hendler I, Blackwell S, Mehta S, Whitty J, Russell E, Sorokin Y, Cotton D. 2005. The levels of leptin, adiponectin, and resistin in normal weight, overweight, and obese pregnant women with and without preeclampsia. Am J Obstet Gynecol 193: 979–983.

Henson MC, Castracane VD. 2006. Leptin in pregnancy: an update. Biol Reprod 74: 218–229.

Henson MC, Swan KF, Edwards DE, Hoyle GW, Purcell J, Castracane VD. 2004. Leptin receptor expression in fetal lung increases in late gestation in the baboon: a model for human pregnancy. Reprod 127: 87–94.

Herbert J. 1993. Peptides in the limbic system: neurochemical codes for co-ordinated adaptive responses to behavioural and physiological demand. Neurobiology 41: 723–791.

Hershey AD, Chase M. 1952. Independent functions of viral protein and nucleic acid in growth of bacteriophage. J General Physiol 36: 39–56.

Hervey GR. 1959. The effects of lesions in the hypothalamus in parabiotic rats. J Physiol 145: 336–352.

Heyland A, Moroz LL. 2005. Cross-kingdom hormonal signaling: an insight from thyroid hormone functions in marine larvae. J Exp Biol 208: 4355–4361.

Hillman LS. 1990. Mineral and vitamin D adequacy in infants fed human milk or formula between 6 and 12 months of age. J Pediatr 117: S134–S142.

Hobolth A, Christensen OF, Mailund T, Schierup MH. 2007. Genomic relationships and speciation times of human, chimpanzee, and gorilla inferred from a coalescent hidden Markov model. PLoS Genetics 3: 294–304.

Hoffman DJ, Wang Z, Gallagher D, Heymsfield SB. 2005. Comparison of visceral adipose tissue mass in adult African Americans ans whites. Obesity Res 13: 66–74.

Hohmann G, Fruth B. 1993. Field observations on meat sharing among bonobos (*Pan paniscus*). Folia Primatologica, 60: 225–229.

Hohmann G, Fruth B. 2003. Intra- and inter-sexual aggression by bonobos in the context of mating. Behav 140 (11–12): 1389–1413.

Holick MF. 1994. Vitamin D: new horizons for the 21st century. Am J Clin Nutr 60: 619–630.

Holick MF. 2004. Vitamin D: importance in the prevention of cancer, type 1 diabetes, heart disease and osteoporosis. Am J Clin Nutr 79: 362–371.

Holliday R. 1990. Mechanisms for the control of gene activity during development. Biol Rev Cambr Philos Soc 65: 431–471.

Holliday R. 2006. Epigenetics: a historical overview. Epigenetics 1: 76–80.

Hosoi T, Kawagishi T, Okuma Y, Tanaka J, Nomura Y. 2002. Brain stem is a direct target for leptin's action in the central nervous system. Endocrinology 143: 3498–3504.

Hossain P, Kawar B, El Nahas M. 2007. Obesity and diabetes in the developing world: a growing challenge. N Eng J Med 356: 213–215.

Houseknecht KL, McGuire MK, Portocarrero CP, McGuire MA, Beerman K. 1997. Leptin is present in human milk and is related to maternal plasma leptin concentration and adiposity. Biochem Biophys Res Comm 240: 742–747.

Howlett J, Ashwell M. 2008. Glycemic response and health: summary of a workshop. Am J Clin Nutr 87 (suppl): 212S–216S.

Hsu F-C, Lenchik L, Nicklas BJ, Lohman K, Register TC, Mychaleckyj J, Langefeld CD, Freedman BI, Bowden DW, Carr JJ. 2005. Heritability of body composition measured by DXA in the Diabetes Heart Study. Obesity Res 13: 312–319.

Huising MO, et al. 2004. Structural characterization of a cyprinid (*Cyprinus carpio* L.) CRH, CRH-BP, and CRH-R1, and the role of these proteins in the acute stress response. J Molecular Endocrinol 32: 627–648.

Huising MO, Flik G. 2005. The remarkable conservation of corticotropin-releasing hormone (CRH)-binding protein in the honeybee (*Apis mellifera*) dates the CRH system to a common ancestor of insects and vertebrates. Endocrinology 146: 2165–2170.

Huising MO, Geven EJ, Kruiswijk CP, Nabuurs SB, Stolte EH, Spanings FAT, Verburg-van Kemnade BMJ, Flik G. 2006. Increased leptin expression in common carp (*Cyprinus carpio*) after food intake but not after fasting or feeding to satiation. Endocrinol 147: 5786–5797.

Hyppönen E, Power C. 2006. Vitamin D status and glucose homeostasis in the 1958 British birth cohort. Diabetes Care 29: 2244–2246.

Iacobellis G, Sharma AM. 2007. Obesity and the heart: redefinition of the relationship. Obesity Rev 8: 35–39.

Irwin M, Thompson J, Miller C, Gillin JC, Ziegler M. 1999. Effects of sleep and sleep deprivation on catecholamine and interleukin-2 levels in humans: clinical implications. JCEM 84: 1979–1985.

Isganaitis E, Lustig RH. 2005. Fast food, central nervous system insulin resistance, and obesity. Arterioscler Thromb Vasc Biol 25: 2451–2462.

Jackson KG, Robertson MD, Fielding BA, Frayn KN, Williams CM. 2002. Olive oil increases the number of triacylglycerol-rich chylomicron particles compared with other oils: an effect retained when a second standard meal is fed. Am J Clin Nutr 76: 942–949.

Jacobson P, Torgenson JS, Sjostrom L, Bouchard C. 2007. Spouse resemblance in body mass index: effects on adult obesity prevalence in the offspring generation. Am J of Epidemiol 165 (1): 101–108.

Jakimiuk AJ, Skalba P, Huterski R, Haczynski J, Magoffin DA. 2003. Leptin messenger ribonucleic acid (mRNA) content in the human placenta at term: relationship to levels of leptin in cord blood and placental weight. Gynecol Endocrinol 17: 311–316.

Jang H-J, Kokrashvili Z, Theodorakis MJ, Carlson OD, Kim B-J, Zhou J, Kim HH, Xu X, Chan SL, Juhaszova M, Bernier M, Mosinger B, Margolskee RF, Egan JM. 2007. Gut-expressed gustducin and taste receptors regulate secretion of glucagon-like peptide-1. PNAS 104: 15069–15074.

Janson CH, Terborgh JW. 1979. Age, sex, and individual specialization in foraging behavior of the brown capuchin (*Cebus apella*). Am J Phys Anthro 50: 452.

Jasienska G, Thune I, Ellison PT. 2006. Fatness at birth predicts adult susceptibility to ovarian suppression: an empirical test of the Predictive Adaptive Response hypothesis. PNAS 103: 12759–12762.

Jasienska G, Ziomkiewicz A, Lipson SF, Thune I, Ellison PT. 2005. High ponderal index at birth predicts high estradiol levels in adult women. Am J Hum Biol 18: 133–140.

Jensen MD. 2006. Is visceral fat involved in the pathogenesis of the metabolic syndrome? Human model. Obesity 14 (suppl): 20S–24S.

Jensen MD, Cryer PE, Johnson CM, Murray MJ. 1996. Effects of epinephrine on regional free fatty acid and energy metabolism in men and women. Ann Rev Physiol 33: 259–264.

Jetter KM, Cassady DL. 2005. The availability and cost of healthier food items. University of California Agricultural Issues Center, AIC Issues Brief 29: 1–6.

Ji H, Friedman MI. 1999. Compensatory hyperphagia after fasting tracks recovery of liver energy status. Physiol Behav 68: 181–186.

Ji H, Friedman MI. 2003. Fasting plasma triglyceride levels and fat oxidation predict dietary obesity in rats. Physiol Behav 78: 767–772.

Johanson D, White T. 1979. A Systematic Assessment of Early African Hominids. Science 202: 321–330.

Johnson, MS, Thomson, SC, Speakman, JR. 2001a. Effects of concurrent pregnancy and lactation in *Mus musculus*. J Exp Bio 204: 1947–1956.

Johnson, MS, Thomson, SC, Speakman, JR. 2001b. Inter-relationships between resting metabolic rate, life-history traits and morphology in *Mus musculus*. J Exp Bio 204: 1937–1946.

Johnson MS, Thomson SC, Speakman JR. 2001c. Lactation in the laboratory mouse *Mus musculus*. J Exp Bio 204: 1925–1935.

Johnson RM, Johnson TM, Londraville RL. 2000. Evidence for leptin expression in fishes. J Exp Zool 286: 718–724.

Jones M. 2007. Feast: Why Humans Share Food. New York: Oxford University Press.

Jorde LB, Wooding SP. 2004. Genetic variation, classification, and "race." Nat Genet 36: 528–533.

Jørgensen ME, Borch-Johnsen K, Bjerregaard P. 2006. Lifestyle modifies obesity-associated risk of cardiovascular disease in a genetically homogeneous population. Am J Clin Nutr 84: 29–36.

Juge-Aubrey CE, Somm E, Giusti V, Pernin A, Chicheportiche R, Verdumo C, Rohner-Jeanrenaud F, Burger D, Dayer J-M, Meier CA. 2003. Adipose tissue is a major source of interleukin-1 receptor antagonist. Diabetes 52: 1104–1110.

Kalkwarf HJ, Specker BL, Bianchi DC, Ranz J, Ho M. 1997. The effect of calcium supplementation on bone density during lactation and after weaning. N Eng J Med 337: 523–528.

Kalliomäki M, Collado MC, Salminen S, Isolauri E. 2008. Early differences in fecal microbiota composition in children may predict overweight. Am J Clin Nutr 87: 534–538.

Kamagai J. 2001. Chronic central infusion of ghrelin increases hypothalamic neuropeptide Y and agouti-related protein mRNA levels and body weight in rats. Diabetes 50 (11): 2438–2443.

Karelis AD, Brochu M, Rabasa-Lhoret R. 2004. Can we identify metabolically healthy but obese individuals (MHO)? Diabetes and Metabol 30: 569–572.

Karelis AD, Faraj M, Bastard JP, St-Pierre DH, Brochu M, Prud'homme D, Rabasa-Lhoret R. 2005. The metabolically healthy but obese individual presents a favorable inflammation profile. JCEM 90: 4145–4150.

Karelis AD, St-Pierre DH, Conus F, Rabasa-Lhoret R, Poehlman ET. 2004. Metabolic and body composition factors in subgroups of obesity: what do we know? JCEM 89: 2569–2575.

Katschinski M. 2000. Nutritional implications of cephalic-phase gastrointestinal responses. Appetite 34: 189–196.

Katschinski M, Dahmen G, Reinshagen M, Beglinger C, Koop H, Nustede R, Adler G. 1992. Cephalic stimulation of gastrointestinal secretory and motor responses in humans. Gastroenterology 103: 383–391.

Kawai K, Sugimoto K, Nakashima K, Miura H, Ninomiya Y. 2000. Leptin as a modulator of sweet taste sensitivities in mice. PNAS 97: 11044–11049.

Keita SOY, Kittles RA, Royal CDM, Bonney GE, Furbert-Harris P, Dunston GM, Rotimi CN. 2004. Conceptualizing human variation. Nat Genet 36: S17–S20.

Kelly K. 1993. Environmental enrichment for captive wildlife through the simulation of gum feeding. Animal Welfare Information Center Newsletter 4 (3): 1–2, 5–10. Accessed at www.nal.usda.gov/awic/newsletters/v4n3/4n3.htm.

Kenagy GJ, Vleck D. 1982. Daily temporal organization of metabolism in small mammals: adaptation and diversity. In Vertebrate Circadian Systems, Aschoff J, Dann S, Groos GA (eds.). Berlin: Springer-Verlag.

Kennedy A, Gettys TW, Watson P, Wallace P, Ganaway E, Pan Q, Garvey WT. 1997. The metabolic significance of leptin in humans: gender-based differences in relationship to adiposity, insulin sensitivity, and energy expenditure. JCEM 82: 1293–1300.

Kennedy GC. 1953. The role of depot fat in the hypothalamic control of food intake in the rat. Proc Royal Soc London 140: 578–592.

Kenny DE, Irlbeck NA, Chen TC, Lu Z, Holick MF. 1999. Determination of vitamins D, A, and E in sera and vitamin D in milk from captive and free-ranging polar bears (Ursus mauritimus), and 7-dehydrocholecterol levels in skin from captive polar bears. Zoo Biol 17: 285–293.

Kershaw EE, Flier JS. 2004. Adipose tissue as an endocrine organ. JCEM 89: 2548–2556.

Keskitalo K, Knaapila A, Kallela M, Palotie A, Wessman M, Sammalisto S, Peltonen L, Tuorila H, Perola M. 2007. Sweet taste preferences are partly genetically determined: identification of a trait locus on chromosome 16. Am J Clin Nutr 86: 55–63.

Kim S, Popkin BM. 2006. Current perspectives on obesity and health: black and white, or shades of grey? Int J Epidemiol 35: 69–71.

Kissileff HR, Pi-Sunyer X, Thornton J, Smith GP. 1981. C-terminal octapeptide of cholecystokinin decreases food intake in man. Am J Clin Nutr 34: 154–160.

Kitano H, Oda K, Matsuoka Y, Csete M, Doyle J, Muramatsu M. 2004. Metabolic syndrome and robustness tradeoffs. Diabetes 53 (suppl 3): S6–S15.

Kleiber M. 1932. The Fire of Life. Huntington, NY: Robert E. Krieger.

Kluger MJ, Rothenburg BA. 1979. Fever and reduced iron: their interaction as a host defense response to bacterial infection. Science 203: 374–376.

Knowler WC, Pettitt DJ, Saad MF, Bennett PH. 1990. Diabetes mellitus in the Pima Indians: incidence, risk factors and pathogenesis. Diabetes Metab Rev 6: 1–27.

Knutson KL, Spiegel K, Penev P, van Cauter E. 2007. The metabolic consequences of sleep deprivation. Sleep Med Rev 11: 163–178.

Kochan Z. 2006. Leptin is synthesized in the liver and adipose tissue of the dunlin (Calidris alpine). Gen Comp Endocrinol 148: 336–339.

Kojima M, Hosoda H, Date Y, Nakazato M, Matsuo H, Kangawa K. 1999. Ghrelin is a growth-hormone-releasing acylated peptide from stomach. Nature 402: 656–660.

Kos K, Harte AL, James S, Snead DR, O'Hare JP, McTernan PG, Kumar S. 2007. Secretion of neuropeptide Y in human adipose tissue and its role in maintenance of adipose tissue mass. Am J Physiol Endocrinol Metab 293: E1335–E1340.

Koska, J, DelParigi, A, de Courten, B, Weyer, C, Tataranni, PA. 2004. Pancreatic polypeptide is involved in the regulation of body weight in Pima Indian male subjects. Diabetes 53: 3091–3096.

Kothapalli KSD, Anthony JC, Pan BS, Hsieh AT, Nathanielsz PW, and Brenna JT. 2007. Differential cerebral cortex transcriptomes of baboon neonates consuming moderate and high docosahexaenoic acid formulas. PLoS One 2 (4): e370.

Kothapalli KSD, Pan BS, Hsieh AT, Anthony JC, Nathanielsz PW, and Brenna JT. 2006. Comprehensive differential transcriptome analysis of cerebral cortex of baboon neonates consuming arachidonic acid and moderate and high docosahexaenoic acid formulas. FASEB J 20: A1347.

Koutsari C, Jensen MD. 2006. Free fatty acid metabolism in human obesity. J Lipid Res 47: 1643–1650.

Kovacs CS, Kronenberg IIM. 1998. Maternal-fetal calcium and bone metabolism during pregnancy, puerperium, and lactation. Endocrine Rev 18: 832–872.

Kovacs P, Harper I, Hanson RI, Infante AM, Bogardus C, Tataranni PA, Baier LJ. 2004. A novel missense substitution (Va11483Ile) in the fatty acid synthase gene (FAS) is associated with percentage of body fat and substrate oxidation rates in nondiabetic Pima Indians. Diabetes 53: 1915–1919.

Kratzsch J, Lammert A, Bottner A, Seidel B, Mueller G, Thiery J, Hebebrand J, Kiess W. 2002. Circulating soluble leptin receptor and free leptin index during childhood, puberty, and adolescence. JCEM 87: 4587–4594.

Kripke D, Simons R, Garfinkel L, Hammond E. 1979. Short and long sleep and sleeping pills. Is increased mortality associated? Arch Gen Psychiatry 36: 103–116.

Kugyelka JG, Rasmussen KM, Frongillo EA. 2004. Maternal obesity is negatively associated with breastfeeding success among Hispanic but not black women. J Nutr 134: 1746–1753.

Kuk JL, Katzmarzyk PT, Nichaman MZ, Church TS, Blair SN, Ross R. 2006. Visceral fat is an independent predictor of all-cause mortality in men. Obesity 14: 336–341.

Kuk JL, Lee SJ, Heymsfield SB, Ross R. 2005. Waist circumference and abdominal adipose tissue distribution: influence of age and sex. Am J Clin Nutr 81: 1330–1334.

Kunz, LH, King, JC. 2007. Impact of maternal nutrition and metabolism on health of the offspring. Seminars in Fetal and Neonatal Med 12: 71–77.

Kuo LE, Kitlinska JB, Tilan JU, Baker SB, Johnson MD, Lee EW, Burnett MS, Fricke ST, Kvetnansky R, Herzog H, Zukowska Z. 2007. Neuropeptide Y acts directly in the periphery on fat tissue and mediates stress-induced obesity and metabolic syndrome. Nat Med 13: 803–811.

Kuzawa CW. 1998. Adipose tissue in human infancy and childhood: an evolutionary perspective. Yrbk Phys Anthropol 41: 177–209.

Kuzawa CW, Quin EA, Adair LS. 2007. Leptin in a lean population of Filipino adolescents. Am J Phys Anthro 132: 642–649.

Laaksonen M, Piha K, Sarlio-Lähteekorva S. 2007. Relative weight and sickness absence. Obesity 15: 465–472.

Laden G, Wrangham R. 2005. The rise of hominids as an adaptive shift in fallback foods: plant underground storage organs (USOs) and australpith origins. J Hum Evol 49: 482–498.

Laird SM, Quinton N, Anstie B, Li TC, Blakemore AIF. 2001. Leptin and leptin binding activity in recurrent miscarriage women: correlation with pregnancy outcome. Human Reproduction 16: 2008–2013.

Lammert A, Kiess W, Glasow A, Bottner A, Kratzsch J. 2001. Different isoforms of the soluble leptin receptor determine the leptin binding activity of human circulating blood. Biochem Biophys Res Commun 283: 982–988.

Lamont LS. 2005. Gender differences in amino acid use during endurance exercise. Nutr Rev 63: 419–422.

Lamont LS, McCullough AJ, Kalhan SC 2001. Gender differences in leucine, but not lysine, kinetics. J Appl Physiol 91: 357–362.

Lamonte MJ, Blair SN. 2006. Physical activity, cardiorespiratory fitness, and adiposity: contributions to disease risk. Curr Opin Clin Nutr Metab Care 9: 540–546.

Laugerette F, Passilly-Degrace P, Patris B, Niot I, Febbraio M, Montmayeur J-P, Besnard P. 2005. CD36 involvement in orosensory detection of dietary lipids, spontaneous fat preference, and digestive secretions. J Clin Invest 115: 3177–3184.

Lê K-A, Tappy L. 2006. Metabolic effects of fructose. Curr Opin Clin Nutr Metab Care 9: 469–475.

Leakey MD, Roe DA (eds.). 1994. Olduvai Gorge. Vol. 5, Excavations in Beds III, IV, and the Masek Beds, 1968–1971. Cambridge: Cambridge University Press.

LeBlanc J, Soucy J, Nadeau A. 1996. Early insulin and glucagon responses to different food items. Horm Metab Res 28: 276–279.

Lee AT, Plump A, DeSimone C, Cerami A, Bucala R. 1995. A role for DNA mutations in diabetes associated teratogenesis in transgenic embryos. Diabetes 44: 20–24.

Lee CD, Blair S, Jackson A. 1999. Cardiorespiratory fitness, body composition,

and all-cause and cardiovascular disease mortality in men. Am J Clin Nutr
69: 373–380.

Lee H-M, Wang G, Englander EW, Kojima M, Greeley GH, Jr. 2002. Ghrelin, a
new gastrointestinal endocrine peptide that stimulates insulin secretion:
enteric distribution, ontogeny, influence of endocrine, and dietary manipulations. Endocrinol 143: 185–190.

Lee JM, Appugliese D, Kaciroti N, Corwyn RF, Bradley RH, Lumeng JC. 2007.
Weight status in young girls and the onset of puberty. Pediatr 119: E624–
E630.

LeGrande EK, Brown CC. 2002. Darwinian medicine: applications of evolutionary biology for veterinarians. Can Vet J 43: 556–559.

Leibowitz SF, Chang G-Q, Dourmashkin JT, Yun R, Julien C, Pamy PP. 2006.
Leptin secretion after a high-fat meal in normal-weight rats: strong predictor
of long-term body fat accrual on a high-fat diet. Am J Physiol Endocrinol
Metab 290: E258–E267.

Leitzmann MF, Park Y, Blair A, Ballard-Barbash R, Mouw T, Hollenbeck AR,
Schatzkin A. 2007. Physical activity recommendations and decreased risk of
mortality. Arch Intern Med 167: 2453–2460.

Lemieux S, Prud'homme D, Bouchard C, Tremblay A, Després J-P. 1993. Sex differences in the relation of visceral adipose tissue accumulation to total body
fatness. Am J Clin Nutr 58: 463–467.

Leonard, WR, Robertson, ML. 1992. Nutritional requirements and human evolution: a bioenergetics model. Am J Hum Biol 4: 179–195.

Leonard, WR, Robertson, ML. 1994. Evolutionary perspectives on human nutrition: the influence of brain and body size on diet and metabolism. Am J Hum
Biol 6: 77–88.

Leonard, WR, Robertson, ML, Snodgrass, JJ, Kuzawa, CW. 2003. Metabolic
correlates of hominid brain evolution. Comp Biochem Physiol A 135: 5–15.

Leonard, WR, Snodgrass, JJ, Robertson, ML. 2007. Effects of brain evolution on
human nutrition and metabolism. Ann Rev Nutr 27: 311–327.

Leperq J, Challier JC, Guerre-Millo M, Cauzac M, Vidal H, Haugel-de Mouzon
S. 2001. Prenatal leptin production: evidence that fetal adipose tissue produces leptin. JCEM 86: 2409–2413.

Lewis K, Li C, Perrin MH, Blount A, Kunitake K, Donaldson C, Vaughan J, Reyes
TM, Gulyas J, Fischer W, Bilezikjian L, Rivier J, Sawchenko PE, Vale WW.
2001. Identification of urocortin III, an additional member of the corticotropin-releasing factor (CRF) family with high affinity for the CRF2 receptor.
PNAS USA 98: 7570–7575.

Ley RE, Turnbaugh PJ, Klein S, Gordon JI. 2006. Microbial ecology: human gut
microbes associated with obesity. Nature 444: 1022–1023.

Li H-j, Ji C-y, Wang W, Hu Y-h. 2005. A twin study for serum leptin, soluble leptin receptor, and free insulin-like growth factor-I in pubertal females. JCEM 90: 3659–3664.

Li X, Li W, Wang H, Bayley DL, Cao J, Reed DR, Bachmanov AA, Huang L, Legrand-Defretin V, Beauchamp GK, Brand JG. 2006. Cats lack a sweet taste receptor. J Nutr 136: 1932S–1934S.

Licinio J, Negrão AB, Mantzoro C, Kaklamani V, Wong M-L, Bongiorno PB, Mulla A, Cearnal L, Veldhuis JD, Flier JS, McCann SM, Gold PW. 1998. Synchronicity of frequently sampled, 24-hr concentrations of circulating leptin, luteinizing hormone, and estradiol in healthy women. PNAS USA 95: 2541–2546.

Lietzmann MF, Park Y, Blair A, Ballard-Barbash R, Mouw T, Hollenbeck AR, Schatzkin A. 2007. Physical activity recommendations and decreased risk of mortality. Arch Intern Med 167: 2453–2460.

Lihn AS, Pedersen SB, Richelsen B. 2005. Adiponectin: action, regulation and association to insulin sensitivity. Obesity Rev 6: 13–21.

Lindeberg S, Cordain L, Eaton SB. 2003. Biological and clinical potential of a Paleolithic diet. J Nutr and Enviro Med 13: 149–160.

Linder K, Arner P, Flores-Morales A, Tollet-Egnell P, Norstedt G. 2004. Differentially expressed genes in visceral or subcutaneous adipose tissue of obese men and women. J Lipid Res 45: 148–154.

List JF, Habener JF. 2003. Defective melanocortin 4 receptors in hyperphagia and morbid obesity. N Eng J Med 348: 1160–1163.

Lostao MP, Urdaneta E, Martinez-Anso E, Barber A, Martinez JA. 1998. Presence of leptin receptors in rat small intestine and leptin effect on sugar absorption. FEBS Lett 423: 302–306.

Lourenço AEP, Santos RV, Orellana JDY, Coimbra CEA. 2008. Nutrition transition in Amazonia: obesity and socioeconomic change in the Suruí Indians from Brazil. Am J Hum Bio 00: 000–000.

Lu GC, Rouse DJ, DuBard M, et al. 2001. The effect of the increasing prevalence of maternal obesity on perinatal morbidity. Am J Obstet Gynecol 185: 845–849.

Ludwig DS. 2000. Dietary glycemic index and obesity. J Nutr 130 (suppl): 280S–283S.

Luscombe-Marsh ND, Smeets AJPG, Westerterp-Plantenga MS. 2008. Taste sensitivity for monosodium glutamate and an increased liking of dietary protein. Br J Nutr 99: 904–908.

Ma L, Hanson RL, Que LN, Cali AMG, Fu M, Mack JL, Infante AM, Kobes S, Bogardus C, Shuldiner AR, Baier LJ. 2007. Variants in *ARHGEF11*, a candidate gene for the linkage to type 2 diabetes on chromosomes 1q, are nominally associated with insulin resistance and type 2 diabetes in Pima Indians. Diabetes 56: 1454–1459.

Ma L, Tataranni PA, Bogardus C, Baier LJ. 2004. Melanocortin 4 receptor gene variation is associated with severe obesity in Pima Indians. Diabetes 53: 2696–2699.

Ma L, Tataranni PA, Hanson RL, Infante AM, Kobes S, Bogardus C, Baier LJ. 2005. Variations in peptide YY and Y2 receptor genes are associated with severe obesity in Pima Indian men. Diabetes 54: 1598–1602.

MacLean PS, Higgins JA, Jackman M, Johnson GC, Fleming-Elder BK, Wyatt H, Melanson EL, Hill JO. 2006. Peripheral metabolic responses to prolonged weight reduction that promote rapid, efficient regain in obesity-prone rats. Am J Physiol Regul Integr Comp Physiol 290: 1577–1588.

Mallon L, Broman JE, Hetta J. 2005. High incidence of diabetes in men with sleep complaints or short sleep duration. Diabetes Care 28: 2762–2767.

Margolskee RF, Dyer J, Kokrashvili Z, Salmon KS, Ilegems E, Daly K, Maillet EI, Ninomiya Y, Mosinger B, Shirazi-Beechy SP. 2007. T1R3 and gustducin in gut sense sugars to regulate expression of Na⁺-glucose cotransporter 1. PNAS 104: 15075–15080.

Mars M, de Graaf C, de Groot L, Kok FJ. 2005. Decreases in fasting leptin and insulin concentrations after acute energy restriction and subsequent compensation in food intake. Am J Clin Nutr 81: 570–577.

Martin RD. 1981. Relative brain size and basal metabolic rate in terrestrial vertebrates. Nature 293: 57–60.

Martin RD. 1983. Human Brain Evolution in an Ecological Context. New York: American Museum of Natural History.

Martin RD. 1996. Scaling of the mammalian brain: the maternal energy hypothesis. News in Physiol Sciences 11: 149–156.

Martínez V, Barrachina MD, Ohning G, Taché Y. 2002. Cephalic phase of acid secretion involves activation of medullary TRH receptor subtype 1 in rats. Am J Physiol Gastrointest Liver Physiol 283: G1310–G1319.

Matkovic V, Ilich JZ, Skugor M, Badenhop NE, Goel P, Clairmont A, Klisovic D, Nahhas RW, Landoll JD. 1997. Leptin is inversely related to age at menarche in human females. JCEM 82: 3239–3245.

Matson CA, Ritter RC. 1999. Long-term CCK-leptin synergy suggests a role for CCK in the regulation of body weight. Am J Physiol Regul Integr Comp Physiol 276: R1038–R1045.

Matter KC, Sinclair SA, Hostetler SG, Xiang H. 2007. A comparison of the characteristics of injuries between obese and non-obese inpatients. Obesity 15: 2384–2390.

Mattes RD. 2002. Oral fat exposure increases the first phase triacylglycerol concentration due to release of stored lipid in humans. J Nutr 132: 3656–3662.

Mattes RD. 2005. Fat taste and lipid metabolism in humans. Physiol Behav 86: 691–697.

Maynard LA, Loosli JK, Hintz HF, Warner RG. 1979. Animal Nutrition. 7th ed. New York: McGraw-Hill.

McDowell MA, Brody DJ, Hughs JP. 2007. Has age at menarche changed? Results from the National Health and Nutrition Examination Survey (NHANES) 1999–2004. J Adolescent Health 40: 227–231.

McEwen BS. 1998. Stress, adaptation, and disease: allostasis and allostatic load. Ann NY Acad Sci 840: 33–44.

McEwen BS. 2000. Allostasis and allostatic load: implications for neuropsychopharmacology. Neuropsychopharmacology 22: 108–124.

McEwen BS. 2005. Stressed or stressed out: what is the difference? J Psychiatry Neurosci 30: 315–318.

McEwen BS. 2007. Physiology and neurobiology of stress and adaptation: central role of the brain. Physiol Rev 87: 873–904.

McEwen BS, Stellar E. 1993. Stress and the individual: mechanisms leading to disease. Arch Int Med 153: 2093–2101.

McGrew WC, Brennan JA, Russell J. 1986. An artificial "gum-tree" for marmosets (*Callithrix j. jacchus*). Zoo Biology 5: 45–50.

McHenry HM, Coffing K. 2000. *Australopithecus* to *Homo*: transformations in body and mind. Ann Rev Anthropol 29: 125–146.

McNab BK, Brown JH. 2002. The Physiological Ecology of Vertebrates: A View from Energetics. Ithaca: Cornell University Press.

Melis AP, Hare B, Tomasello M. 2006. Engineering cooperation in chimpanzees: tolerance constraints on cooperation. Anim Behav 72: 275–286.

Mendel G. 1865. Experiments in plant hybridization. Meetings of the Brunn Nat Hist Soc, Brno, current Czech Republic. February 8 and March 8, 1865.

Merchant JL. 2007. Tales from the crypts: regulatory peptides and cytokines in gastrointestinal homeostasis and disease. J Clin Invest 117: 6–12.

Miescher F. 1871 Der physiologische Process der Athmung. Akademische Habilitationsrede 1871. In Die Histochemischen und Physiologischen Arbeiten von Friedrich Miescher--A, Arbeiten von F. Miescher. W His et al. (eds.), 35–54. Vol. 2. Leipzig: FCW Vogel.

Miles R. 2008. Neighborhood disorder, perceived safety, and readiness to encourage use of local playgrounds. Am J Prev Med 34: 275–281.

Milligan LA. 2005. Concentration of sIgA in the milk of *Macaca mulatta* [abstract]. Am J of Phys Anthropol Annual Meeting Issue: 153.

Milligan LA. 2008. Nonhuman primate milk composition: relationship to phylogeny, ontogeny and, ecology. PhD diss., University of Arizona.

Milligan LA, Rapoport SI, Cranfield MR, Dittus W, Glander KE, Oftedal OT, Power ML, Whittier CA, Bazinet RP. 2008. Fatty acid composition of wild anthropoid primate milks. Comp Biochem Physiol Pt B 149: 74–82.

Millikan GC, Bowman RI. 1967. Observations of Galapagos tool-using finches in captivity. Living Bird 6: 23–41.

Milton K. 1987. Primate diets and gut morphology: implications for hominid evolution. In Food and Evolution: Toward a Theory of Food Habits, Harris M, Ross EB (eds.), 93–115. Philadelphia: Temple University Press.

Milton K. 1988. Foraging behavior and the evolution of primate cognition. In Machiavellian Intelligence: Social Expertise and the Evolution of Intellect in Monkeys, Apes, and Humans, Whiten A and Byrne R (eds.), 285–305. New York: Oxford University Press.

Milton K. 1999a. Nutritional characteristics of wild Primate foods: do the natural diets of our closest living relatives have lessons for us? Nutrition 15: 488–498.

Milton K. 1999b. A hypothesis to explain the role of meat-eating in human evolution. Evol Anthropol 8: 11–21.

Milton K, Demment MW. 1988. Digestion and passage kinetics of chimpanzees fed high and low fiber diets and comparison with human data. J Nutr 118: 1082–1088.

Mistry AM, Swick A, Romsos DR. 1999. Leptin alters metabolic rates before acquisition of its anorectic effect in developing neonatal mice. Am J Physiol Regul Integr Comp Physiol 277: R742–R747.

Mitani JC. 2006. Demographic influences on the behavior of chimpanzees. Primates 47: 6–13.

Mitani JC, Watts DP. 1999. Demographic influences on the hunting behavior of chimpanzees. Am J Phys Anthropol 109: 439–454.

Mittendorfer B. 2003. Sexual dimorphism in human lipid metabolism. J Nutr 135: 681–686.

Mizuno TM, Bergen H, Funabashi T, Kleopoulos SP, Zhong YG, Bauman WA, Mobbs CV. 1996. Obese gene expression: reduction by fasting and stimulation by insulin and glucose in lean mice, and persistent elevation in acquired (diet-induced) and genetic (yellow *agouti*) obesity. PNAS 93: 3434–3438.

Mock CN, Grossman DC, Kaufman RP, Mack CD, Rivara FP. 2002. The relationship between body weight and risk of death and serious injury in motor vehicle crashes. Accid Anal Prev 34: 221–228.

Mojtabai R. 2004. Body mass index and serum folate in childbearing women. Eur J Epidemiol 19: 1029–1036.

Monro JA, Shaw M. 2008. Glycemic impact, glycemic glucose equivalents, glycemic index, and glycemic load: definitions, distinctions, and implications. Am J Clin Nutr 87 (suppl): 237S–243S.

Montecucchi PC, Henschen A. 1981. Amino acid composition and sequence analysis of sauvagine, a new active peptide from the skin of *Phyllomedusa sauvagei*. Int J Pept Protein Res. 18: 113–120.

Monteiro CA, Conde WL, Popkin BM. 2004. The burden of disease from undernutrition and overnutrition in countries undergoing rapid nutrition transition: a view from Brazil. Am J Pub Health 94: 433–434.

Moore TR. 2004. Adolescent and adult obesity in women: a tidal wave just beginning. Clin Obstet Gynecol 47: 884–9.

Moore-Ede MC. 1986. Physiology of the circadian timing system: predictive versus reactive homeostasis. Am J Physiol 250: R737–752.

Moran TH, Kinzig KP. 2004. Gastrointestinal satiety signals II. Cholecystokinin. Am J Physiol Gastrointest Liver Physiol 286: G183–G188.

Morris JG. 1999. Ineffective vitamin D synthesis in cats is reversed by an inhibitor of 7-dehydrocholesterol-Δ^7-reductase. J Nutr 129: 903–908.

Morris KL, Zemel MB. 2005. 1,25-dihydroxyvitamin D3 modulation of adipocyte glucocorticoid function. Obesity Res 13: 670–677.

Morton NM, Emilsson V, Liu YL, Cawthorne MA. 1998. Leptin action in intestinal cells. J Biol Chem 273: 26194–26201.

Mountain JL, Risch N. 2004. Assessing genetic contributions to phenotypic differences among "racial" and "ethnic" groups. Nat Genet 36: S48–S53.

Mrosovsky N. 1990. Rheostasis: The Physiology of Change. New York: Oxford University Press.

Muglia LJ. 2000. Genetic analysis of fetal development and parturition control in the mouse. Pediatr Res 47: 437–443.

Narayan KMV, Boyle JP, Thompson TJ, Gregg EW, Williamson DF. 2007. Effect of BMI on lifetime risk for diabetes in the U.S. Diabetes Care 30: 1562–1566.

Natalucci G, Reidl S, Gleiss A, Zidek T, Frisch H. 2005. Spontaneous 24-h ghrelin secretion pattern in fasting subjects: maintenance of a meal-related pattern. Eur L Endocrinol 152: 845–850.

National Academy of Sciences. 2006. Assessing fitness for military enlistment: physical, medical, and mental health standards. Committee on Youth Population and Military Recruitment: Physical, Medical, and Mental Health Standards, National Research Council.

National Center for Health Statistics. 2005. Quick stats: percentage of adults who reported an average of $\geq$ 6 hours of sleep per 24-hour period, by sex and age group—United States, 1985 and 2004. JAMA 294: 2692.

Nead KG, Halterman JS, Kaczorowski JM, Auinger P, Weitzman M. 2004. Over-weight children and adolescents: a risk group for iron deficiency. Pediatrics 114: 104–108.

Nedergaard J, Bengtsson T, Cannon B. 2007. Unexpected evidence for active brown adipose tissue in adult humans. Am J Physiol--Endocrinol and Metabol 293: E444–E452.

Neel JV. 1962. Diabetes mellitus: a "thrifty" genotype rendered detrimental by "progress"? Am J Hum Genet 14: 353–362.

Nesse RM, Berridge KC. 1997. Psycoactive drug use in evolutionary perspective. Science 278: 63–66.

NHLBI press release. 2008. For safety, NHLBI changes intensive blood sugar treatment in trial of diabetes and cardiovascular disease. February 6. Accessed at www.nhlbi.nih.gov/health/prof/heart/other/accord/.

Nicholls DG. 2001. A history of UCP1. Biochem Soc Trans 29: 751–755.

Nicholls DG, Rial E. 1999. A history of the first uncoupling protein, UCP1. J Bioenergetics Biomembranes 31: 399–406.

Nielsen S, Guo ZK, Albu JB, Klein S, O'Brien PC, Jensen MD. 2003. Energy expenditure, sex, and endogenous fuel availability in humans. J Clin Invest 111: 981–988.

Nielson S, Guo ZK, Johnson M, Hensrud DD, Jensen MD. 2004. Splanchic lipolysis in human obesity. J Clin Invest 113: 1582–1588.

Niijima A, Togiyama T, Adachi A. 1990. Cephalic-phase insulin release induced by taste stimulus of monosodium glutamate (umami) taste. Physiol Behav 48: 905–908.

Nilsson PM, Rööst M, Engström G, Hedblad B, Berglund G. 2004. Incidence of diabetes in middle-aged men is related to sleep disturbances. Diabetes Care 27: 2464–2469.

Norgan NG. 1990. Body mass index and body energy stores in developing countries. Euro J Clin Nutr 44: 79–84.

Norgan NG, Ferro-Luzzi A. 1982. Weight-height indices as estimators of fatness in men. Human Nutr–Clin Nutr 36: 363–372.

Norgren R. 1995. Gustatory system. In The Rat Nervous System, Pazinos G (ed.). New York: Academic Press.

Oddy DJ. 1970. Food in nineteenth-century England: nutrition in the first urban society. Proc of the Nutr Soc 29: 150–157.

Oftedal OT. 1984. Milk composition, milk yield, and energy output at peak lactation: a comparative review. Symp Zool Soc Lon 51: 33–85.

Oftedal OT. 1993. The adaptation of milk secretion to the constraints of fasting in bears, seals, and baleen whales. J of Dairy Sci 76: 3234–3246.

Oftedal OT, Alt GL, Widdowson EM, Jakubasz MR. 1993. Nutrition and growth of suckling black bears (*Ursus americanus*) during their mothers' winter fast. Brit J Nutr 70: 59–79.

Ogden CL, Carrol MD, Curtin LR, McDowell MA, Tabak CJ, Flegal KM. 2006. Prevalence of overweight and obesity in the United States, 1999–2004. JAMA 295: 1549–1555.

Ogden CL, Fryar CD, Carroll MD, Flegal KM. 2004. Mean body weight, height, and body mass index, United States, 1960–2002. Advance Data from Vital and Health Statistics 347: 1–18. Accessed at www.cdc.gov/nchs/data/ad/ad347.pdf.

Oguma Y, Sesso HD, Paffenbarger RS, Lee IM. 2002. Physical activity and all cause mortality in women: a review of the evidence. Br J Sports Med 36: 162–172.

Ohara I, Otsuka S, Yugari Y. 1988. Cephalic-phase response of pancreatic exocrine secretion in conscious dogs. Am J Physiol Gastrointest Liver Physiol 254: G424–G428.

Okawara Y, Morley SD, Burzio LO, Zwiers H, Lederis K, Richter D. 1988 Cloning and sequence analysis of cDNA for corticotropin-releasing factor precursor from the teleost fish *Catostomus commersoni*. PNAS 85: 8439–8443.

O'Keefe JH, Cordain L. 2004. Cardiovascular disease resulting from a diet and lifestyle at odds with our Paleolithic genome: how to become a 21st-century hunter-gatherer. Mayo Clin Proc 79: 101–108.

Olivereau M, Olivereau J. 1988. Localization of CRF-like immunoreactivity in the brain and pituitary of teleost fish. Peptides 9: 13–21.

O'Reardon JP, Ringel BL, Dinges DF, Allison KC, Rogers NL, Martino NS, Stunkard AJ. 2004. Circadian eating and sleeping patterns in the night eating syndrome. Obesity Res 12: 1789–1796.

Østbye T, Dement JM, Krause KM. 2007. Obesity and workers' compensation. Arch Intern Med 167: 766–773.

Ostlund RE, Yang JW, Klein S, Gingerich R. 1996. Relation between plasma leptin concentration and body fat, gender, diet, age, and metabolic covariates. JCEM 81: 3909–3913.

O'Sullivan AJ, Kriketos AD, Martin A, Brown MA. 2006. Serum adiponectin levels in normal and hypertensive pregnancy. Hypertension in Pregnancy 25: 193–203.

Paczoska-Eliasiewicz HE, Gertler A, Proszkowiec M, Proudman J, Hrabia A, Sechman A, Mika M, Jacek T, Cassy S, Raver N, Rzasa J. 2003. Attenuation by leptin of the effects of fasting on ovarian function in hens (*Gallus domesticus*). Reproduction 126 (6): 739–751.

Paczoska-Eliasiewicz HE, Proszkowiec-Weglarz M, Proudman J, Jacek T, Mika

M, Sechman A, Rzasa J, Gertler A. 2006. Exogenous leptin advances puberty in domestic hen. Domestic Animal Endocrinol 31: 211–226.

Pannacciulli N, Le DS, Salbe AD, Chen K, Reiman EM, Tataranni PA, Krakoff J. 2007. Postprandial glucagon-like peptide-1 (GLP-1) response is positively associated with changes in neuronal activity of brain areas implicated in satiety and food intake regulation in humans. Neuroimage 35: 511–517.

Papas MA, Alberg AJ, Ewing R, Helzlsouer KJ, Gary TL, Klassen AC. 2007. The built environment and obesity. Epidemiol Rev 29: 129–143.

Park Y-W, Allison DB, Heymsfield SB, Gallagher D. 2001. Larger amounts of visceral adipose tissue in Asian Americans. Obesity Res 9: 381–387.

Parra R. 1978. Comparison of foregut and hindgut fermentation in herbivores. In Ecology of Arboreal Folivores, Montgomery GG (ed.), 205–229. Washington, DC: Smithsonian Institution Press.

Parsons TJ, Power C, Manor O. 2001. Fetal and early life growth and body mass index from birth to early adulthood in 1958 British cohort: longitudinal study. BMJ 323: 1331–1335.

Pasquali R, Cantobelli S, Casimirri F, Capelli M, Bortoluzzi L, Flamia R, Labate AMM, Barbara L. 1993. The hypothalamic-pituitary-adrenal axis in obese women with different patterns of body fat distribution. JCEM 77: 341–346.

Pasquali R, Gambineri A, Pagotto U. 2006. The impact of obesity on reproduction in women with polycystic ovary syndrome. BJOG 113: 1148–1159.

Pasquali R, Pelusi C, Genghini S, Cacciari M, Gambineri A. 2003. Obesity and reproductive disorders in women. Hum Repro Update 9: 359–372.

Patel MS, Srinivasan M. 2002. Metabolic programming: causes and consequences. J Biol Chem 277: 1629–1632.

Paulsen IT, Press CM, Ravel J, Kobayashi DY, Myers GA, Mavrod DV, Deboy RT, Seshadri R, Ren Q, Madupu R, Dodson RJ, Durkin AS, Brinkac LM, Daugherty SC, Sullivan SA, Rosovitz MJ, Gwinn ML, Zhou L, Nelson WC, Weidman J, Watkins K, Tran K, Khouri H, Pierson EA, Pierson III LS, Thomashow LS, Loper JE. 2005. Complete genome sequence of the plant commensal pseudomonas fluorescens pf-5: insights into the biological control of plant disease. Nature Biotech 23: 873–878.

Pavlov IP. 1902. The Work of the Digestive Glands. London: Charles Griffin.

Peciña S, Schulkin J, Berridge KC. 2006. Nucleus accumbens corticotropin-releasing factor increases cue-triggered motivation for sucrose reward: paradoxical positive incentive effects in stress? BMC Biology 4: 8. doi: 10.1186/1741-7007-4-8

Pedersen SB, Kristensen K, Hermann PA, Katzenellenbogen JA, Richelsen B. 2004. Estrogen controls lipolysis by up-regulating α2A-adrenergic receptors directly in human adipose tissue through the estrogen receptor α. Implications for the female fat distribution. JCEM 89: 1869–1878.

Perreault L, Lavely JM, Kittleson JM, Horton TJ. 2004. Gender differences in lipoprotein lipase activity after acute exercise. Obesity Res 12: 241–249.

Perry GH, Dominy NJ, Claw KG, Lee AS, Fiegler H, Redon R, Werner J, Villanea FA, Mountain JL, Misra R, Carter NP, Lee C, Stone AC. 2007. Diet and the evolution of human amylase gene copy number variation. Nat Genet 39: 1256–1260.

Peters JC, Wyatt HR, Donahoo WT, Hill JO. 2002. From instinct to intellect: the challenge of maintaining healthy weight in the modern world. Obesity Rev 3: 69–74.

Peters JH, Karpiel AB, Ritter RC, Simasko SM. 2004. Cooperative activation of cultured vagal afferent neurons by leptin and cholecystokinin. Endocrinol 145: 3652–3657.

Peters JH, McKay BM, Simasko SM, Ritter RC. 2005. Leptin-induced satiation mediated by abdominal vagal afferents. Am J Physiol Regul Integr Comp Physiol 288: R879–R884.

Picó C, Oliver P, Sánchez J, Palou A. 2003. Gastric leptin: a putative role in the short-term regulation of food intake. Br J Nutr 90: 735–741.

Place AR. 1992. Comparative aspects of lipid digestion and absorption: physiological correlates of wax ester digestion. Am J Physiol Regul Integr Comp Physiol 263: R464–R471.

Plummer TW, Stanford CB. 2000. Analysis of a bone assemblage made by chimpanzees at Gombe National Park, Tanzania. J Hum Evol 39 (3): 345–365.

Pobiner BL, DeSilva J, Sanders WJ, Mitani JC. 2007. Taphonomic analysis of skeletal remains from chimpanzee hunts at Ngogo, Kibale National Park, Uganda. J Hum Evol 52: 614–636.

Poitout V. 2003. The ins and outs of fatty acids on the pancreatic β cell. Trends Endocrinol Metab 14: 201–203.

Popkin BM. 2001. The nutrition transition and obesity in the developing world. J Nutr 131: 871S–873S.

Popkin BM. 2002. An overview on the nutrition transition and its health implications: the Bellagio meeting. Public Health Nutr 5: 93–103.

Porte D, Jr., Baskin DG, Schwartz MW. 2005. Insulin signaling in the central nervous system: a critical role in metabolic homeostasis and disease from C. elegans to humans. Diabetes 54: 1264–1276.

Power ML. 1991. Digestive function, energy intake, and the response to dietary gum in captive callitrichids. Ph.D. diss., University of California at Berkeley. 235.

Power ML. 2004. Viability as opposed to stability: an evolutionary perspective on physiological regulation. In Allostasis, Homeostasis, and the Costs of Adaptation, Schulkin J (ed.), 343–364. Cambridge: Cambridge University Press.

Power ML, Heaney RP, Kalkwarf HJ, Pitkin RM, Repke JT, Tsang RC, Schulkin J. 1999. The role of calcium in health and disease. Am J Obstet Gynecol 181: 1560–1569.

Power ML, Oftedal OT, Tardif SD. 2002. Does the milk of callitrichid monkeys differ from that of larger anthropoids? Am J Primatol 56: 117–127.

Power ML, Schulkin J. 2006. Functions of corticotropin-releasing hormone in anthropoid primates: from brain to placenta. Am J Hum Biol 18: 431–447.

Power ML, Tardif SD, Power RA, Layne DG. 2003. Resting energy metabolism of Goeldi's monkey (*Callimico goeldii*) is similar to that of other callitrichids. Am J Primatol 60: 57–67.

Power RA, Power ML, Layne DG, Jaquish CE, Oftedal OT, Tardif SD. 2001. Relations among measures of body composition, age, and sex in the common marmoset monkey (*Callithrix jacchus*). Comp Med 51: 218–223.

Powley TL. 1977. The ventralmedial hypothalamic syndrome, satiety and a cephalic-phase hypothesis. Psychol Rev 84: 89–126.

Powley TL. 2000. Vagal circuitry mediating cephalic-phase responses to food. Appetite 34: 184–188.

Powley TL, Berthoud H-R. 1985. Diet and cephalic-phase insulin responses. Am J Clin Nutr 42: 991–1002.

Prentice A, Jarjou LM, Cole TJ, Stirling DM, Dibba B, Fairweather-Tait S. 1995. Calcium requirements of lactating Gambian mothers: effects of a calcium supplement on breast-milk calcium concentration, maternal bone mineral content, and urinary calcium excretion. Am J Clin Nutr 62: 58–67.

Prentice A, Jebb S. 2004. Energy intake/physical activity interactions in the homeostasis of body weight regulation. Nutr Rev 62: S98–S104.

Prentice AM. 2005. The emerging epidemic of obesity in developing countries. Int J Epidemiol 1–7.

Prentice AM, Rayco-Solon P, Moore SE. 2005. Insights from the developing world: thrifty genotypes and thrifty phenotypes. Proc Nutr Soc 64: 153–161.

Preshaw RM, Cooke AR, Grossman MI. 1966. Sham-feeding and pancreatic secretion in the dog. Gastroenterology 50: 171–178.

Proulx K, Richard D, Walker C-D. 2002. Leptin regulates appetite-related neuro-peptides in the hypothalamus of developing rats without affecting food intake. Endocrinol 143: 4683–4692.

Pruetz JD, Bertolani P. 2007. Savanna chimpanzees, *Pan troglodytes verus,* hunt with tools. Curr Biol 17 (5): 412–417.

Pryer J. 1993. Body mass index and work-disabling morbidity: results from a Bangladeshi case study. Eur J Clin Nutr 47: 653–7.

Racette SB, Hagberg JM, Evans EM, Holloszy JO, Weiss EP. 2006. Abdominal obesity is a stronger predictor of insulin resistance than fitness among 50–95 year olds. Diabetes Care 29: 673–678.

Ramsay JE, Ferrell WR, Crawford L, Wallace AM, Greer IA, Sattar N. 2002. Maternal obesity is associated with dysregulation of metabolic, vascular, and inflammatory pathways. JCEM 87: 4231–4237.

Rask E, Olsson T, Söderber S, Andrew R, Livingstone DEW, Johnson O, Walker BR. 2001. Tissue-specific dysregulation of cortisol metabolism in human obesity. JCEM 86: 1418–1421.

Rask E, Walker BR, Söderber S, Livingstone DEW, Eliasson M, Johnson O, Andrew R, Olsson T. 2002. Tissue-specific changes in peripheral cortisol metabolism in obese women: increased adipose 11β-hydroxysteroid dehydrogenase type 1 activity. JCEM 87: 3330–3336.

Ray JG, Wyatt PR, Vermeulen MJ, Meir C, Cole DE. 2005. Greater maternal weight and the ongoing risk of neural tube defects after folic acid flour fortification. Obstet Gynecol 105: 261–265.

Rechtschaffen A, Gilliland MA, Bergmann BM, Winter JB. 1983. Physiological correlates of prolonged sleep deprivation in rats. Science 221: 182–184.

Reed DR, Lawler MP, Tordoff MG. 2008. Reduced body weight is a common effect of gene knockout in mice. BMC Genetics 9: 4.

Renehan AG, Tyson M, Egger M, Heller RF, Zwahlen M. 2008. Body-mass index and incidence of cancer: a systematic review and meta-analysis of prospective observational studies. Lancet 371: 569–578.

Resnick HE, Redline S, Shahar E, Gilpin A, Newman A, Walter R, Ewy GA, Howard BV, Punjabi NM. 2003. Diabetes and sleep disturbances. Diabetes Care 26: 702–709.

Rice T, Perusse L, Bouchard C, Rao DC. 1999. Familial aggregation of body mass index and subcutaneous fat measures in the longitudinal Quebec family study. Genet Epidemiol 16: 316–334.

Richelsen B. 1986. Increased a 2- but similar b-adrenergic receptor activities in subcutaneous gluteal adipocytes from females compared with males. Eur J Clin Invest 16: 302–309.

Richter CP. 1936. Increased salt appetite in adrenalectomized rats. Am J Physiol 115: 155–161.

Richter CP. 1953. Experimentally produced reactions to food poisoning in wild and domesticated rats. Ann NY Acad Sci 56: 225–239.

Robson SL. 2004. Breast milk, diet, and large human brains. Curr Anthropol 45: 419–425.

Rodríguez G, Samper MP, Olivares JL, Ventura P, Moreno LA, Pérez-González JM. 2005. Skinfold measurements at birth: sex and anthropometric influence. Arch Dis Child Fetal Neonatal Ed 90: F273–F275.

Rodríguez-Cuenca S, Monjo M, Proenza AM, Roca P. 2005. Depot differences in steroid receptor expression in adipose tissue: possible role of the local steroid milieu. Am J Physiol Endocrinol Metab 288: E200–E207.

Rolls BJ, Roe LS, Meengs JS. 2006. Reductions in portion size and energy density of foods are addictive and lead to sustained decreases in energy intake. Am J Clin Nutr 83: 11–17.

Rolls BJ, Roe LS, Meengs JS. 2007. The effect of large portion sizes on energy intake is sustained for 11 days. Obesity 15: 1535–1543.

Rosati A, Stevens J, Hare B, Hauser M. 2007. The evolutionary origins of human patience: temporal preferences in chimpanzees, bonobos, and human adults. Curr Biol 17: 1663–1668.

Rosenbaum M, Nicolson M, Hirsch J, Heymsfield SB, Gallagher D, Chu F, Leibel RL. 1996. Effects of gender, body composition, and menopause on plasma concentrations of leptin. JCEM 81: 3424–3427.

Ross N. 1997. Effects of diet- and exercise-induced weight loss on visceral adipose tissue in men and women. Sports Med 24: 55–64.

Roth J, Qiang X, Marbán SL, Redelt H, Lowell BC. 2004a. The obesity pandemic: where have we been and where are we going? Obesity Res 12: 88S–101S.

Roth J, Volek JS, Jacobson M, Hickey J, Stein DT, Klein S, Feinman R, Schwartz GJ, Segal-Isaacson CJ. 2004b. Paradigm shifts in obesity research and treatment: roundtable discussion. Obesity Res 12: 145S–148S.

Royal CDM, Dunston GM. 2004. Changing the paradigm from "race" to human genome variation. Nat Genet 35: S5–S7.

Rozin P. 1976. The selection of food by rats, humans, and other animals. In Advances in the Study of Behavior, Rosenlatt JS, Hinde RA, Shaw E, Beer C (eds.). Vol. 6. New York: Academic Press.

Rozin P. 2005. The meaning of food in our lives: a cross-cultural perspective on eating and well-being. J Nutr Educ Behav 37: S107–S112.

Rozin P, Schulkin J. 1990. Food selection. In Handbook of Behavioral Neurobiology, Stricker EM (ed.). New York: Plenum Press.

Ruff CB, Trinkaus E, Holliday TW. 1997. Body mass and encephalization in Pleistocene Homo. Nature 387: 173–176.

Russell JA, Leng G. 1998. Sex, parturition, and motherhood without oxytocin? J Endocrinol 157: 343–359.

Saad MF, Damani S, Gingerich RL, Riad-Gabriel MG, Khan A, Boyadjian R, Jinagouda SD, El-Tawil K, Rude RK, Kamdar V. 1997. Sexual dimorphism in plasma leptin concentration. JCEM 82: 579–584.

Saguy AC, Riley KW. 2005. Weighing both sides: morality, mortality, and framing contests over obesity. J Health Politics Policy Law 30: 869–921.

Sahu A. 2004. Minireview: a hypothalamic role in energy balance with special emphasis on leptin. Endocrinol 145: 2613–2620.

Sallis JF, Glanz K. 2006. The role of built environments in physical activity, eating, and obesity in childhood. The Future of Children 16: 89–108.

Samaras K, Spector TD, Nguten TV, Baan K, Campbell LV, Kelly PJ. 1997. Genetic factors determine the amount and distribution of fat in women after the menopause. J Clin Epidemiol Metab 82: 781–785.

Sapolsky RM. 2001. Physiological and pathophysiological implications of social stress in mammals. In Coping with the Environment: Neural and Endocrine Mechanisms, McEwen BS, Goodman HM (eds.). New York: Oxford University Press.

Sarich VM. Wilson AC. 1973. Generation time and genomic evolution in primates. Science 179: 1144–1147.

Schlundt DG, Briggs NC, Miller ST, Arthur CM, Goldzweig IA. 2007. BMI and seatbelt use. Obesity 15: 2541–2545.

Schmid SM, Hallschmid M, Jauch-Chara K, Bandorf N, Born J, Schultes B. 2007. Sleep loss alters basal metabolic hormone secretion and modulates the dynamic counterregulatory response to hypoglycemia. JCEM 92: 3044–3051.

Schmidt-Nielsen K. Animal Physiology: Adaptation and Environment. 1994. Cambridge: Cambridge University Press.

Schrauwen P, Hesselink MKC. 2004. Oxidative capacity, lipotoxicity, and mitochondrial damage in type 2 diabetes. Diabetes 53: 1412–1417.

Schulkin J. 1991. Sodium Hunger. Cambridge: Cambridge University Press.

Schulkin J. 1999. Corticotropin-releasing hormone signals adversity in both the placenta and the brain: regulation by glucocorticoids and allostatic overload. J Endocrinol 161: 349–356.

Schulkin J. 2001. Calcium Hunger: Behavioral and Biological Regulation. Cambridge: Cambridge University Press.

Schulkin J. 2003. Rethinking Homeostasis: Allostatic Regulation in Physiology and Pathophysiology. Cambridge: MIT Press.

Schulz LO, Bennet PH, Ravussin E, Kidd JR, Kidd KK, Esparza J, Valencia ME. 2006. Effects of traditional and western environments on prevalence of type 2 diabetes in Pima Indians in Mexico and the U.S. Diabetes Care 29: 1866–1871.

Schulze MB, Manson JE, Ludwig DS, et al. 2004. Sugar-sweetened beverages, weight gain, and incidence of type 2 diabetes in young and middle-aged women. JAMA 292: 927–934.

Schwartz GJ, Moran TH. 1996. Sub-diaphragmatic vagal afferent integration of meal-related gastrointestinal signals. Neurosci Behav Rev 20: 47–56.

Schwartz MW, Woods SC, Porte D, Jr., Seeley RJ, Baskin DG. 2000. Central nervous system control of food intake. Nature 404: 661–671.

Schwartz MW, Woods SC, Seeley RJ, Barsh GS, Baskin DG, Leibel RL. 2003. Is the energy homeostasis inherently biased toward weight gain? Diabetes 52: 232–238.

Schweitzer MH, Suo Z, Avci R, Asara JM, Allen MA, Arce FT, Horner JR. 2007. Analyses of soft tissue from *Tyrannosaurus rex* suggest the presence of protein. Science 316: 277–280.

Scott EM, Grant PJ. 2006. Neel revisited: the adipocyte, seasonality, and type 2 diabetes. Diabetologia 49: 1462–1466.

Seasholtz AF, Valverde RA, Denver RJ. 2002. Corticotropin-releasing hormone-binding protein: biochemistry and function from fishes to mammals. J Endocrinol 175: 89–97.

Seidell JC, Pérusse L, Després J-P, Bouchard C. 2001. Waist and hip circumferences have independent and opposite effects on cardiovascular disease risk factors: the Quebec family study. Am J Clin Nutr 74: 315–321.

Senut B, Pickford M, Gommery D, Mein P, Cheboi K, Coppens Y. 2001. First hominid from the Miocene (Lukeino Formation, Kenya). Comptes Rendus de l'Academie des Sciences, Series IIA—Earth and Planetary Sci 332, 2: 137–144.

Seppälä-Lindroos A, Vehkavaara S, Häkkinen AM, Goto T, Westerbacka J, Sovijärvi A, Halavaara J, Yki-Jarvinen H. 2002. Fat accumulation in the liver is associated with defects in insulin suppression of glucose production and serum free fatty acids independent of obesity in normal men. JCEM 87: 3023–3028.

Sharrock KCB, Kuzawa CW, Leonard WR, Tanner S, Reyes-Garcia VE, Vadez V, Huanca T, McDade TW. 2008. Developmental changes in the relationship between leptin and adiposity among Tsimané children and adolescents. Am J Hum Bio oo: oo–oo.

Shi H, Dirienzo D, Zemel MB. 2001. Effects of dietary calcium on adipocyte lipid metabolism and body weight regulation in energy-restricted aP2-agouti transgenic mice. FASEB J 15: 291–293.

Shipman P, Walker A. 1989. The costs of becoming a predator. Am Anthropol 88: 26–43.

Short L, Horne J. 2002. Toucans, Barbets, and Honeyguides. New York: Oxford University Press.

Sierra-Johnson J, Johnson BD, Bailey KR, Turner ST. 2004. Relationships between insulin sensitivity and measures of body fat in asymptomatic men and women. Obesity Res 12: 2070–2077.

Singh R, Artaza JN, Taylor WE, Braga M, Yuan X, Gonzalez-Cadavid NF, Bhasin S. 2006. Testosterone inhibits adipogenic differentiation in 3T3-L1 cells: nuclear translocation of androgen receptor complex with beta-catenin and T-cell factor 4 may bypass canonical Wnt signaling to down-regulate adipogenic transcription factors. Endocrinology 147: 141–154.

Škopková M, Penesová A, Sell H, Rádiková Ž, Vlček M, Imrich R, Koška J, Ukropec J, Eckel J, Klimeš I, Gašperíková D. 2007. Protein array reveals differentially expressed proteins in subcutaneous adipose tissue in obesity. Obesity 15: 2396–2406.

Slawik M, Vidal-Puig AJ. 2006. Lipotoxicity, overnutrition, and energy metabolism in aging. Ageing Res Rev 5: 144–164.

Smeets AJ, Westerterp-Plantenga MS. 2006. Oral exposure and sensory-specific satiety. Physiol and Behavior 89: 281–286.

Smith GP. 1995. Pavlov and appetite. Int Physiol Behav Sci 30: 169–174.

Smith GP. 2000. The controls of eating: a shift from nutritional homeostasis to behavioural neuroscience. Nutrition 16: 814–820.

Smith SR, de Jonge L, Pellymounter M, Nguyen T, Harris R, York D, Redmann S, Rood J, Bray GA. 2001. Peripheral administration of human corticotropin-releasing hormone: a novel method to increase energy expenditure and fat oxidation in man. JCEM 86: 1991–1998.

Smith-Kirwin SM, O'Connor DM, De Johnston J, Lancey ED, Hassink SG, Funanage VL. 1998. Leptin expression in human mammary epithelial cells and breast milk. JCEM 83: 1810–1813.

Snih SA, Ottenbacher KJ, Markides KS, Kuo Y-F, Eschbach K, Goodwin JS. 2007. The effect of obesity on disability vs. mortality in older Americans. Arch Intern Med 167: 774–780.

Snijder MB, Dekker JM, Visser M, Bouter LM, Stehouwer CDA, Kostense PJ, Yudkin JS, Heine RJ, Nijpels G, Seidell JC. 2003. Associations of hip and thigh circumferences independent of waist circumference with the incidence of type 2 diabetes: the Hoorn Study. Am J Clin Nutr 77: 1192–1197.

Sobhani I, Buyse M, Goiot H, Weber N, Laigneau JP, Henin D, Soul JC, Bado A. 2002. Vagal stimulation rapidly increases leptin secretion in human stomach. Gastroenterology 122: 259–263.

Sookoian S, Gemma C, Garcfa SI, Gianotti TF, Dieuzeide G, Roussos A, Tonietti M, Trifone L, Kanevsky D, González CD, Pirola CJ. 2007. Short allele of serotonin transporter gene promoter is a risk factor for obesity in adolescents. Obesity 15: 271–276.

Sooranna SR, Ward S, Bajoria R. 2001. Fetal leptin influences birth weight in twins with discordant growth. Pediatr Res 49: 667–672.

Soucy J, LeBlanc J. 1999. Protein meals and postprandial thermogenesis. Physiol Behav 65: 705–709.

Spanovich S, Niewiarowski PH, Londraville RL. 2006. Seasonal effects on circulating leptin in the lizard *Sceloporus undulatus* from two populations. Comp Biochem Physiol Pt B, Biochem and Molecular Biol 143: 507–513.

Speakman JR. 2006. Thrifty genes for obesity and the metabolic syndrome-time to call off the search? Diab Vasc Dis Res 3: 7–11.

Speakman JR. 2007. A nonadaptive scenario explaining the genetic predisposition to obesity: the "predation release" hypothesis. Cell Metab 6: 5–12.

Speakman JR, Djafarian K, Stewart J, Jackson DM. 2007. Assortative mating for obesity. Am J Clin Nutr 86: 316–323.

Speakman JR, Ergon T, Cavanagh R, Reid K, Scantlebury DM, Lambin X. 2003. Resting and daily energy expenditures of free-living field voles are positively correlated but reflect extrinsic rather than intrinsic effects. PNAS 100: 14057–14062.

Speakman JR, Gidney A, Bett J, Mitchell IP, Johnson MS. 2001. Effect of variation in food quality on lactating mice *Mus musculus*. J Exp Bio 204: 1957–1965.

Speiser PW, Rudolf MCJ, Anhalt H, Camacho-Hubner C, Chiarelli F, Eliakim A, Freemark M, Gruters A, Hershkovitz E, Iughetti L, Krude H, Latzer Y, Lustig RH, Pescovitz OH, Pinhas-Hamiel O, Rogol AD, Shalitan S, Sultan C, Stein D, Vardi P, Werther GA, Zadik Z, Zuckerman-Levin N, Hochberg Z. 2005. Consensus statement: childhood obesity. JCEM 90: 1871–1887.

Spiegel D, Sephton S. 2002. Re: night shift work, light at night, and risk of breast cancer. J Nat Cancer Institute 94: 530.

Spiegel K, Knutson K, Leproult R, Tasali E, van Cauter E. 2005. Sleep loss: a novel risk factor for insulin resistance and type 2 diabetes. J Appl Physiol 99: 2008–2019.

Spiegel K, Leproult R, L'Hermite-Balériaux M, Copinnschi G, Penev PD, Van Couter E. 2004. Leptin levels are dependent on sleep duration: relationships with sympathovagal balance, carbohydrate regulation, cortisol, and thyrotropin. JCEM 89: 5762–5771.

Spoor F, Leakey MG, Gathogo PN, Brown FH, Antón SC, McDougall I, Kiarie C, Manthi FK, Leakey LN. 2007. Implications of new early Homo fossils from Ileret, east of Lake Turkana, Kenya. Nature 448: 688–691.

Stanford CB. 2001. The ape's gift: meat-eating, meat-sharing, and human evolution. In Tree of Origin, de Waal FBM (ed.). Cambridge: Harvard University Press.

Stanford CB, Wallis J, Matama H, Goodall J. 1994. Patterns of predation by chimpanzees on red colobus monkeys in Gombe National Park, 1982–1991. Am J Phys Anthropol 94: 213–228.

Stein CJ, Colditz GA. 2004. The epidemic of obesity. JCEM 89: 2522–2525.

Stellar E. 1954. The physiology of motivation. Psychol Rev 61: 5–22.

Stenzel-Poore MP, Heldwein KA, Stenzel P, Lee S, Vale WW. 1992. Characterization of the genomic corticotropin-releasing factor (CRF) gene from *Xenopus laevis:* two members of the CRF family exist in amphibians. Mol Endocrinol 6: 1716–1724.

Sterling P. 2004. Principles of allostasis: optimal design, predictive regulation, pathophysiology, and rational therapeutics. In Allostasis, Homeostasis, and the Costs of Adaptation, Schulkin J (ed.), 17–64. Cambridge: Cambridge University Press.

Sterling P, Eyer J. 1988. Allostasis: a new paradigm to explain arousal pathology. In Handbook of Life Stress, Cognition, and Health, Fisher S, Reason J (eds.). New York: John Wiley.

Stewart PM, Boulton A, Kumar S, Clark PMS, Shakleton CHL. 1999. Cortisol metabolism in human obesity: impaired cortisone to cortisol conversion in subjects with central obesity. JCEM 84: 1022–1027.

Stiner MC. 1993. Modern human origins: faunal perspectives. Ann Rev Anthropol 22: 55–82.

Stiner MC. 2002. Carnivory, coevolution, and the geographic spread of the genus *Homo*. J Archaeological Res 10: 1–63.

Straif K, Baan R, Grosse Y, Secretan B, El Ghissassi F, Bouvard V, Altieri A, Benbrahim-Tallaa L, Cogliano V. 2007 Carcinogenicity of shift-work, painting, and fire-fighting. Lancet Oncol 8: 1065–1066.

Strum SC. 1975. Primate predation: interim report on the development of a tradition in a troop of olive baboons. Science 187: 755–757.

Strum SC. 2001. Almost Human: a journey into the world of baboons. Chicago: University of Chicago Press.

Stubbs RJ, Tolkamp BJ. 2006. Control of energy balance in relation to energy intake and energy expenditure in animals and man: an ecological perspective. Br J Nutr 95: 657–676.

Stunkard AJ. 1988. The Salmon lecture. Some perspective on human obesity: its causes. Bull NY Acad of Med 64 (8): 902–923.

Stunkard AJ, Grace WJ, Wolff HG. 1955. The night-eating syndrome: a pattern of food intake among certain obese patients. Am J Med 19: 78–86.

Stunkard AJ, Harris JR, Pedersen NL, McClearn GE. 1990. The body-mass index of twins who have been reared apart. N Eng J Med 322: 1483–1487.

Stunkard AJ, Sørensen TI, Hanis C, Teasdale TW, Chakraborty R, Schull WJ,

Schulsinger F. 1986. An adoption study of human obesity. N Eng J Med 314: 193–198.

Subar AF, Krebs-Smith SM, Cook A, Kahle LL. 1998. Dietary sources of nutrients among US children, 1989–1991. Pediatrics 102 (4 Pt 1): 913–923.

Sui X, LaMonte MJ, Blair SN. 2007. Cardiorespiratory fitness as a predictor of nonfatal cardiovascular events in asymptomatic women and men. Am J Epidemiol 165: 1413–1423.

Sui X, LaMonte MJ, Laditka JN, Hardin JW, Chase N, Hooker SP, Blair SN. 2007. Cardiorespiratory fitness and adiposity as mortality predictors in older adults. JAMA 298: 2507–2516.

Sumner AE, Farmer NM, Tulloch-Reid MK, Sebring NG, Yanovski JA, Reynolds JC, Boston RC, Premkumar A. 2002. Sex differences in visceral adipose tissue volume among African Americans. Am J Clin Nutr 76: 975–979.

Sun X, Zemel MB. 2004. Role of uncoupling protein 2 (UCP2) expression and 1alpha, 25-dihydroxyvitamin D3 in modulating adipocyte apoptosis. FASEB J 18: 1430–1432.

Sun X, Zemel MB. 2007. Calcium and 1,25-dihydroxyvitamin D3 regulation of adipokine expression. Obesity 15 (2): 340–348.

Sun Y, Ahmed S, Smith RG. 2003. Deletion of Ghrelin impairs neither growth nor appetite. Molecular Cellular Biol 23: 7973–7981.

Suter KJ, Pohl CR, Wilson ME. 2000. Circulating concentrations of nocturnal leptin, growth hormone, and insulin-like growth factor-I increase before the onset of puberty in agonadal male monkeys: potential signals for the initiation of puberty. JCEM 85: 808–814.

Swanson LW, Simmons DM. 1989. Differential steroid hormone and neural influences on peptide mRNA levels in CRH cells of the paraventricular nucleus: a hybridization histochemical study in the rat. J Comp Neurol 285: 413–435.

Taché Y, Perdue MH. 2004. Role of peripheral CRF signaling pathways in stress-related alterations of gut motility and mucosal function. Neurogastroenterol Motil 16 (suppl): 137–142.

Taheri S, Lin L, Austin D, Young T, Mignot E. 2004. Short sleep duration is associated with reduced leptin, elevated ghrelin, and increased body mass index. PLoS Medicine/Public Library of Science 1: e62.

Takaya K, Ariyasu H, Kanamoto N, Iwakura H, Yoshimoto A, Harada M, Mori K, Komatsu Y, Usui T, Shimatsu A, Ogawa Y, Hosoda K, Akamizu T, Kojima M, Kangawa K, Nakao K. 2000. Ghrelin strongly stimulates growth hormone (GH) release in humans. JCEM 85: 1169–1174.

Tam CS, de Zegher F, Garnett SP, Baur LA, Cowell CT. 2006. Opposing influences of prenatal and postnatal growth on the timing of menarche. JCEM 91: 4369–4373.

Taouis M, Chen J-W, Daviaud C, Dupont J, Derouet M, Simon J. 1998. Cloning the chicken leptin gene. Gene 208: 239–242.

Tardif SD, Power M, Oftedal OT, Power RA, Layne DG. 2001. Lactation, maternal behavior, and infant growth in common marmoset monkeys (*Callithrix jacchus*): effects of maternal size and litter size. Behav Ecol Sociobiol 51: 17–25.

Tchernof A, Desmeules A, Richard C, Laberge P, Daris M, Mailloux J, Rheaume C, Dupont P. 2004. Ovarian hormone status and abdominal visceral adipose tissue metabolism. JCEM 89: 3425–3430.

Tebbich S, Taborsky M, Fessl B, Dvorak M. 2002. The ecology of tool use in the woodpecker finch (*Cactospiza pallida*). Ecology Letters 5: 656–664.

Teff KL. 2000. Nutritional implications of the cephalic-phase reflexes: endocrine responses. Appetite 34: 206–213.

Teff KL, Devine, J, Engelman, K. 1995. Sweet taste: effect on cephalic phase insulin release in men. Physiol and Behav 57: 1089–1095.

Teff KL, Elliott SS, Tschöp M, Kieffer TJ, Rader D, Heiman M, Townsend RR, Keim NL, D'Alessio D, Havel PJ. 2004. Dietary fructose reduces circulating insulin and leptin, attenuates postprandial suppression of ghrelin, and increases triglycerides in women. JCEM 89: 2963–2972.

Teff KL, Engelman K. 1996. Oral sensory stimulation improves glucose tolerance in humans: effects on insulin, C-peptide, and glucagon. Am J Physiol 270: R1371–R1379.

Teff KL, Mattes RD, Engelman K. 1991. Cephalic-phase insulin release in normal weight males: verification and reliability. Am J Physiol 261: E430–E436.

Teff KL, Townsend RR. 1999. Early-phase insulin infusion and muscarinic blockade in obese and lean subjects. Am J Physiol 277: R198–R208.

Teleki G. 1973. The Predatory Behavior of Wild Chimpanzees. Lewisburg, PA: Bucknell University Press.

Temple JL, Legierski CM, Giacomelli AM, Salvy SJ, Epstein LH. 2008. Overweight children find food more reinforcing and consume more energy than do nonoverweight children. Am J Clin Nutr 87: 1121–1127.

Terborgh J. 1984. Five New World Primates. Princeton: Princeton University Press.

Thomas DE, Elliott EJ, Baur L. 2007. Low glycaemic index or low glycaemic load diets for overweight and obesity. Cochrane Database Syst Rev 3: CD005105.

Thompson SD, Power ML, Rutledge CE, Kleiman DG. 1994. Energy metabolism and thermoregulation in the golden lion tamarin (*Leontopithecus rosalia*). Folia Primatol 63: 131–143.

Thouless CR, Fanshawe JH, Bertram CR. 1989. Egyptian vultures *Neophron percnopterus* and ostrich *Struthio camelus* eggs: the origin of stone-throwing behavior. Ibis 131: 9–15.

Tittelbach TJ, Berman DM, Nicklas BJ, Ryan AS, Goldberg AP. 2004. Racial differences in adipocyte size and relationship to the metabolic syndrome in obese women. Obesity Res 12: 990–998.

Tittelbach TJ, Mattes RD. 2001. Oral stimulation influences postprandial triacylglycerol concentrations in humans: nutrient specificity. J Am Coll Nutr 20: 485–493.

Todes DP. 2002. Pavlov's Physiology Factory: Experiment, Interpretation, Laboratory Enterprise. Baltimore: Johns Hopkins University Press.

Tomasetto C, Karam SM, Ribieras S, Masson R, Lefebvre O, Staub A, Alexander G, Chenard MP, Rio MC. 2000. Identification and characterization of a novel gastric peptide hormone: the motilin-related peptide. Gastroenterology 119: 395–405.

Tordoff MG, Friedman, MI. 1989. Drinking saccharin increases food intake and preference—IV. Cephalic phase and metabolic factors. Appetite 12: 37–56.

Travers JB, Travers SP, Norgren R. 1987. Gustatory neural processing in the hindbrain. Annual Rev of Neuroscience 10: 595–632.

Trayhurn P, Bing C, Wood IS. 2006. Adipose tissue and adipokines—energy regulation from the human perspective. J Nutr 136: 1935S–1939S.

Trevathan WR, Smith EO, McKenna JJ. 1999. Evolutionary Medicine. New York: Oxford University Press.

Trevathan WR, Smith EO, McKenna JJ. 2007. Evolutionary Medicine and Health: New Perspectives. New York: Oxford University Press.

Trifiletti LB, Shields W, Bishai D, McDonald E, Reynaud F, Gielen A. 2006. Tipping the scales: obese children and child safety seats. Pediatr 117: 1197–1202.

Tritos NA, Kokkotou EG. 2006. The physiology and potential clinical applications of ghrelin, a novel peptide hormone. Mayo Clin Proc 81: 653–660.

Trujillo ME, Scerer PE. 2005. Adiponectin—journey from an adipocyte secretory protein to biomarker of the metabolic syndrome. J Int Med 257: 167–175.

Tschop M, Smiley DL, Heiman ML. 2000. Ghrelin induces adiposity in rodents. Nature 407: 908–913.

Tso P, Liu M. 2004. Apolipoprotein A-IV, food intake, and obesity. Physiol Behav 83: 631–643.

Turnbaugh PJ, Ley RE, Mahowald MA, Magrini V, Mardis ER, Gordon JI. 2006.

An obesity-associated gut microbiome with increased capacity for energy harvest. Nature 444: 1027–1031.

Uppot RN, Sahani DV, Hahn PF, Gervais D, Mueller PR. 2007. Impact of obesity on Medical imaging and image-guided intervention. AJR 188: 433–440.

Vale W, Spiess J, Rivier C, Rivier J. 1981. Characterization of a 41-residue ovine hypothalamic peptide that stimulates secretion of corticotropin and β-endorphin. Science 78: 1394–1397.

van Dam RM, Wilett WC, Manson JE, Hu FB. 2006. The relationship between overweight in adolescence and premature death in women. Ann Intern Med 145: 91–97.

Van der Merwe M-T, Pepper MS. 2006. Obesity in South Africa. Obesity Rev 7: 315–322.

Van Pelt RE, Evans EM, Schechtman KB, Ehsani AA, Kohrt WM. 2002. Contributions of total and regional fat mass to risk for cardiovascular disease in older women. Am J Physiol Endocrinol Metab 282: E1023–E1028.

Vasilakopoulou A, le Roux CW. 2007. Could a virus contribute to weight gain? Int J Obes 31: 1350–1356.

Vasudevan S, Tong Y, Steitz JA. 2007. Switching from repression to activation: microRNAs can up-regulate translation. Science 318: 1931–1934.

Votruba SB, Jensen MD. 2006. Sex-specific differences in leg fat uptake are revealed with a high-fat meal. Am J Physiol Endocrinol Metab 291: E1115–E1123.

Waddington CH. 1942. Canalization of development and the inheritance of acquired characters. Nature 150: 563–565.

Wade GN, Jones JE. 2004. Neuroendocrinology of nutritional infertility. Am J Physiol Regul Integr Comp Physiol 287: R1277–R1296.

Waga IC, Dacier AK, Pinha PS, Tavares MCH. 2006. Spontaneous tool use by wild capuchin monkeys (Cebus libidinosus) in the Cerrado. Folia Primatol 77: 337–344.

Wallace B, Cesarini D, Lichtenstein P, Johannesson M. 2007. Heritability of ultimate game responder behavior. PNAS 104: 15631–15634.

Waller DK, Shaw GM, Rasmussen SA, Hobbs CA, Canfield MA, Siega-Riz AM, Gallaway MS, Correa A. 2007. Prepregnancy obesity as a risk factor for structural birth defects. Arch Pediatr Adolesc Med 161: 745–750.

Wang JX, Davies MJ, Norman RJ. 2002. Obesity increases the risk of spontaneous abortion during infertility treatment. Obesity Res 10: 551–554.

Waterland RA, Jirtle RL. 2003. Transposable elements: targets for early nutritional effects on epigenetic gene regulation. Molecular and Cellular Biol 23: 5293–5300.

Watson JD, Crick FHC. 1953. Molecular structure of nucleic acids: a structure for the deoxyribose nucleic acid. Nature 171: 737–738.

Watts DP, Mitani JC. 2002. Hunting behavior of Chimpanzees at Ngogo, Kibale National Park, Uganda. Int J Primatol 23: 1–28.

Weedman K. 2005. Gender and stone tools: an ethnographic study of the Konso and Gmao hideworkers of southern Ethiopia. In Gender and Hide Production, Frink L, Weedman K (eds.), 175–196. Walnut Creek, CA: AltaMira Press.

Weigle DS, Duell PB, Conner WE, Steiner RA, Soules MR, Kuijper JL. 1997. Effect of fasting, refeeding, and dietary fat restriction on plasma leptin levels. JCEM 82: 561–565.

Weingarten HP, Powley TL. 1980. Ventromedial hypothalamic lesions elevate basal and cephalic-phase gastric acid output. Am J Physiol 239: G221–G229.

Weisberg SP, McCann D, Desai M, Rosenbaum M, Leibel RL, Ferrante AW, Jr. 2003. Obesity is associated with macrophage accumulation in adipose tissue. J Clin Invest 112: 1796–1808.

Wellen KE, Hotamisligil GS. Inflammation, stress, and diabetes. 2005. J Clin Invest 115: 1111–1119.

West DB, Fey D, Woods SC. 1984. Cholecystokinin persistently suppresses meal size but not food intake in free-feeding rats. Am J Physiol Regul Integr Comp Physiol 246: R776–R787.

White FJ, Wood KD. 2007. Female feeding priority in bonobos, Pan paniscus, and the question of female dominance. Am J of Primatol 69: 837–850.

White TD, Suwa G, Asfaw B. 1994. Australopithecus ramidus, a new species of early hominid from Ethiopia. Nature 371: 306–312.

Wicks D, Wright J, Rayment P, Spiller R. 2005. Impact of bitter taste on gastric motility. Eur J Gastroenterol Hepatol 17: 961–965.

Wild S, Roglic G, Green A, Sicree R, King H. 2004. Global prevalence of diabetes: estimates for the year 2000 and projections for 2030. Diabetes Care 27: 1047–1053.

Wilkins MHF, Stokes AR, Wilson HR. 1953. Molecular structure of nucleic acids: molecular structure of deoxypentose nucleic acids. Nature 171: 738–740.

Williams CM. 2004. Lipid metabolism in women. Proc Nutr Soc 63: 153–160.

Williams GW, Nesse RM. 1991. The dawn of Darwinian medicine. Quart Rev of Biol 66: 1–22.

Williams LS, Rotich J, Qi R, Fineberg N, Espay A, Bruno A, Fineberg SE, Tierney WR. 2002. Effects of admission hyperglycemia on mortality and costs in acute ischemic stroke. Neurology 59: 67–71.

Wimmer R, Kirsch S, Rappold GA, Schempp W. 2002. Direct evidence for the Homo-Pan clade. Chromosome Res 10: 55–61.

Wingfield JC. 2004. Allostatic load and life cycles: implications for neuroendocrine control mechanisms. In Allostasis, Homeostasis, and the Costs of Adaptation, Schulkin J (ed.), 302–342. Cambridge: Cambridge University Press.

Won Y-J, Hey J. 2005. Divergence population genetics of chimpanzees. Molecular Biol Evol 22: 297–307.

Wong SNP, Sicotte P. 2007. Activity budget and ranging patterns of *Colobus vellerosus* in forest fragments in central Ghana. Folia Primatol 78: 245–254.

Wood B, Collard M. 1999. The human genus. Science 284: 65–71.

Wood B, Richmond BG. 2000. Human evolution: taxonomy and paleobiology. J of Anat 197: 19–60.

Woodhouse LJ, Gupta N, Bhasin M, Singh AB, Ross R, Phillips J, Bhasin S. 2004. Dose-dependent effects of testosterone on regional adipose tissue distribution in healthy young men. JCEM 89: 718–726.

Woods SC. 1991. The eating paradox. How we tolerate food. Psychol Rev 98: 488–505.

Woods SC. 2006. Dietary synergies in appetite control: distal gastrointestinal tract. Obesity 14: 171S–178S.

Woods SC, Gotoh K, Clegg DJ. 2003. Gender differences in the control of energy homeostasis. Exp Biol Med 228: 1175–1180.

Woods SC, Hutton RA, Makous W. 1970. Conditioned insulin secretion in the albino rat. Proc Soc Exp Biol Med 133: 965–968.

Woods SC, Seeley RJ, Porte D, Jr., Schwartz MW. 1998. Signals that regulate food intake and energy homeostasis. Science 280: 1378–1383.

Woods SC, Vasselli JR, Kaestner E, Szakmary GA, Milburn GA, Vitiello MV. 1977. Conditioned insulin secretion and meal feeding in rats. J Cop Physiol Psychol 91: 128–133.

Wortsman J, Matsuoka LY, Chen TC, Lu Z, Holick MF. 2000. Decreased bioavailability of vitamin D in obesity. Am J Clin Nutr 72: 690–693.

Wraith A, Törnsten A, Chardon P, Harbitz I, Chowdhary BP, Andersson L, Lundin L-G, Larhammar D. 2000. Evolution of the neuropeptide Y receptor family: gene and chromosome duplications deduced from the cloning and mapping of the five receptor subtype genes in pig. Genome Res 3: 302–310.

Wrangham RW. 2001. Out of the *Pan*, into the fire: from ape to human. In Tree of Origin, de Waal FBM (ed.). Cambridge: Harvard University Press.

Wrangham RW, Conklin-Brittain NL. 2003. The biological significance of cooking in human evolution. Comp Biochem Physiol Pt A 136: 35–46.

Wrangham RW, Jones JH, Laden G, Pilbeam D, Conklin-Brittain NL. 1999. The raw and the stolen: cooking and the ecology of human origins. Curr Anthropol 40: 567–594.

Wrangham RW, Peterson D. 1996. Demonic Males: Apes and the Origins of Human Violence. Boston: Houghton Mifflin.

Wren AM, Seal LJ, Cohen MA, Byrnes AE, Frost GS, Murphy KG, Dhillo WS, Ghatei MA, Bloom SR. 2001b. Ghrelin enhances appetite and increases food intake in humans. JCEM 86: 5992.

Wren AM, Small CJ, Abbott CR, Dhillo WS, Seal LJ, Cohen MA, Batterham RL, Taheri S, Stanley SA, Ghatei MA, Bloom SR. 2001a. Ghrelin causes hyperphagia and obesity in rats. Diabetes 50: 2540–2547.

Xiang H, Smith GA, Wilkins JR, Chen G, Hostetler SG, Stallones L. 2005. Obesity and risk of nonfatal unintentional injuries. Am J Prev Med 29: 41–45.

Xu H, Barnes GT, Yang Q, Tan G, Yang D, Chou CJ, Sole J, Nichols A, Ross JS, Tartaglia LA, Chen H. 2003. Chronic inflammation in fat plays a crucial role in the development of obesity-related insulin resistance. J Clin Invest 112: 1821–1830.

Yaggi HK, Araujo AB, McKinlay JB. 2006. Sleep duration as a risk factor for the development of type 2 diabetes. Diabetes Care 29: 657–661.

Yajnik CS. 2004. Early life origins of insulin resistance and type 2 diabetes in India and other Asian countries. J Nutr 134: 205–210.

Yan LL, Daviglus ML, Liu K, Stamler J, Wang R, Pirzada A, Garside DB, Dyer AR, Van Horn L, Liao Y, Fries JF, Greenland P. 2006. Midlife body mass index and hospitalization and mortality in older age. JAMA 295: 190–198.

Young TK, Bjerregaard P, Dewailly E, Risica PM, Jorgensen ME, Ebbesson SEO. 2007. Prevalence of obesity and its metabolic correlates among the circumpolar Inuit in 3 countries. Am J of Pub Health 97: 691–695.

Young WS, Shepard E, Amico J, Hennighausen L, Wagner K-U, La Marca ME, McKinney C, Ginns EI. 1996. Deficiency in mouse oxytocin prevents milk ejection, but not fertility and parturition. J Neuroendocrinol 8: 847–853.

Zafra MA, Molina F, Puerto A. 2006. The neural/cephalic-phase reflexes in the physiology of nutrition. Neurosci Biobehav Rev 30: 1032–1044.

Zellner DA, Loaiza S, Gonzalez Z, Pita J, Morales J, Pecora D, Wolf A. 2006. Food selection changes under stress. Physiol and Behav 87: 789–793.

Zemel MB. 2002. Regulation of adiposity and obesity risk by dietary calcium: mechanisms and implications. J Am Coll Nutr 21: 146S–151S.

Zemel MB. 2004. Role of calcium and dairy products in energy partitioning and weight management. Am J Clin Nutr 79: 907S–912S.

Zhang JV, Ren P-G, Avsian-Kretchmer O, Luo C-W, Rauch R, Klein C, Hseuh A. 2005. Obestatin, a peptide encoded by the ghrelin gene, opposes ghrelin's effects on food intake. Science 310: 996–999.

Zhang Y, Proenca R, Maffei M, Baron M, Leopold L, Friedman JM. 1994. Positional cloning of the mouse Obese gene and its human analog. Nature 372: 425–531.

Zhu S, Layde, PM, Guse, CE, Laud, PW, Pintar, F, Nirula, R, Hargarten, S. 2006. Obesity and risk for death due to motor vehicle crashes. Am J Pub Health 96: 734–739.

Zhu X, Barch Lee C. 2008. Walkability and safety around elementary schools: economic and ethnic disparities. Am J Prev Med 34: 282–290.

Zigman JM, Elmquist JK. 2003. Minireview: from anorexia to obesity: the yin and yang of body weight control. Endocrinol 144: 3749–3756.

Zuberbuhler K, Jenny D. 2002. Leopard predation and primate evolution. J Hum Evol 43: 873–886.

Index

···

Australopithecines, 45–49, 59, 74, 83, 86

Basal metabolic rate (BMR), 40, 42, 64, 121, 146–50, 152; during activity vs. sleep, 149–50; body size and, 147–49, 160–61; caloric intake and, 147, 161; food restriction and, 151; measurement of, 152
Binge eating, 292
Biological energy, 138
Biology, 4–10, 317; of fat, 7–8, 17, 244–64; thermodynamics and, 15, 136, 140–43
Bipedalism, 45–47
Birth defects, 287–88
Birth weight, 5, 17, 18, 41, 285; heart disease risk and, 300; macrosomia, 41, 288–89, 307; metabolic syndrome risk and, 303; neonatal fatness, 5, 17–18, 41, 279–83, 290, 307, 319–21; obesity risk and, 305
Bitter taste, 218, 219
Blood-brain barrier, 171–72, 184, 186, 191, 195, 222, 272
Body mass index (BMI), 24–26, 28–29, 324; calcium intake and, 100; cardiovascular fitness and, 325; extreme obesity and, 26, 30; heritability of, 292, 313; maternal, macrosomia and, 41; menarche and, 284, 285; mortality and, 34–36, 37–38; predation pressure and, 83–84; race/ethnicity and, 26, 29, 34, 263, 293, 310; visceral fat and, 262, 263; work absenteeism, illness and, 114–15
Body morph type, 313–14
Body size: advantages of larger, 50–52, 66; appetite and, 192–93; basal metabolic rate and, 147–49, 160–61; energy intake and, 50–51, 160–61; evolution and, 47–50, 66
Bonobos, 68–69, 80–83
Brain: appetite, satiety and, 194–200; cephalic-phase responses and, 220–23; expensive tissue hypothesis of evolution of, 64–66, 151–53; fatty acids and development of, 107–8; gut-brain peptides, 16, 173–84, 189–92, 206;

hierarchical organization of, 198–99, 205; leptin and, 178–79, 183; neural circuits of feeding behavior, 194–99, 205, 242; postnatal growth of, 280–81; shift of energy expenditure from gut to, 65–66, 152
Brain size, 319–22; dietary fat and, 106–7, 320; evolution of, 47–50, 66, 79, 319; meals and, 74–78, 79, 89; metabolic rate and, 64–65, 152–53, 321; social complexity theory of, 78
Brain stem, 196–99, 205, 223
Breast milk, 19, 321; fat content of, 108–9, 276–77, 279; immune-function molecules in, 281–82, 321; leptin in, 182; vitamin D in, 251
Brown fat, 245, 320
Built environment, 111, 123–26, 135

Calcium: balance of, 156, 157; deficiency of, 187; intake of, 72, 100, 103, 132–33, 156, 252; during pregnancy and lactation, 277–79
Calcium ions, 141, 142
Caloric theory of heat, 137
Caloric value of foods, 121, 153, 154
Cancer, 34, 127
Carbohydrates, 153–54, 200, 203, 229; metabolizable energy value for, 158, 159
Cardiovascular disease, 28, 34, 44, 261, 270, 300–301, 303
Cardiovascular fitness, 123, 325
Catabolism, 139, 145, 154
Cause of obesity, 41–43, 154
CD36, 219–20
Central nervous system (CNS), 170, 172, 194, 207–8, 242; cephalic-phase responses and, 220–23. See also Brain
Cephalic-phase responses, 16, 188, 206–25, 227–28, 230; in appetite, 232–36; CNS and, 220–23; concept of, 208–10, 225; conditioned, 220–21, 227; demonstrated by sham feeding, 214, 215, 217; elimination of, 222; evidence for, 213–17; in feeding biology, 211–13; of ghrelin secretion, 234; insulin response, 214, 216–17, 219,

221, 223–25, 230; Pavlov's studies of, 206–9, 213, 215; vs. reactive responses, 215–16; in regulatory physiology, 210–11; salivation, 209; in satiety, 231, 236–37; stimulation of, 213–15; to taste, 214, 217–20
Cerebral cortex, 197–99, 205
Chemical environment, 125–26, 304
Children: birth defects in, 287–88; car safety seats for, 37; diet of, 42; early-life poverty, nutrition, and heart disease risk, 300–301; maternal BMI and adult weight of, 41; neonatal fatness, 5, 17–18, 41, 279–83, 290, 307, 319–21; obesity among, 31; physical activity of, 123, 124; television viewing by, 122–23
Chimpanzee-human speciation, 46–47, 79
Chimpanzees, 68–69, 71–74, 80–86
Cholecystokinin (CCK), 166, 175, 183, 189–90, 196, 237; appetite and, 189–90, 196, 237; leptin and, 237; receptors for, 189, 237
Circadian cycle, 211; appetite trained to, 235–36; of energy balance, 161; of ghrelin secretion, 233–35; of insulin secretion, 221, 233; of leptin levels, 128, 180, 203, 211, 233; night shift work and, 126–27
Circumventricular organs, 195–96
Comfort foods, 239
Concentration gradient, 141
Conditional responses, 209
Conservation of: energy, 136–38, 139, 140
Cooking, 58–59, 69
Cooperation, 87–89; efficiency and, 84–86; fairness and, 82–83; tolerance and, 78–79, 87
Corticotropin-releasing hormone (CRH), 166–69, 175, 239–41, 254, 263, 270
Cortisol, 166, 211, 249, 253, 254, 263, 270, 303
Cushing's syndrome, 263, 270
Cytokines, 17, 91, 249, 254
Cytosine triphosphate (CTP), 146, 157

Deoxyribonucleic acid (DNA), 146, 164, 177, 294–98, 302, 316
Development: early-life poverty, nutrition, and heart disease risk, 300–301; epigenetic factors in, 301–4; in utero programming of physiology, 18, 41, 119–20, 299–300, 302–3, 305–6
Diabetes, type 2, 28, 31, 32–34, 44, 117, 131, 261, 270, 329; gestational, 288; in Pima Indians, 311–13; race and, 308–10; sleep and, 127, 128
Diet: adaptation and, 52–54; of ancestral *Homo* species, 52, 54–56, 59, 86; of children, 42; digestion of starch in, 61–63; digestive machinery and, 63–64; fat in, 66, 106–7, 153–54, 229, 246, 314–15, 320; fiber in, 55, 58, 59–60, 63; food-restricted, 6, 151; latitude and fat in, 314–15; liquid calories in, 116–18; meat in, 54–55, 68–69, 72–73, 111; nutrient deficiencies in, 99–100, 103, 112, 132–34, 186–87; paleolithic, 99–101; poverty, heart disease risk and, 300–301; in pregnancy, 119, 305–6; quality of, 59; vegetables in, 42; vitamin C in, 56; Western, 130. *See also* Food(s)
Digestible energy, 154
Digestive process, 228–31
Digestive system, 57–59; endocrine, 172–73; functions of, 172–73; gas formation in, 62; gut-brain peptides, 16, 173–75, 189–92, 206; gut flora of, 134–45; gut kinetics, 59–61; gut taste sensors, 173; leptin receptors in, 177, 182; modern diet and, 63–64; Pavlov's studies of, 206–9, 213, 215; starch digestion in, 61–63
Diseases, obesity-related, 13, 14, 17, 18, 20, 34–36, 41, 44, 131, 211, 261–63, 270, 293, 324–26, 329
Dyslipidemia, 261, 262, 263, 270, 303

Eating. *See* Feeding/eating behavior
Eating disorders, 188
Eating out, 120, 125, 227
Efficient feeding, 84–86
Endocrine-disrupting chemicals, 304

Endocrine system, 170–73, 247–48; adipose tissue and endocrine function, 248–50; digestive, 172–73, 248

Energy, 136–37; biological, 138; conservation of, 136–38, 139, 140; digestible, 154; gross, 154; metabolizable, 154, 158, 159

Energy balance, 7–8, 11, 65, 154–55, 201, 226–27, 327; appetite, satiety and, 239–42; circadian cycle of, 161; energy stores and, 158–60; metabolic balance trials, 155–57; negative, 155, 158, 161, 227, 241; positive, 136, 155, 161, 227, 241; proxy measures of, 226; regulation of, 241

Energy-dense foods, 42, 66, 113, 114, 115, 130, 153–54

Energy expenditure, 6, 8, 12, 15, 16, 19, 40, 42–43, 64–65, 140, 146–50, 245; additive theory of, 152; calories for, 121; components of, 146, 147; heat as, 143–44; of human baby, 280; of lactating mice, 150–51; metabolic rate and, 65, 147–50, 152; physical activity and, 121, 150–51; in pregnancy, 266; regulation of, 150–51; shift from gut to brain, 65–66, 152; total, 65, 146, 147, 150–51

Energy intake, 6, 12–13, 15, 16, 40, 140; body size and, 50–51; measures of, 153–54; undernutrition and, 113–14

Energy metabolism, 8, 12, 15, 16, 136–62, 200

Energy storage, 6, 7, 67, 136, 155, 158–61, 187, 246

Entropy, 140–41; eating and, 145–46

Environment: built, 111, 123–26, 135; chemical, 125–26, 304; food-restricted, 6, 151; heat transfer to, 143–44; interactions with biology, 5, 18; mismatch between adaptive responses and, 15, 23, 90–110, 289–90, 291, 327; modification of, 19–20; obesogenic, 15, 111–35; pathogens in, 19, 282, 321–22

Enzymes, 139, 145

Epidemiology, 21–24

Epigenetic factors, 301–4

Erectile dysfunction, 286, 291

Estrogens, 126, 166, 244, 252, 253, 258, 269, 270, 273, 283, 286

Evolution, 9–11, 14, 40–41, 163–66, 317; adaptation and, 23, 40, 91, 92–93; common ancestry, 163–64; diet and, 6, 54–56; genetics and, 164–66, 168, 176–77, 185, 294–95, 297; information molecules and, 166–67, 238–39, 318–19; of meals, 14–15, 68–89, 212; mismatch between environment and, 15, 23, 90–110, 289–90, 327; obesity and, 10–11, 14, 40–41, 319–22; of our early ancestors, 45–67, 69–70; pharmacology and, 328–30; physical activity and, 326–27; reproduction and, 266, 289–90

Evolutionary medicine, 90–92

Expensive tissue hypothesis, 64–66, 151–53

Fairness, 82–83

Fast foods, 125, 133

Fat, body: abdominal, 261–63, 267–70, 274; adaptive purpose of, 5, 7; biology of, 7–8, 17, 244–64; excessive, 7, 13, 17, 155; metabolizable energy value for, 158, 159, 245; mortality and, 5; neonatal, 5, 17–18, 279–83, 290, 307, 319–21; reproduction and, 266–67, 276–79, 283, 321; sex differences in, 17, 18, 265, 267–70, 290, 291; steroid hormones and, 252–53, 263, 270–71; storage of, 5, 6, 7, 10–11, 67, 113, 136, 155, 157, 158, 246, 291; visceral, 261–64, 268–70, 271, 272, 290, 291, 310. *See also* Adipose tissue

Fat, dietary, 66, 153–54, 229, 246; brain size and, 106–7, 320; latitude and, 314–15

Fat metabolism, 200, 201, 204, 205, 246, 294; racial differences in, 310–11; sex differences in, 265, 274–75

Fat taste, 218, 219–20

Fatty acids, 16, 154, 157, 158, 244; brain and, 107–8; long-chain polyunsaturated, 107–8; in milk, 108–9; receptors for, 219; sex differences in uptake and release of, 274, 290, 291; as signaling molecules, 119; in visceral fat, 262–63, 270

Feeding/eating behavior: anticipatory responses in, 16, 188, 206–25, 227–28, 230; brain size and, 74–78; of chimpanzees, 68–69, 71–74; constraints on, 212, 230, 242; cooperation, efficiency and, 84–86; cooperation, fairness and, 82–83; cooperation, tolerance and, 78–79, 87; entropy and, 145–46; food-choice behavior, 187, 205; grazing, 64; meals, 14–15, 68–89, 212; motivation for, 186, 200, 232, 239; neural circuits of, 194–99, 205, 242; night eating syndrome, 128; paradox of, 226–43; patience and, 84–85; phases of, 228–29; predation and, 83–84; regulation of, 6, 188, 194–200; self-medication hypothesis of, 239–40; social function of, 69, 70, 71, 86–87, 120, 200

Fertility, 17, 266–67, 286; leptin and, 283, 284; male, 286, 291. *See also* Reproduction

Fiber, 55, 58, 59–60, 63

Folate, 133–34, 305

Food(s): absorption of, 229–31; acquisition of, 40, 42, 226; in ancestral vs. developed-nation environment, 42, 43; anticipatory responses to, 16, 188, 206–25; availability of, 6, 199–200; caloric value of, 153, 154; comfort, 239; cooking of, 58–59, 69; cost of, 6–7, 42, 116, 131; energy-dense, 42, 66, 113, 114, 115, 130, 153–54; fast, 125, 133; high glycemic, 56, 63–64, 67, 118–19; modern, 111, 113–16; mouth feel of, 219; portion size of, 121, 227; sharing of, 69, 71, 73–74, 76, 86; as signaling molecules, 119–20; sweet, 66, 104–6; taste of, 101–4, 113, 217–20; value related to scarcity of, 10, 101–4. *See also* Diet

Food insecurity, 10, 290

Food intake, regulation of, 6, 12–13, 16, 19, 188, 201–202, 205, 211, 225; appetite, satiety and, 186–205, 239–

42; cephalic-phase responses in, 16, 188, 206–25, 227–28, 230; leptin in, 16, 17, 98, 118, 166, 177–84, 226, 236–37; to maintain homeostasis, 6, 11–12, 15, 91, 93–96, 228, 231, 242; paradox of feeding, 226–42; taste and, 217–20; time scale and, 227

Friendships and obesity, 134

Fructose, 104, 118, 231, 232, 240

Fuel oxidation theory of appetite, 232

Gastric acid, 173, 209, 213, 236

Genetics, 292–316; amylase gene number, 61–63, 164–66; assortative mating, 294, 313–14; DNA, 146, 164, 177, 294–98, 302, 316; epigenetic factors, 301–4; gene duplications, 164–66, 168, 176–77, 185; gene polymorphisms, 18, 19, 38, 292, 293, 298–99, 311–13, 316; genetic drift hypothesis, 265, 291; in utero programming of physiology, 18, 41, 119–20, 299–300, 302–3, 305–6; old vs. new, 294–98; paralogous genes, 166; race/ethnicity and, 308–9; thrifty genotypes and phenotypes, 10, 41, 293, 303–7; vulnerability to obesity and, 17–18, 19, 22–23, 38–39, 44, 292–93, 308–9, 316, 319

Genome, 46, 295–96, 308, 315–16

Ghrelin, 118, 128, 174–75, 191–92, 211, 232–35, 238, 296

Global obesity, 29–34, 44, 128–30

Glucocorticoids, 244, 249, 252–53

Glucose, 16, 63–64, 118–19, 127, 154, 157–59, 201, 224, 231, 263, 276, 303, 329–30

Glycemic index, 56, 63–64, 67, 118–19

Glycogen, 157–60, 193, 201

Grazing behavior, 64

Gross energy, 154

Guanosine triphosphate (GTP), 146, 157

Gut-brain peptides, 16, 173–84, 189–92, 206

Gut flora, 134–35

Gut kinetics, 59–61

Gut taste sensors, 173

Obesity *(cont.)*
180, 183–84, 203–4; malnutrition
and, 29, 34, 111, 114–16, 132–34;
meals and, 86–89; measurement of,
24–26; metabolism and, 201–4; non-
health consequences of, 37–38; pharmacological interventions for, 20,
328–30; poverty and, 111, 129, 130;
prevalence of, 1, 5, 7, 13–14, 21, 44,
324; race/ethnicity and, 10, 34, 308–
9; reproduction and, 265–91; resistance to, 44; vitamin D and, 252; vulnerability to, 10, 17–18, 34, 43–44
Obesogenic environment, 15, 111–35;
built environment, 111, 123–26, 135;
eating out, 120, 125, 227; food as
signaling molecules, 119–20; high-
fructose foods, 118; high glycemic
foods, 56, 63–64, 67, 118–19; liquid
calories, 116–18; modern food, 111,
113–16; nutrition transition, 128–31;
physical inactivity, 121–23; portion
size, 121, 227; sleep pattern, 126–28
Obestatin, 238, 296
Orexins, 127
Orthologs, 164; of CRH, 167–68; of
leptin, 181–83
Osteoarthritis, 17, 28
Oxytocin, 9

Pancreatic polypeptide-fold family, 175–
77, 214, 311
Paradox of feeding, 226–43
Paralogs, 166; of CRH, 168
Parathyroid hormone, 100
Paraventricular nucleus, 196
Pathogen exposure, 19, 282, 321–22
Pavlov, Ivan, 206–9, 213, 215
Peptides, 16, 96, 163, 166–72; in diges-
tive system, 172–73; gut-brain, 16,
173–84, 189–92, 206; paralogous, 168
Peptide YY, 175–77, 189, 311
Pharmacological interventions, 20,
328–30
Phosphorus intake, 100
Phylogeny, 52–53, 163–64, 302
Physical activity, 20, 42–43, 121–22,
124–25, 325; aging and, 325; BMI,
disease risk and, 325; cardiovascular

fitness and, 123, 325; of children and
adolescents, 123, 124; energy expenditure and, 121, 150–51; evolution
and, 326–27; lack of, 122–23, 325–
26; vs. resting, 98–99; sex differences
in fat oxidation during, 274–75
Pima Indians, 311–13, 316, 326
Polycystic ovary syndrome, 253
Polymorphisms, genetic, 18, 19, 38,
292, 293, 298–99, 311–13, 316
Portion size, 121, 227
Poverty, 111, 129, 130–31; nutrition,
heart disease risk and, 300–301
Predation pressure, 83–84
Pregnancy and lactation: calcium in,
277–79; diet in, 119, 305–6; epigenetic factors and, 302–3; in fasting
animals, 276–77; folate in, 133–34,
305; in utero programming of physiology, 18, 41, 119–20, 299–300, 302–
3, 305–6; leptin in, 255–56; neonatal
fatness, 5, 17–18, 41, 279–83, 290,
307, 319–21; nutritional costs of,
266; obesity and outcome of, 286–89;
sleep and weight retention after, 128.
See also Reproduction
Progesterone, 252
Proopiomelanocortin, 196–97
Protein, 72, 153–54, 156, 200, 203,
229; metabolizable energy value for,
158, 159
Puberty, 284–86

Race/ethnicity: BMI and, 26, 29, 34,
263, 293, 310; diabetes risk and,
308–13; fat deposition and metabolism and, 263, 310–11; latitude,
dietary fat and, 314–15; metabolic
flexibility and, 204; metabolic syndrome and, 263; vulnerability to obesity and, 10, 34, 308–9
Receptors, 166; for cholecystokinin,
189, 237; for CRH, 167, 168; for
fatty acids, 219; for ghrelin, 238; for
hormones, 247, 248; for leptin, 177–
80, 182, 197, 236, 256, 285, 295; for
melanocortin 4, 292, 313; for pancreatic polypeptide-fold family, 175–76;
for sex hormones, 271

Recreational facilities, 124–25
Renin-angiotensin system, 172, 186
Reproduction, 9, 17, 18, 146, 265–91;
fat, leptin and, 180–81, 238, 254,
255–56, 266–67, 283, 290; fat advantages for, 276–79, 290–91, 321; and
fat in women, 283; male fertility, 286,
291; neonatal fatness, 5, 17–18, 41,
279–83, 290, 307, 319–21; obesity
and fertility, 286. *See also* Pregnancy
and lactation
Restaurant meals, 120, 125, 227
Rheostasis, 94, 96
Ribonucleic acid (RNA), 146, 182,
295–97

Safety issues, 37–38
Salivation, 209, 213
Satiation, 188–89; gut-brain peptides
and, 189–91
Satiety, 16–17, 188–92; appetite, energy
balance and, 239–42; brain, appetite
and, 194–200; cephalic-phase responses
in, 231, 236–37; function of, 242;
short-term signals of, 227, 231
Sedentary lifestyle, 42, 43, 122–23
Self-medication hypothesis of eating,
239–40
Sex differences: in adiponectin level,
258; in body fat, 17, 18, 265, 267–70,
290; in bonobos and chimpanzees,
80–81; in fat metabolism and storage,
265, 274–75, 291, 310; in leptin levels, 273–74; in metabolic flexibility,
204; in obesity prevalence, 30–31; in
obesity-related mortality, 34–36; in
response to central insulin and leptin,
272–73
Sham-feeding studies, 214, 215, 234
Single nucleotide polymorphisms, 298–
99, 311, 316
Skeletal muscle contractions, 141, 142
Sleep, 126–28; basal metabolic rate during, 149–50; diabetes and, 127, 128;
energy balance during, 161; leptin
during, 128, 180
Social behavior: of bonobos and chimpanzees, 81; eating as, 69, 70, 71, 77,
86–87, 120, 200

Sodas, 116–18, 132
Sodium: appetite for, 186–87, 194, 219;
deficiency of, 186–87, 219; intake of,
100; regulation of, 172, 186
Sour taste, 218
South Pacific islanders, 28, 29, 31, 32,
44, 128
Spermatogenesis, 286
Starch digestion, 61–63
Steroid hormones, 16, 96, 163, 166,
171, 195, 244, 249, 325; fat and,
252–53, 263, 270–71
Sweet foods, 66, 173; cephalic-phase
responses to, 214, 216–17; high-
fructose, 104, 118, 231, 232, 240;
honey, 104–6; liquids, 116–18

Taste, 217–20; bitter, 218, 219;
cephalic-phase responses to, 214,
217–19; fat, 218, 219–20; gut taste
sensors, 173; for MSG, 218–219, 219;
preferences and aversions to, 101–4,
113, 220–21; salty, 219; sour, 218;
sweet, 66, 104–6, 116–18, 173, 214,
216–217; umami, 218–19
Television viewing, 122–23
Testosterone, 166, 244, 252, 253, 269,
270–71, 299
Thermodynamics, 15, 136, 140–43
Thermogenesis, 231, 245, 320
Thermoregulation, 65, 123, 143–44,
146, 150, 245
Thiamine deficiency, 187
Thrifty genotypes and phenotypes, 10,
41, 293, 303–5; criticism of
hypothesis, 306–7
Tolerance, 78–79, 87
Tool use, 72, 74–76, 85, 86, 104
Toxins, 125–26, 304
Transportation, 123–24
Triglycerides, 245, 263, 270
Tumor necrosis factor α, 254, 257,
262
Twin studies, 292

Umami taste, 218–19
Undernutrition, 113–14
Uracil triphosphate (UTP), 146, 157
Urban growth, 130, 131